The Social Brain

A Developmental Perspective

The Social Brain

A Developmental Perspective

edited by Jean Decety

The MIT Press
Cambridge, Massachusetts
London, England

This book was set in Stone Serif by Westchester Publishing Services. Printed and bound in the United States of America.

Library of Congress Cataloging-in-Publication Data

Names: Decety, Jean, editor.
Title: The social brain : a developmental perspective / edited by Jean Decety.
Description: Cambridge, Massachusetts : MIT Press, [2020] | Includes bibliographical references and index.
Identifiers: LCCN 2019046415 | ISBN 9780262044141 (hardcover)
Subjects: LCSH: Social perception. | Developmental psychology. | Evolutionary psychology. | Neurosciences.
Classification: LCC BF323.S63 .S624 2020 | DDC 153—dc23
LC record available at https://lccn.loc.gov/2019046415

10 9 8 7 6 5 4 3 2 1

Contents

Preface

Social cognition encompasses all the information-processing mechanisms that underlie how people capture, process, store, and apply information about others to navigate social situations. It focuses on the importance of cognitive and emotional processes in our social interactions. The way we think about others plays a major role in how we think, feel, and interact with the world around us.

The past decade has seen an increase in the study of the developmental origins of the social mind. While evolutionary theory defines how social competencies have been shaped by natural selection, developmental psychologists have come up with ingenious ways to test the social abilities of infants and young children, focusing on what they do rather than on what they say. Similarly, neuroscientists have begun to examine the neurobiological mechanisms that implement and guide early social cognition. This new knowledge supports the view that social cognition is present early in infancy and childhood in surprisingly sophisticated forms. Exploring this dynamic relationship between biology and social cognition, from infancy through childhood, allows us to examine key processes in typical and atypical development. From this research, a new picture of childhood and human nature has emerged. Far from being mere unfinished adults, babies and young children are exquisitely designed by evolution to capture relevant social information and to learn and explore their social environment.

This new volume brings together a range of empirical and theoretical views from both developmental psychology and developmental neuroscience, and covers a core set of questions and topics that concern the development of the social mind. The basic topics about the origins, development, and biological bases of the human social mind include voice and face recognition, language, theory of mind, group dynamics, morality, prosocial behavior, and social decision-making. Contributions from evolutionary psychology, developmental psychology, cognitive neuroscience, and behavioral economics inform these included themes.

Structure of the Book

The book is divided into five parts. The first part provides scientific advances in early social perception and cognition. Evidence about the emergence and development of social information is presented by Marie-Hélène Grosbras and Pascal Belin. The authors highlight similarities with the development of face processing and discuss the implications and future directions for research. In the second chapter, Fabrice Damon, David Méary, Michelle de Haan, and Olivier Pascalis review early social perceptual biases associated with face processing. They argue that there is a continuity in social cognitive development in the capacities specific to the social domain and linked to later theory of mind. In chapter 3, Fransisca Ting, Melody Buyukozer Dawkins, Maayan Stavans, and Renée Baillargeon review the evidence for four moral principles that guide infants' moral reasoning. They demonstrate that very early moral cognition is remarkably sophisticated and provides a rich foundation for infants' adaptation to their social world. Next, Tobias Grossmann discusses the emerging body of infant neuroimaging studies, focusing on the role of the medial prefrontal cortex in social information processing. Although imitation plays a fundamental role in social learning, Virginia Slaughter challenges the view that human infants possess from birth an innate mechanism to imitate other people's actions. In the sixth chapter, Diane Goldenberg, Narcis Marshall, Sofia Cardenas, and Darby Saxbe review the empirical research on how parents are represented in a child's brain and how a child is represented in the parents' brains, providing insight into potential mechanisms that support the co-construction of neural representations over time.

The second section of this volume explores recent research on early infant competences for language and theory of mind. In chapter 7, Ghislaine Dehaene-Lambertz, Ana Fló, and Marcela Peña propose that human cognition has been boosted beyond the codevelopment of social cognition through language and symbolic thinking, which can be observed from the first months of life on. Dora Kampis, Frances Buttelmann, and Ágnes Melinda Kovács review recent findings regarding theory of mind abilities in infants and children and explore the ability to understand that the same reality may be represented under different subjective descriptions by different people. Markus Paulus, in chapter 9, presents a developmental account on how young children become moral agents. He makes the case that a full understanding of moral development requires taking into account the dynamic interplay between biological bases, the social environment, and the active role of the child. In chapter 10, Caitlin Hudac and Jessica Sommerville identify how electrophysiology can contribute to identifying the psychological processes that underlie early theory of mind abilities and early social moral concerns. In chapter 11, Xiao Pan Ding and Kang Lee focus on verbal deception and how theory of mind and executive functions play a functional role in the development of spontaneous lying. They also provide neuroimaging studies to understand the neural mechanisms of lying.

The third section of the volume focuses on the origins and development of prosocial behavior. In chapter 12, Jean Decety and Nikolaus Steinbeis review evidence showing that natural selection has equipped the human brain with a set of innate predispositions that motivate us to be social, cooperative, and altruistic. They show how the mechanisms involved in prosociality mature and interact with social and cultural environments. In chapter 13, Valerie Kuhlmeier, Tara Karasewich, and Kristen Dunfield present the partner choice model to account for the advantage of being selective with our prosocial behavior. In chapter 14, Lior Abramson, Florina Uzefovsky, and Ariel Knafo-Noam focus on how empathy is related to genetic variation across people and how this relationship varies through development and is specific to different components of empathy. Next, Hagit Sabato and Tehila Kogut, in chapter 15, discuss the development of children's sharing behavior and highlight the importance of the interaction between the prospective helper's and the recipient's (or the situation's) characteristics in our understanding of the development of prosociality in children.

The fourth section includes two contributions about how young children make social categories. In chapter 16, Susan Gelman and Rachel Fine provide an up-to-date overview of research on psychological essentialism in children's categories. They discuss how essentialism can distort children's social judgments of others. Then, in chapter 17, Marjorie Rhodes considers three myths and three developmental lessons about the development and nature of social categorization. She provides new insight into how to ameliorate their negative consequences and improve intergroup relations.

The final section of the volume is dedicated to atypical social cognition. In chapter 18, Kevin Pelphrey, Jennifer Frey, and Michael Crowley review developmental social neuroscience evidence about autism spectrum disorder biomarkers. In chapter 19, Essi Viding, Eamon McCrory, and Ruth Roberts focus on the developmental origins of psychopathy and show the progress that has been made in charting brain abnormalities associated with several neurocognitive hallmarks of the disorder. In the final chapter, Valerie Reyna and Christos Panagiotopoulos present the fuzzy trace theory to explain how cognitive representations of moral and monetary decisions, along with reward motivation and social values, are crucial for understanding the adaptive social brain as well as developmental atypicalities such as autism and psychopathy.

Overall, this volume brings a fresh perspective to the development of social cognition and shows the value of bringing together developmental psychology, behavioral economics, and cognitive neuroscience to illuminate our understanding of the origins, mechanism, functions, and development of the many capacities that have evolved to facilitate and regulate a wide variety of behaviors and motivations, fine-tuned to group living.

Jean Decety
Chicago, IL

I Early Social Perception and Cognition

1 Development of Voice Perception in the Human Brain

Marie-Hélène Grosbras and Pascal Belin

Overview

Routinely and effortlessly we extract a wealth of socially relevant information from voices. This goes far beyond speech comprehension and allows us to recognize a variety of attributes from the speaker such as identity or affect. Here we review evidence showing that this expertise starts developing very early on in infancy and continues to refine throughout childhood and part of adolescence. We also examine the maturation of dedicated voice processing mechanisms in the brain. We highlight similarities with the development of face processing and discuss implications and future directions for research.

Introduction

Identifying relevant features from voices is paramount not only for developing linguistic competences, but also for building sophisticated social interactions throughout the life span. Indeed, humans exhibit impressive skills for extracting subtle information from a voice, including identity, age, gender, and minute emotional or intentional expressions. An extensive corpus of studies has described the perceptual and neurophysiological processes specific to voices as compared to other sounds. This involves voice-sensitive brain regions, which are more active when listening to conspecific vocalizations than when listening to environmental sounds, centered primarily around the temporal voice areas (TVA) along the superior temporal cortex (see figure 1.1B), and including also prefrontal, limbic, and subcortical regions. These processes are probably phylogenetically ancient as homologous processes are present in other species. Their ontogeny is less known.

Yet the ability to perceive, categorize, and appraise complex stimuli such as voices in a vast range of contexts necessitates advanced expertise, which must be built as an individual grows up. The consolidation of this expertise is facilitated by the bias that even young infants show to attend to social cues, including faces or voices, more than other stimuli in their environment (Valenza, Simion, Cassia, & Umiltà, 1996; Vouloumanos, Hauser, Werker, & Martin, 2010). Being able to process vocal stimuli in a sophisticated way is obviously important for language acquisition. However, in addition—and often overlooked in the developmental literature—this ability is also fundamental for learning to interact with others. In this regard, voice has sometimes been compared to an "auditory face" (Belin, Fecteau, & Bedard, 2004), conveying cues about an individual traits and states of mind. Indeed, both voices and faces carry important social information; they are present early in the infant's environment, and their perception also is modulated by context. One might thus expect a similar developmental trajectory for the perception of both channels of social interaction.

Extensive research has shown that face perception skills (including identity, emotion, and intention recognition) are present very early on in infancy (e.g., Walker-Andrews, 1997) and continue to improve during adolescence (Scherf, Behrmann, & Dahl, 2012). Likewise, brain activity in the fusiform cortex sensitive to face stimuli, which is present in young infants (Halit, Csibra, Volein, & Johnson, 2004), continues to mature during adolescence (Scherf et al., 2012). In comparison, the development of voice processing has been less investigated, and these studies have focused principally on infancy and early childhood. Yet, as with faces, expertise for voices continues to mature throughout childhood and adolescence, alongside improvement and complexification of verbal but also nonverbal communication. Here we present an overview of the main findings, at the behavioral and neural levels, concerning voice processing in infants, children, and adolescents.

Development of Voice Perception Skills

Infancy and Toddlerhood
Using measures of heart rate variability, several studies have provided evidence that near-term fetuses in utero are sensitive to familiar voices (Lee & Kisilevsky, 2014). At birth, infants prefer to orient toward their mother's

voice (DeCasper & Fifer, 1980). This sensitivity to voice identity appears earlier than recognition of speech-related features and may facilitate language acquisition. Indeed, it is only from two months onward that infants can discriminate between different phonemes (Friederici, 2005), and their recognition improves if the phonemes are spoken by their mother than an unfamiliar speaker (Barker & Newman, 2004).

Early on, infants also can extract categorical identity attributes from voices. After using habituation paradigms, Miller (1983) reported that six-month-old infants, but not two-month-old infants, could distinguish male from female voices. Around the same period, infants show responses that appear different for adults' approving or prohibitive vocalizations, even if they are spoken in another language (Fernald, 1993); infants can discriminate happiness, anger, and sadness from the tone of voice (Flom & Bahrick, 2007). In the same study five-month-olds were found to achieve chance level for recognizing facial expressions. In fact, several studies have concluded that during the first year of life voice is the primary channel for recognizing emotion, even though infants demonstrate some knowledge of the matching between facial expressions and the corresponding affective vocal expressions (Walker-Andrews & Lennon, 1991). Accordingly, infants adapt their behavior primarily in response to the affect expressed by their mother's voice rather than face. For instance, Mumme, Fernald, and Herrera (1996) showed that twelve-month-olds interacted more with a novel toy if their mother orally expressed a positive rather than negative emotion. By contrast, facial expressions did not influence the child's behavior.

The earliest development of voice recognition skills compared to face recognition could be explained by several factors. First, during fetal development the auditory system is stimulated earlier than the visual system, giving a head start to vocal perception skills. Second, in the context of an immature orienting system, auditory processing becomes more relevant than visual processing in many contexts, when the interacting person is not directly in the field of view. Lastly, the evolutionary importance of expression of emotion from voice likely contributes to why voice processing or at least affective voice processing develops earlier than face processing in ontogeny.

Childhood and Adolescence

Although babies and toddlers can extract identity, gender, and affective information from voice above chance level, they are still not as proficient as adults. Considerable improvement takes place during childhood and adolescence. Research on this topic is somewhat disjoint from that on infant development because the methods used to study preverbal individuals differ from those used in children, which are more similar to investigations in adults. Although voice is less researched than face perception, developmental studies have indicated that, similar to faces, the development of voice perception is different from the development of perception of other, nonsocial stimuli.

Also, children perform well at tasks directly testing voice recognition. In an early study by Bartholomeus (1973), four-year-old children were at 57 percent correct at recognizing (free naming) their kindergarten classmates from hearing their voice (compared with 68 percent for the teachers). They performed far better with face recognition (97 percent for children, 100 percent for adults). Not much was done in this research domain in the following years, when efforts focused on the nature and development of children's linguistic representations. The implicit conclusion was that little change took place after age four with regards to processing of paralinguistic information from voices.

Yet studies using more controlled designs and populations have demonstrated that identity recognition based on voices continues to improve during school years. Creel and Jimenez (2012) observed that children aged three to six years were better than chance when discriminating between speakers, even when the speakers were close in terms of their voice acoustic parameters, but the children were still far worse than adults (60 percent versus 90 percent correct responses). Mann, Diamond, and Carey (1979) showed that the recognition of recently learned voices improved from age six to ten, followed by a dip in performance during early adolescence before attainment of adult capacity by age fourteen. This was confirmed with a similar task using monosyllabic words instead of sentences. Levi and Schwartz (2013) showed that children (aged seven to nine years) performed worse than early adolescents (aged ten to twelve years), who were themselves not as good as adults.

The emerging picture is thus that preschoolers but also school-aged children are less proficient than adults at mapping fine-grained voice features to individuals. Nevertheless, for more salient distinctions—such as male versus female speakers, or child versus adult voices—adult-level performance is achieved much earlier (Creel & Jimenez, 2012). Interestingly, no own-age bias was reported, similar to what has been described for faces in this age range (Macchi Cassia, Pisacane, & Gava, 2012).

Affective state recognition from voice also improves considerably during childhood. The recognition of basic emotions from the tone of voice in sentences improves between five and ten years of age (Boucher, Lewis, & Collis, 2000; Cohen, Prather, Town, & Hynd, 1990; Friend, 2000; Sauter, Panattoni, & Happé, 2013; Van Lancker, Cornelius, & Kreiman, 1989). This is also the case when using meaningless sentences, single words (Matsumoto & Kishimoto, 1983), or nonverbal interjections or vocal bursts (Allgood & Heaton, 2015; Grosbras, Ross, & Belin, 2018; Sauter et al., 2013) as stimuli. This development is delayed compared with recognition of affective intent from music and from development of linguistic comprehension. This suggests a maturational trajectory distinct from language and from general emotion comprehension development. Rather, it seems related to a specific tuning of vocal cues decoding with age. It continues also during adolescence, until about fifteen years of age, with parallel trajectories for boys and girls (figure 1.1A) (Grosbras et al., 2018; Chronaki, Hadwin, Garner, Maurage, & Sonuga-Barke, 2015; Morningstar, Ly, Feldman, & Dirks, 2018).

This developmental time line is comparable to what has been described from the literature on face perception (Herba & Phillips, 2004; Wade, Lawrence, Mandy, & Skuse, 2006). Also, studies directly comparing the two modalities in children have suggested—contrary to what has been observed in infants—an advantage of the face channel for recognizing identity (Bartholomeus, 1973) or emotion (Gil, Aguert, Le Bigot, Lacroix, & Laval, 2014; Nowicki & Mitchell, 1998; Zupan, 2015), similar to that observed in adults. Furthermore, if lexical (Friend, 2000) or contextual information including facial expression (Gil et al., 2014) is present, children aged three to nine years rely more on this information than on prosody to decipher the affective state of people.

In summary, the ability to extract information about the speaker's identity or emotion continues to mature until adolescence, after a significant amount of exposure to both familiar and novel voices. This resembles what has been described for the ability to extract the same information from faces. This late refinement of social perception is in line with social

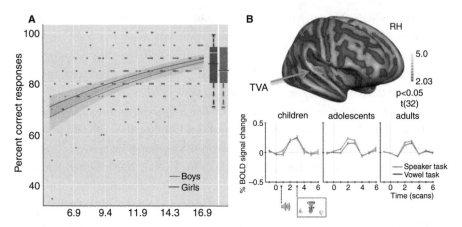

Figure 1.1
A. Percentage correct in a four-alternative forced choice for emotion from nonlinguistic utterances from ages four to seventeen. The lines represent the predicted performances for males and females separately. Shaded areas represent confidence intervals. Boxplots represent median and 75 percentile in adults' data. (Reprinted with permission from Grosbras et al., 2018.) B. Functional magnetic resonance imaging response in the right TVA when children, adolescents, or adults judged the speaker's identity (light grey) or categorized the vowel pronounced from short utterances (dark grey). Note the highest signal for the speaker task for the adolescents and adults but not the children. TVA: temporal voice area. (Reprinted with permission from Bonte et al., 2016.)

challenges during the transition into adulthood: as social relationships become more complex, individuals need to be able to detect and categorize subtle social cues more accurately.

Life history theories posit that the adaptive value of adolescence as a specific stage in the human life span is to allow for better social skills building and that this is related to increased brain size and protracted development of the brain networks involved in social perception (Bogin, 1994; Joffe, 1997). This account raises the question of how brain development may support development in vocal decoding abilities. Indeed, regions of the temporal lobes, including the TVA, are among the latest to reach an adult level of structural maturity (Mills, Lalonde, Clasen, Giedd, & Blakemore, 2014). In the next section, we examine evidence of the maturation of functional brain activity related to voice perception throughout infancy, childhood, and adolescence.

Development of Voice Perception in the Brain

Infancy and Toddlerhood

Brain responses to vocal stimuli have been studied in infants and toddlers by use of electroencephalography (EEG), functional near-infrared spectroscopy (fNIRS), and more recently functional magnetic resonance imaging (fMRI). Although many studies have aimed at finding neural correlates of speech, a few have focused on voice perception per se, thereby using nonlinguistic stimuli or at least stimuli without any semantic content.

Beauchemin and colleagues (2011) measured EEG responses in newborns evoked by either the mother's voice or another familiar voice in contrast to unfamiliar voices, all uttering the vowel /a/ (mismatch-negativity paradigm). Differences in the response to the mother and stranger voices emerged 100 milliseconds after stimulus presentation, peaking at 200 milliseconds, with the higher signal and stronger left hemisphere dominance appearing for the mother's voice. Using a similar technique, Purhonen and colleagues (2004) observed that four-month-old infants displayed an increased negativity at 300 milliseconds after they heard their mother's as compared to a stranger's voice. Both fNIRS studies in neonates (Saito et al., 2007) and fMRI studies in two- to three-month-old infants (Dehaene-Lambertz et al., 2010) also showed changes in metabolic signal in the frontal cortex for the mothers compared with strangers' voices. Although it is difficult to say to what extent these observations are related to perceptual, attentional, or emotional effects, they do demonstrate that from the very first weeks of life, at an early stage of cortical auditory processing the timing, the intensity and topology of brain activity reflect some information about the speaker's identity, sufficient to differentiate the mother's voice from other voices.

Other brain imaging studies have shown sensitivity to prosody or emotional content of voice as well. For instance, Grossmann, Striano, and Friederici (2005) observed in seven-month-old infants a slow-wave event-related brain potential (ERP) over the temporal cortex from 500 milliseconds to 1,000 milliseconds after the onset of a word uttered with a happy or angry but not a neutral prosody. This was confirmed by a fNIRS study showing modulation of hemodynamic response in the right temporal and right prefrontal cortex when seven-month-olds listened to sentences spoken happily or angrily as compared to neutral prosody (Grossmann, Oberecker,

Koch, & Friederici, 2010). This suggests enhanced sensory processing for the affectively loaded auditory stimuli.

Using the mismatched negativity technique, which measures ERP signal in response to a "deviant" stimulus in a stream of stimuli form another category, Cheng and colleagues (2012) showed that even neonates display a response recorded from frontal-central electrodes sensitive to nonlinguistic happy and fearful prosody as compared to neutral sounds. In a follow-up study Zhang and colleagues (2017) showed increased hemodynamic response over the right middle and superior temporal cortex when neonates listened passively to emotional prosody (fearful angry or happy prosody in pseudo sentences—that is, sentences devoid of semantic meaning). Blasi and colleagues (2011), who used fMRI in seven-month-old infants, did not observe any significant difference between happy and neutral nonspeech vocalizations. Sad vocalizations, by contrast, elicited specific activation, but over the insula and orbitofrontal cortex and not over the superior temporal cortex.

Thus, at birth and in the first months, the brain is certainly sensitive to the affective content of a voice in addition to its familiarity, but it remains unclear exactly which emotions elicit the most differential signals. At any rate, this fits with the behavioral observations described in the previous section.

Yet although these studies show that speakers' attributes such as identity or affective states are processed differently in the infant brain very early on, they do not show whether there is something specific about voice per se. The same responses could be observed when contrasting two musical instruments, for instance, or other auditory objects with different familiarity or symbolic value. Also, depending on the study, the brain responses are observed over the temporal or frontal cortex.

To address this question, Cheng and colleagues (2012) contrasted happy and angry vocalizations to acoustically matched synthetic sounds, and they observed a mismatch negativity at the electrodes compatible with the location of the TVA that was specific to vocal sounds, thus demonstrating higher temporal cortex activity for vocal as compared to nonvocal synthetic stimuli in neonates. Yet in an fNIRS study contrasting nonverbal vocalizations to natural environmental sounds (taken from the voice localizer commonly used in adults and including animal vocalizations), Grossmann and colleagues (2010) observed voice-specific activity in seven-month-olds but not in four-month-olds, albeit in a location more posterior than what would be expected from direct comparison with data from adults (Belin & Grosbras,

2010). Using fMRI and contrasting familiar nonlinguistic vocalizations and environmental sounds Blasi and colleagues (2011) reported activity in several foci along the superior temporal sulcus in three- to seven-month-old infants. These different results might be due to differences in methodology, but they also may reflect the fact that cortical voice sensitivity in early life might be less robust and less specific. In line with this hypothesis, Lloyd-Fox and colleagues (2012) reported voice-selective hemodynamics in the same location as well as in a slightly more anterior channel, with a linear increase between three and seven months of age.

Furthermore a fNIRS study found that four-month-old infants showed similarly strong responses in the temporal cortex on hearing monkey calls as they did when hearing human speech; this indicates that at this young age infants are still sensitive to vocalizations from both species (Minagawa-Kawai et al., 2011). This could also possibly explain why Grossman and colleagues (2010) did not observe any voice selective activity in four-month-olds, because their control stimuli included animal vocalizations. However, given the small number of brain imaging studies in infants, when exactly voice-sensitive or voice-selective responses appear in the cortex remains to be confirmed. What is certain is that specialized processes exist in the infant brain before the end of the first year that are dedicated to processing voice stimuli.

Childhood and Adolescence

The research on voice brain processing in children and adolescents is even sparser. Using ERP, Rogier and colleagues (2010) observed that four- to five-year-old children displayed frontal evoked activity over frontotemporal channels between 200 and 300 milliseconds after stimulus onset specifically when they listened to voices, but not other auditory stimuli. This response is similar to the "frontal-temporal positivity to voice" described in adults (Charest et al., 2009). This result was replicated in four- to twelve-year-old children as well as in a group of children with cochlear implants, albeit in this case with a slightly later latency (Bakhos, Galvin, Roux, Lescanne, & Bruneau, 2018).

Using fMRI, Abrams and colleagues (2016) described voice-specific activity (while contrasting voices to environmental sounds) along the superior temporal sulcus in children aged seven to twelve years. Interestingly the anterior foci were active only when children listened to their mother's voice and not to unfamiliar voices, raising the possibility that at this age the activity is driven by the social or affective relevance of the vocal source (see also Yamasaki et al., 2013), or that the anterior temporal lobe stores memory for familiar voices. The study made no direct comparison with adults, however.

Using fMRI in children (aged eight to nine years), adolescents (aged fourteen to fifteen years), and young adults, Bonte et al. (2013) looked at the developmental trajectories of both the function and morphology of the superior temporal cortex region. They observed voice-selective activity in the three age-groups along the superior temporal sulcus and gyrus, corresponding to the TVAs. As in infants and adults, the activity was higher in the right hemisphere. Also this voice response in the superior temporal cortex changed from being less selective (lower t values) and more spatially diffuse (larger numbers of voxels) in children toward being highly selective and focal in adults. In a match-to-sample task on vowels uttered by children or adults, focusing either on identifying the vowel spoken or the speaker, the same authors observed higher activity for the speaker task compared to the vowel (speech) task in the left TVA in children adolescents and adults. This effect increased with age (figure 1.1A) (Bonte, Ley, Scharke, & Formisano, 2016).

Thus, the specialization of the temporal cortex for vocal stimuli occurs independent of the development of speech processing. Moreover, compared to adults and adolescents, children recruited additional regions in the cingulate and frontal cortex. Although they should be tested with connectivity analyses, these findings indicate that the functional network for processing voices becomes more focal with age.

In summary, the few studies investigating brain correlates of the paralinguistic aspects of voice processing show that they continue to mature in childhood and adolescence. Similarly to what has been described for other cognitive or perceptual processes, notably face perception, the voice cortical network might develop from a more diffuse, less specialized configuration to a more focal, more specialized one.

Implications and Conclusions

The early presence of specialized processes for voice perception and their protracted refinement to reach adult expertise have implications for the maturation of how an individual interacts with her environment at different moments of life. First, voice processing is one important channel for nonverbal communication, which has different constraints at different life stages. Preverbal infants are entirely dependent on caregivers and thus on social interactions. From birth, it is thus vital to learn quickly how to decode, adapt to, and send social signals. It makes sense that dedicated mechanisms exist in the brain very early on in development. Also, the intrinsic incentive salience and behavioral relevance of stimuli like faces and voices probably help to promote experience-dependent plasticity (Poremba, Bigelow, & Rossi, 2013) in specialized neural substrates, thereby increasing the specificity of response to these stimuli.

Furthermore, in view of the complexity of these signals and the subtlety of the information to be extracted, it is also not surprising that the refinement of social perception—and in particular voice perception—takes over a decade to complete. This notion fits with life history theories of evolution that propose that the emergence of adolescence as a distinct stage in ontogeny, between childhood and adulthood, serves the purpose of developing sophisticated social skills and thus promotes group cohesion (Bogin, 1994). As such, adolescence would have conferred significant reproductive advantages on our evolutionary ancestors, in part by allowing individuals to learn and master adult relationship skills before mating and reproducing. From this evolutionary perspective, the development of voice perception in adolescence could be related to the role of voice in courtship in many species.

The adaptive values of building advanced expertise in decoding social signals is exemplified by the correlation between performance in recognition tests of identity or emotion and social competencies, including popularity, engagement in conflict, conflict resolution, communication, and social interaction in preschool-aged children (Nowicki & Mitchell, 1998) and in children and adolescents (Abrams et al., 2016; Skuk, Palermo, Broemer, & Schweinberger, 2019), as well as academic achievement. In studies that also tested facial expression recognition, social competence was associated more strongly with vocal than facial emotion recognition (Collins & Nowicki, 2001). This might relate to the fact that, as in adults, in children and adolescents performance in identity and emotion recognition becomes more challenging with voices than faces (Stevenage et al., 2013). This contrasts with early life when independent movement is limited and vision still not well focused; infants rely more on audition, and thus on voice of their caregivers, than on vision to pick up social signals.

Beyond the discrepancies in the relative timing of the maturation and strength of representation of vocal and facial perception, the evidence presented here underscores the striking similarities between the two processes, which can be related to these evolutionary considerations. These commonalities include early development, progressive refinement as social environments change, and cortical representation becoming more specialized and less diffuse, which conforms with the idea of common organizing principles for representation of face and voice (Yovel & Belin, 2013).

In addition, voice is important as the carrier of speech. The studies we have reviewed often highlight that the maturation of processing of paralinguistic information from voices is not directly related to measures of language maturation (Creel & Jimenez, 2012; Doherty, Fitzsimons, Asenbauer, & Staunton, 1999). Also, children with specific language impairment are not necessarily impaired in tasks that necessitate the processing of nonlinguistic vocal information (Boucher et al., 2000; Levi & Schwartz, 2013). Nevertheless, the development of voice perception interacts with the development of linguistic abilities. As in adults, infants and young children are better at processing language if the speaker's voice is familiar to them (Schmale, Cristia, & Seidl, 2012; van Heugten & Johnson, 2017). In consequence, early development of voice perception skills may help scaffold the building of language.

Conversely, speech recognition may also interact with voice perception development. In adults, bottom-up acoustic analysis and top-down influence from phonological processing collaboratively govern voice recognition (Schmale et al., 2012). Generally we recognize better the identity and emotion, and memorize better a person's voice (Scherer, Banse, & Wallbott, 2001) if she speaks the same language as us, even in conditions when the intelligibility is degraded (Fleming, Giordano, Caldara, & Belin, 2014). This "language familiarity effect" has been described across developmental stages. Infants and preschool-aged children prefer interacting with children and adults of same language (Kinzler, Dupoux, & Spelke, 2007). Seven-month-old infants can distinguish two voices better if the two talkers speak their native language. The strength of the difference in performance in voice recognition for native compared to unfamiliar languages seems to increase during childhood (Levi & Schwartz, 2013). Although this deserves further investigation, these results point toward an intertwining between the development of paralinguistic and linguistic aspects of voice signal processing.

Thus, the development of voice perception is related to the development of social competencies and to the development of language. This is unlikely to be the whole story, though. A fact that is rarely discussed in studies that tend to focus on averaged age-effect is the large interindividual variability. For instance, in Barthomoleus's study of speaker's recognition in children, the mean performance in four-year-olds was 57 percent correct responses, but it ranged between 11 percent and 95 percent in their sample (Bartholomeus, 1973). In our test of emotion recognition from voice we observed that this interindividual variability decreased from childhood to late adolescence. Although it echoes developmental findings in other domains, the origin of this interindividual variability has yet to be well characterized. It could be related partly to cultural factors (Matsumoto & Kishimoto, 1983).

Interindividual variability is also reflected by the association between voice perception skills and neurodevelopmental pathologies. For instance, children on the autistic spectrum are impaired at recognizing voice identity (Klin, 1991; Schelinski, Riedel, & von Kriegstein, 2014) and emotion (Lindner & Rosen, 2006) whereas the perception of vocal speech is relatively intact, at least when speech is presented with a good signal-to-noise ratio (deGelder, Vroomen, & van der Heide, 1991). In addition, the functional organization of voice-selective cortex in children with autism differs from children with typical development (Yoshimura et al., 2017). Even in children without an autism diagnosis there is a correlation between autistic traits and activity in voice selective cortex (Abrams et al., 2016). Moreover, a deficit in voice perception in adolescents has been associated with socioemotional pathology that emerges in adolescence such as anxiety and depression (McClure & Nowicki, 2001).

Conclusion

The development of voice processing is special, nonlinear, and relates to social development and communication abilities. It relies on a network of regions in the temporal and frontal cortex, the developmental trajectories of which remain to be better described. Understanding this development in relation to the constraints of different ontogenic stages is important in the study of neurodevelopmental disorders such as autism. In view of the increased use of mediated interactions, especially in the adolescent population, greater understanding is also important for the design of future interacting artificial agents.

References

Abrams, D. A., Chen, T., Odriozola, P., Cheng, K. M., Baker, A. E., Padmanabhan, A., ... Menon, V. (2016). Neural circuits underlying mother's voice perception predict social communication abilities in children. *Proceedings of the National Academy of Sciences of the United States of America, 113*(22), 6295–6300.

Allgood, R., & Heaton, P. (2015). Developmental change and cross-domain links in vocal and musical emotion recognition performance in childhood. *British Journal of Developmental Psychology, 33*(3), 398–403.

Bakhos, D., Galvin, J., Roux, S., Lescanne, E., & Bruneau, N. (2018). Cortical processing of vocal and nonvocal sounds in cochlear-implanted children: An electrophysiological study. *Ear and Hearing, 39*(1), 150–160.

Barker, B. A., & Newman, R. S. (2004). Listen to your mother! The role of talker familiarity in infant streaming. *Cognition, 94*(2), B45–53.

Bartholomeus, B. (1973). Voice identification by nursery school children. *Canadian Journal of Psychology, 27*(4), 464–472.

Beauchemin, M., Gonzalez-Frankenberger, B., Tremblay, J., Vannasing, P., Martinez-Montes, E., Belin, P., ... Lassonde, M. (2011). Mother and stranger: An electrophysiological study of voice processing in newborns. *Cerebral Cortex, 21*(8), 1705–1711.

Belin, P., Fecteau, S., & Bedard, C. (2004). Thinking the voice: Neural correlates of voice perception. *Trends in Cognitive Sciences, 8*(3), 129–135.

Belin, P., & Grosbras, M. H. (2010). Before speech: Cerebral voice processing in infants. *Neuron, 65*(6), 733–735.

Blasi, A., Mercure, E., Lloyd-Fox, S., Thomson, A., Brammer, M., Sauter, D., ... Murphy, D. G. (2011). Early specialization for voice and emotion processing in the infant brain. *Current Biology, 21*(14), 1220–1224.

Bogin, B. (1994). Adolescence in evolutionary perspective. *Acta Paediatrica Supplement, 406*, 29–35; discussion 36.

Bonte, M., Frost, M. A., Rutten, S., Ley, A., Formisano, E., & Goebel, R. (2013). Development from childhood to adulthood increases morphological and functional inter-individual variability in the right superior temporal cortex. *Neuroimage, 83*, 739–750.

Bonte, M., Ley, A., Scharke, W., & Formisano, E. (2016). Developmental refinement of cortical systems for speech and voice processing. *Neuroimage, 128*, 373–384.

Boucher, J., Lewis, V., & Collis, G. M. (2000). Voice processing abilities in children with autism, children with specific language impairments, and young typically developing children. *Journal of Child Psychology and Psychiatry, and Allied Disciplines, 41*(7), 847–857.

Charest, I., Pernet, C. R., Rousselet, G. A., Quinones, I., Latinus, M., Fillion-Bilodeau, S., ... Belin, P. (2009). Electrophysiological evidence for an early processing of human voices. *BMC Neuroscience, 10*, 127.

Cheng, Y., Lee, S. Y., Chen, H. Y., Wang, P. Y., & Decety, J. (2012). Voice and emotion processing in the human neonatal brain. *Journal of Cognitive Neuroscience, 24*(6), 1411–1419.

Chronaki, G., Hadwin, J. A., Garner, M., Maurage, P., & Sonuga-Barke, E. J. (2015). The development of emotion recognition from facial expressions and non-linguistic vocalizations during childhood. *British Journal of Developmental Psychology, 33*(2), 218–236.

Cohen, M., Prather, A., Town, P., & Hynd, G. (1990). Neurodevelopmental differences in emotional prosody in normal children and children with left and right temporal lobe epilepsy. *Brain and Language, 38*(1), 122–134.

Collins, M., & Nowicki S. (2001). African American children's ability to identify emotion in facial expressions and tones of voice of European Americans. *Journal of Genetic Psychology, 162*(3), 334–346.

Creel, S. C., & Jimenez, S. R. (2012). Differences in talker recognition by preschoolers and adults. *Journal of Experimental Child Psychology, 113*(4), 487–509.

DeCasper, A. J., & Fifer, W. P. (1980). Of human bonding: Newborns prefer their mothers' voices. *Science, 208*(4448), 1174–1176.

deGelder, B. , Vroomen, J., & van der Heide, L. (1991). Face recognition and lip-reading in autism. *European Journal of Cognitive Psychology, 3*(1), 69–86.

Dehaene-Lambertz, G., Montavont, A., Jobert, A., Allirol, L., Dubois, J., Hertz-Pannier, L., & Dehaene, S. (2010). Language or music, mother or Mozart? Structural and environmental influences on infants' language networks. *Brain and Language, 114*(2), 53–65.

Doherty, C. P., Fitzsimons, M., Asenbauer, B., & Staunton, H. (1999). Discrimination of prosody and music by normal children. *European Journal of Neurology, 6*(2), 221–226.

Fernald, A. (1993). Approval and disapproval: Infant responsiveness to vocal affect in familiar and unfamiliar languages. *Child Development, 64*(3), 657–674.

Fleming, D., Giordano, B., Caldara, R., & Belin, P. (2014). A language-familiarity effect for speaker discrimination without comprehension. *Proceedings of the National Academy of Sciences of the United States of America, 111*(38), 13795–13798.

Flom, R., & Bahrick, L. E. (2007). The development of infant discrimination of affect in multimodal and unimodal stimulation: The role of intersensory redundancy. *Developmental Psychology, 43*(1), 238–252.

Friederici, A. D. (2005). Neurophysiological markers of early language acquisition: From syllables to sentences. *Trends in Cognitive Science, 9*(10), 481–488.

Friend, M. (2000). Developmental changes in sensitivity to vocal paralanguage. *Developmental Science, 3*(2), 148–162.

Gil, S., Aguert, M., Le Bigot, L., Lacroix, A., & Laval, V. (2014). Children's understanding of others' emotional states: Inferences from extralinguistic or paralinguistic cues? *International Journal of Behavioral Development, 38*(6), 10.

Grosbras, M. H., Ross, P. D., & Belin, P. (2018). Categorical emotion recognition from voice improves during childhood and adolescence. *Scientific Reports, 8*(1), 14791.

Grossmann, T., Oberecker, R., Koch, S. P., & Friederici, A. D. (2010). The developmental origins of voice processing in the human brain. *Neuron, 65*(6), 852–858.

Grossmann, T., Striano, T., & Friederici, A. D. (2005). Infants' electric brain responses to emotional prosody. *NeuroReport, 16*(16), 1825–1828.

Halit, H., Csibra, G., Volein, A., & Johnson, M. H. (2004). Face-sensitive cortical processing in early infancy. *Journal of Child Psychology and Psychiatry, and Allied Disciplines, 45*(7), 1228–1234.

Herba, C., & Phillips, M. (2004). Annotation: Development of facial expression recognition from childhood to adolescence: Behavioural and neurological perspectives. *Journal of Child Psychology and Psychiatry, and Allied Disciplines, 45*(7), 1185–1198.

Joffe, T. H. (1997). Social pressures have selected for an extended juvenile period in primates. *Journal of Human Evolution, 32*(6), 593–605.

Kinzler, K. D., Dupoux, E., & Spelke, E. S. (2007). The native language of social cognition. *Proceedings of the National Academy of Sciences of the United States of America, 104*(30), 12577–12580.

Klin, A. (1991). Young autistic children's listening preferences in regard to speech: A possible characterization of the symptom of social withdrawal. *Journal of Autism and Developmental Disorders, 21*(1), 29–42.

Lee, G. Y., & Kisilevsky, B. S. (2014). Fetuses respond to father's voice but prefer mother's voice after birth. *Developmental Psychobiology, 56*(1), 1–11.

Levi, S. V., & Schwartz, R. G. (2013). The development of language-specific and language-independent talker processing. *Journal of Speech, Language, and Hearing Research, 56*(3), 913–920.

Lindner, J. L., & Rosen, L. A. (2006). Decoding of emotion through facial expression, prosody and verbal content in children and adolescents with Asperger's syndrome. *Journal of Autism and Developmental Disorders, 36*(6), 769–777.

Lloyd-Fox, S., Blasi, A., Mercure, E., Elwell, C., & Johnson, M. (2012). The emergence of cerebral specialization for the human voice over the first months of life. *Social Neuroscience, 7*(3), 317–330.

Macchi Cassia, V., Pisacane, A., & Gava, L. (2012). No own-age bias in 3-year-old children: More evidence for the role of early experience in building face-processing biases. *Journal of Experimental Child Psychology, 113*(3), 372–382.

Mann, V. A., Diamond, R., & Carey, S. (1979). Development of voice recognition: Parallels with face recognition. *Journal of Experimental Child Psychology, 27*(1), 153–165.

Matsumoto, D., & Kishimoto, H. (1983). Developmental characteristics in judgments of emotion from nonverbal vocal cues. *International Journal of Intercultural Relations, 7*(4), 415–424.

McClure, E. B., & Nowicki Jr, S. (2001). Associations between social anxiety and nonverbal processing skill in preadolescent boys and girls. *Journal of Nonverbal Behavior, 25*(1), 3–19.

Miller, C. L. (1983). Developmental changes in male/female voice classification by infants. *Infant Behavior and Development, 6*(2–3), 313–330.

Mills, K. L., Lalonde, F., Clasen, L. S., Giedd, J. N., & Blakemore, S. J. (2014). Developmental changes in the structure of the social brain in late childhood and adolescence. *Social Cognitive and Affective Neuroscience, 9*(1), 123–131.

Minagawa-Kawai, Y., van der Lely, H., Ramus, F., Sato, Y., Mazuka, R., & Dupoux, E. (2011). Optical brain imaging reveals general auditory and language-specific processing in early infant development. *Cerebral Cortex, 21*(2), 254–261.

Morningstar, M., Ly, V. Y., Feldman, L., & Dirks, M. A. (2018). Mid-adolescents' and adults' recognition of vocal cues of emotion and social intent: Differences by expression and speaker age. *Journal of Nonverbal behavior, 42*(2), 237–251.

Mumme, D. L., Fernald, A., & Herrera, C. (1996). Infants' responses to facial and vocal emotional signals in a social referencing paradigm. *Child Development, 67*(6), 3219–3237.

Nowicki, S., Jr., & Mitchell, J. (1998). Accuracy in identifying affect in child and adult faces and voices and social competence in preschool children. *Genetic, Social, and General Psychology Monographs, 124*(1), 39–59.

Poremba, A., Bigelow, J., & Rossi, B. (2013). Processing of communication sounds: Contributions of learning, memory, and experience. *Hearing Research, 305*, 31–44.

Purhonen, M., Kilpelainen-Lees, R., Valkonen-Korhonen, M., Karhu, J., & Lehtonen, J. (2004). Cerebral processing of mother's voice compared to unfamiliar voice in 4-month-old infants. *International Journal of Psychophysiology, 52*(3), 257–266.

Rogier, O., Roux, S., Belin, P., Bonnet-Brilhault, F., & Bruneau, N. (2010). An electrophysiological correlate of voice processing in 4- to 5-year-old children. *International Journal of Psychophysiology, 75*(1), 44–47.

Saito, Y., Kondo, T., Aoyama, S., Fukumoto, R., Konishi, N., Nakamura, K.,…
Toshima, T. (2007). The function of the frontal lobe in neonates for response to a
prosodic voice. *Early Human Development, 83*(4), 225–230.

Sauter, D. A., Panattoni, C., & Happé, F. (2013). Children's recognition of emotions
from vocal cues. *British Journal of Developmental Psychology, 31*(1), 97–113.

Schelinski S, Riedel P, & von Kriegstein K (2014). Visual abilities are important for
auditory-only speech recognition: Evidence from autism spectrum disorder. *Neuropsychologia, 65*, 1–11.

Scherer, K. R., Banse, R., & Wallbott, H. G. (2001). Emotion inferences from vocal
expression correlate across languages and cultures. *Journal of Cross-Cultural Psychology, 32*(1), 76–92.

Scherf, K. S., Behrmann, M., & Dahl, R. E. (2012). Facing changes and changing
faces in adolescence: A new model for investigating adolescent-specific interactions
between pubertal, brain and behavioral development. *Developmental Cognitive Neuroscience, 2*(2), 199–219.

Schmale, R., Cristia, A., & Seidl, A. (2012). Toddlers recognize words in an unfamiliar
accent after brief exposure. *Developmental Science, 15*(6), 732–738.

Skuk, V. G., Palermo, R., Broemer, L., & Schweinberger, S. R. (2019). Autistic traits
are linked to individual differences in familiar voice identification. *Journal of Autism
and Developmental Disorders, 49*(7), 2747–2767.

Stevenage, S. V., Neil, G. J., Barlow, J., Dyson, A., Eaton-Brown, C., & Parsons, B.
(2013). The effect of distraction on face and voice recognition. *Psychological Research,
77*(2), 167–175.

Valenza, E., Simion, F., Cassia, V. M., & Umiltà, C. (1996). Face preference at birth.
Journal of Experimental Psychology: Human Perception and Performance, 22(4), 892–903.

Van Heugten, M., & Johnson, E. K. (2017). Input matters: Multi-accent language
exposure affects word form recognition in infancy. *Journal of the Acoustical Society of
America, 142*(2), El196.

Van Lancker, D., Cornelius, C., & Kreiman, J. (1989). Recognition of emotional-
prosodic meanings in speech by autistic, schizophrenic, and normal children. *Developmental Neuropsychology, 5*(2–3), 207–226.

Vouloumanos, A., Hauser, M. D., Werker, J. F., & Martin, A. (2010). The tuning of
human neonates' preference for speech. *Child Development, 81*(2), 517–527.

Wade, A. M., Lawrence, K., Mandy, W., & Skuse, D. (2006). Charting the development
of emotion recognition from 6 years of age. *Journal of Applied Statistics, 33*(3), 297–315.

Walker-Andrews, A. (1997). Infants' perception of expressive behaviors: Differentiation of multimodal information. *Psychological Bulletin, 121*(3), 437–456.

Walker-Andrews, A., & Lennon, E. (1991). Infants' discrimination of vocal expressions: Contributions of auditory and visual information. *Infant Behavior and Development, 14,* 131–142.

Yamasaki, T., Ogata, K., Maekawa, T., Ijichi, I., Katagiri, M., Mitsudo, T., ... Tobimatsu, S. (2013). Rapid maturation of voice and linguistic processing systems in preschool children: A near-infrared spectroscopic study. *Experimental Neurology, 250,* 313–320.

Yoshimura, Y., Kikuchi, M., Hayashi, N., Hiraishi, H., Hasegawa, C., Takahashi, T., ... Minabe, Y. (2017). Altered human voice processing in the frontal cortex and a developmental language delay in 3- to 5-year-old children with autism spectrum disorder. *Scientific Reports, 7*(1), 17116.

Yovel, G., & Belin, P. (2013). A unified coding strategy for processing faces and voices. *Trends in Cognitive Sciences, 17*(6), 263–271.

Zhang, D., Zhou, Y., Hou, X., Cui, Y., & Zhou, C. (2017). Discrimination of emotional prosodies in human neonates: A pilot fNIRS study. *Neuroscience Letters, 658,* 62–66.

Zupan, B. (2015). Recognition of high and low intensity facial and vocal expressions of emotion by children and adults. *Journal of Social Sciences and Humanities, 1,* 332–344.

2   Building a Face-Space for Social Cognition

Fabrice Damon, David Méary, Michelle de Haan, and Olivier Pascalis

Overview

Human infants begin the journey of social cognition armed with a strong bias for attending to faces and to social stimuli in general. We review our current knowledge and understanding of the neurocognitive development of face processing, and we examine how these early social-perceptual biases serve to direct subsequent experience, shaping the face-space to optimize the recognition of the individuals present in the infant's environment, thus progressively adjusting infants to their social groups. We summarize knowledge about social biases as revealed by behavioral measures, eye-tracking measures, and neural measures of the development of the social brain network for face processing. We will argue that there is continuity in social cognitive development in that capacities specific to the social domain in infancy are linked to later theory of mind skills.

Introduction

Humans live in social groups, and their social lives require relationships with other group members and communication between individuals. Social communication has used different perceptual and cognitive systems for processing information including faces, facial expressions, gestures, vocalizations, sounds, and oral language, which each would have emerged at different evolutionary times. These systems are the basis of social cognition. The development of face processing is a prominent facet of the sociocognitive skills encompassed by social cognition (Grossmann, 2015). There is consensus that humans become "face experts" and that the neonatal face processing system develops with experience (Lee, Anzures, Quinn, Pascalis, &

Slater, 2011). Faces provide a large, complex, and dynamic set of physical, emotional, and social information to observers that essentially determines how people will interact, hence the importance of perceiving facial details. From facial information, humans will also form rapid first-impression judgments such as trustworthiness or attractiveness (Todorov, 2017). The ability to process faces effectively is crucial for individuals to succeed in everyday social interactions, and the development of this ability has been studied intensively for the last thirty years. The aim of this chapter is to review our current knowledge and understanding of the neurocognitive development of face processing.

Multidimensional Representation of Faces: The Face-Space Model

Faces are multidimensional stimuli that provide rich sources of visual information with social significance, including all the elements allowing identification of social partners, such as their identity, age, and sex. "Face-space" is an influential framework introduced to account for many of the basic perceptual features of face recognition (Valentine, Lewis, & Hills, 2016). In this model, each face is represented by a location in a psychological similarity space composed of multiple perceptual dimensions. These dimensions correspond to physiognomic variations that serve to discriminate faces; they may be specific parameters (e.g., size of the eyebrow or distance between the eyes) or global face properties (e.g., sex, race, age, and identity). Similarity between faces is thus an analogue of distance in the real world and can be determined by calculating the distance in face-space. Accordingly, faces that are similar to each other tend to be close together in face-space because they have similar values on face-space dimensions. Faces that are dissimilar are separated by a larger distance.

Face-space begins as the central tendency of all these dimensions and can be thought of as the average or prototype of all faces experienced by an individual. Therefore, typical faces are located near the center of the face-space and distinctive (atypical) faces are farther away in the face-space spectrum. The number of dimensions present in the face-space is not theoretically specified, but one assumption of the model is that dimensions are selected and tuned to optimize discrimination among the most frequently encountered faces (typically, but not necessarily, own-race faces), which may be inappropriate to distinguish infrequently experienced face type (e.g., other-race faces).

As predicted by the model, other-race faces are poorly recognized and remembered compared to own-race faces (Rossion & Michel, 2012). When represented in the face-space, other-race faces are relegated in a densely packed clump away from the center of the face-space (see figure 2.1A). In other words, other-race faces are found very similar to each other (densely packed); hence they are hard to discriminate and are distinctive from own-race faces (located away from the center). As a result, the activation of an other-race face is more likely to also activate neighboring faces, making them prone to misidentifications.

Further than face discrimination, categorization is also affected. Because other-race faces are located in a small, high-density cluster, the activation of nearby individuals is faster than the activation of a population of faces more widely distributed in the face-space as are own-race faces. Consequently, the activation of the whole category is easier for other-race compared with own-race faces. This explains a somewhat paradoxical result:

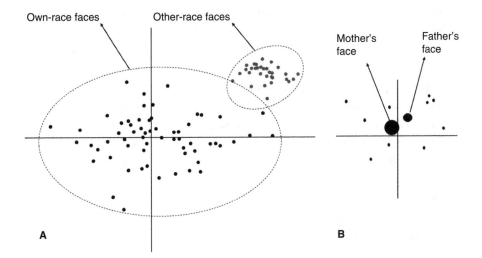

Figure 2.1
A. Own-race and other-race faces schematic representation (black and grey dots, respectively) in a face-space of two unspecified dimensions (for illustrative purposes).
B. Representation of an infant face-space, populated with fewer individuals than an adult face-space. The size of the mother's and father's faces are bigger than the other dots, illustrating their weight on the face-space, according to the frequency of exposure to these faces in the first year of life. Thus, the center of the face-space likely reflects the characteristics of the primary caregiver.

other-race faces are less accurately recognized than own-race faces but are classified *faster* by race (Rossion & Michel, 2012).

The Shaping of the Face-Space in Early Infancy

The infant face-space is probably broad and coarsely tuned for visual stimuli bearing some facial structural properties, and progressively builds up a specific representation of human faces commensurate with the infant's exposure to this kind of visual stimuli (Slater et al., 2010). Due to their limited experience, infants could initially possess a face-space constituted by a small number of dimensions (de Haan, Humphreys, & Johnson, 2002; Humphreys & Johnson, 2007). This would allow a basic comparison of the degree of perceptual similarity of a particular face with a representation of faces based on initial biases (Simion & Di Giorgio, 2015) and early experiences with faces.

How are the face-space dimensions defined when experience with faces is scarce? Despite their lack of experience, even newborns show a predisposition to attend to stimuli that possess the structure of the face over other non-face-like stimuli (for reviews see Reynolds & Roth, 2018; Simion & Di Giorgio, 2015). The origin of this early bias for faces has been a hotly debated topic, notably the question of whether these preferences reflect domain-specific or domain-general mechanisms (Karmiloff-Smith, 2015). The domain-specific view proposes that newborns possess a mechanism that biases their visual attention toward faces, bootstrapping their learning and specialization for faces (de Haan, Humphreys, & Johnson, 2002; Johnson, Senju, & Tomalski, 2015). Another view proposed is that the early face preferences derive from the functional properties of the immature newborn's visual system, generating attentional biases for general visual features. For example, newborns are sensitive to "top-heavy" patterns (i.e., composed of more elements in the upper than in the lower part of the configuration) bounded by a circular area (Simion & Di Giorgio, 2015; for another competing model of face preference in neonates, see also Wilkinson, Paikan, Gredebäck, Rea, & Metta, 2014). Whatever the reason, these biases functionally ensure that newborns attend to conspecifics and subsequently develop face expertise. These biases can be viewed as a set of primitive dimensions that constitute the initial constraints weighting on the emerging face-space, along with the other nonvisual cues that may accompany the exposure to faces (e.g., voice, maternal odor, and tactile stimulation).

Infants' experiences with faces are limited, so the infant face-space is likely populated with very few exemplars (Sugden & Moulson, 2018). As a consequence, the characteristics of the few included faces should have a strong influence on the face-space. In particular, the face of the primary caregiver probably weighs heavily in the central tendency of the face-space and the shaping of the dimensions (see figure 2.1B). This could explain why neonates show longer looking times (interpreted as a preference) for their mother's face compared with the face of a stranger (e.g., Bushnell, 2001). Three months later, infants still present visual preferences that closely match their everyday perceptual experience with faces (Rennels & Davis, 2008): they look longer at female over male faces, at own-race compared with other-race faces, and at adult over infant faces (for a review, see Quinn, Lee, & Pascalis, 2019). Furthermore, Caucasian twelve-month-olds present a preference for faces close to an average face composed of Caucasian female adult faces (Damon, Méary, et al., 2017), reflecting the facial characteristics of their primary caregivers (Sugden & Moulson, 2018).

Successful social communication implies a correct identification of the people with whom the infants interact, especially the primary caregivers, and these visual biases will guide an infant's attention toward the appropriate faces. During the first year, infants lose the ability to discriminate among individual faces of unfrequently encountered categories, while their ability to distinguish faces from frequently encountered categories is maintained and gradually refined with experience (Maurer & Werker, 2014; for a meta-analysis, see Sugden & Marquis, 2017). More specifically, three-month-olds recognized well both own-race and other-race faces while nine-month-olds only recognized faces from their own race, a pattern of findings that further extends to the processing of nonhuman faces such as monkey or sheep faces (for a recent review, see Quinn et al., 2019).

This specialization remains highly malleable. First, the infant's brain exhibits a high degree of neural plasticity during the first year of life. Second, the properties of the infant face-space are favorable to reorganization and changes, as a sparse face-space is far more strongly impacted by each new face experience compared with a face-space populated by many exemplars (Balas, 2012). Thus, although nine-month-olds typically show a loss of ability to discriminate between faces from other species or other races, it is possible to forestall this loss by exposing infants to these infrequent categories of faces (Heron-Delaney et al., 2011; Pascalis et al., 2005).

Interestingly, social interactions seem to play a critical role. In these training studies the infants were exposed to the faces using procedures that involved social interaction; for example, the parents presented the faces in a friendly way and used labels (or names) to depict the faces. Learning to individuate appeared to be a crucial factor because exposure associated with individual labeling led the infants to maintain the ability to discriminate monkey faces, whereas category learning (faces presented associated with a category label similar for all the faces) did not (Scott & Monesson, 2009). These findings provide convergent evidence that the face-space is shaped in response to social needs and is tuned to facilitate a rapid learning of the appropriate way of communicating and interacting within the social group.

The face-processing specialization is hereby presented through the lens of the face-space framework, but it may actually be seen as embedded in the broader context of the developmental mechanism known as "perceptual narrowing." This phenomenon can be defined as a progression whereby infants maintain the ability to discriminate stimuli to which they are exposed while losing the ability to discriminate stimuli to which they are not exposed (Maurer & Werker, 2014). It has been argued that narrowing might occur not only in face processing but in every domain pertaining to social communication—from speech processing to emotion or gesture perception (Pascalis et al., 2014)—in a process that gradually adapts the infants to their native social group.

From Face-Space Scaling to Early Racial Biases

As previously reviewed, as infants grow older, the boundaries between own-race and other-race face categories are increasingly sharpened as dimensions are scaled to discriminate relevant faces. For example, by nine months of age, infants treat other-race faces from multiple races as if they are of the same kind, clearly apart from own-race faces (Quinn et al., 2019). Far from trivial, those perceptual biases have downstream social consequences. They might notably contribute to the emergence of implicit racial biases, and they have been thought of as precursors of an initial in-group–out-group partitioning of faces.

These implicit perceptual biases readily go beyond mere visual cues. In a study examining the social relevance of own-race faces, Xiao, Quinn, et al. (2018) reported that by nine months of age infants spontaneously associated a positive valence to own-race faces and a negative valence other-race faces. Positive and negative valences were operationalized with happy and sad music. On the "congruent" trials, infants viewed a series of own-race faces sequentially paired with happy musical excerpts and a series of other-race faces sequentially paired with sad musical excerpts. On the "incongruent" trials, the associations between the faces and musical excerpts were reversed. They found that nine-month-old infants, but not younger, were more inclined to associate own-race faces with positive valences and other-race faces with negative valences, suggesting the emergence of an "implicit" race bias around nine months of age.

Furthermore, own-race individuals are judged to be reliable. Seven-month-olds preferentially rely on own-race individuals to guide their social interaction and learning, especially under uncertainty (Xiao, Wu, et al., 2018). Infants were presented with a video in which an adult's (either own-race or other-race) gaze direction predicted a toy's appearance 25, 50, or 100 percent of the time. Under uncertainty (50 percent reliability) the infants only followed the gaze of an own-race but not an other-race adult to anticipate the toy's appearance. Infants were, however, pragmatic: they followed both adults equally when the gaze predicted the event's occurrence with 100 percent reliability, and they followed neither when the gaze predicted the event with 25 percent (chance) reliability.

Social Biases Revealed by Face Scanning: Eye-Tracking Measures

The evidence previously reviewed suggests that the first-year of life is a crucial period in the development of a functional face-space for social cognition (Peltola, Yrttiaho, & Leppänen, 2018). Unusual reactions in social situations and poor interest in social interactions are hallmarks of autism spectrum disorder (ASD). Jones and Klin (2013) followed a cohort of infants at high risk for ASD (i.e., who had a sibling with ASD). Testing happened from birth to thirty-six months of age, when the presence of autistic traits can be diagnosed. Infants were shown videos of an actress playing the role of a caregiver while their eye movements were measured. Attention to eyes, which was initially present in all infants, declined markedly between two and six months in infants later diagnosed with autism. Atypical gaze behavior, including poor eye contact and poor attention to faces, is part of the ASD diagnosis (Lord, Rutter, DiLavore, & Risi, 2008). Atypical attention to eyes in faces emerged during the first year of life and was a marker of risk of ASD at three years. However, using a pop-out face paradigm (i.e., measures of the delay before infants detect a face in an image array of multiple items), Elsabbagh et al. (2013) found no differences between typically developing infants and infants later diagnosed with ASD, suggesting intact automatic attention to faces at six months. Face detection may be preserved while other features of face processing such as face scanning and in particular attention to eyes turn out to be atypical. Although more work is needed to characterize visual processing of social scenes, most of the eye-tracking studies with infants later diagnosed with ASD suggest that face processing undergoes profound changes between two and nine months of age (Klin, Shultz, & Jones, 2015).

Neural Correlates of the Adult Face-Processing Networks

Although experience has an important influence on the development of skills such as face recognition and speech perception (Lee et al., 2011; Werker & Hensch, 2015), these skills and their development must ultimately be instantiated in the brain. In adults, face processing is mediated by a distributed neural network involving subcortical areas involved in detecting faces and directing visual attention to them and core cortical areas involved in the detailed visual-perceptual analysis of faces (for reviews, see Bernstein & Yovel, 2015; Duchaine & Yovel, 2015). Both of these components interact with an extended cortical-subcortical system involved in further processing of faces such as the conscious processing of emotional intentions of others (Gobbini & Haxby, 2007; Haxby, Hoffman, & Gobbini, 2000; Johnson, 2005).

In the *subcortical pathway,* information travels from the retina directly to the superior colliculus, then to the pulvinar and on to the amygdala, with projections to the dorsal visual cortex (Johnson et al., 2015). This route is believed to process facial information quickly and automatically and to rely primarily on low spatial frequency information (Johnson, 2005).

The *core system* for visual analysis of faces receives input from the retina via the geniculostriate pathway and can be divided into two separate but interacting streams (see figure 2.2): a ventral stream, which extracts form information from faces, and a dorsal stream, which is specialized for processing dynamic information from faces (Bernstein & Yovel, 2015; Duchaine & Yovel, 2015). The ventral pathway preferentially processes the structure and surface properties of the face, providing the means to

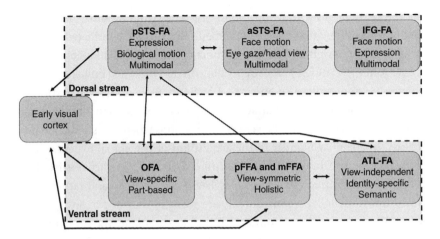

Figure 2.2

Theoretical framework for roles and connections of face-selective areas proposed by Duchaine and Yovel (2015), developed from the model of Haxby et al. (2000). Note that multiple pathways can convey face representations into the network, and not only the occipital face area. ATL-FA, anterior temporal lobe face area; aSTS-FA, anterior superior temporal sulcus face area; FFA, fusiform face area; IFG-FA, inferior frontal gyrus face area; mFFA, middle fusiform face area; pFFA, posterior fusiform face area; OFA, occipital face area; pSTS-FA, posterior superior temporal sulcus face area.

represent invariant features such as identity, age, or sex, but also contributing to facial expression processing. It includes the occipital face area, the fusiform face area, and the anterior temporal lobe face area. The dorsal face pathway plays a role in ongoing social interactions, especially the extraction of changing information from moving faces (e.g., expression, gaze, and mouth movements), but also responds to identity information conveyed by dynamic faces and to biological motion, and is implicated in the multimodal integration of person-related information (e.g., face/voice processing) (Duchaine & Yovel, 2015). This stream is composed of the posterior superior temporal sulcus face area, the anterior superior temporal sulcus face area, and the inferior frontal gyrus face area.

Development of the Social Brain: Infant Face-Processing Network

At least three theoretical frameworks can be put forward to understand social brain development (see Johnson, 2011). The "maturational" account relates the maturation of particular regions of the brain to newly emerging sensory, motor, and cognitive functions, following a one-to-one static mapping of brain–behavior relation over development. Regions are considered matured when they achieve the adult state of functioning, and different components come "online" at different ages depending on the combined influence of gene expression and experience on cortical circuitry. By contrast, the "expertise-based" account proposes that accrued experience with social stimuli could drive the specialization of the very cortical regions where the stimuli are processed, in a similar manner to the acquisition of perceptual expertise for a novel visual category in adults. The third account, "interactive specialization," postulates that brain areas may begin with a broad and rather undefined functionality and consequently are partially activated in a wide range of different contexts and stimuli. During development, the areas become increasingly tuned to respond to particular dimensions of social stimuli, with more specialized and more focal responses. Interactions between regions are fundamental to the development of the specialized function of a specific brain area. The patterns of activity and the patterns of connectivity to other regions both drive the response properties of that specific brain region. In sum, new computational abilities emerge from the interaction and competition between several regions rather than the onset of activity in one region.

How does specialization for brain areas involved in face processing arise? According to the maturational view, we expect to see the gradual addition of new brain modules relating to aspects of face-processing abilities. From an expertise-based perspective, the fusiform face area should become active as the perceptual skill of face processing is acquired. From an interactive specialization view, we expect to observe increasing neural specialization and more restricted localization of face processing with development. With these different predictions in mind, let us review some of the evidence from developmental cognitive neuroscience from birth and during early infancy.

As discussed earlier, newborn infants present a visual bias for faces and face-like stimuli over nonsocial stimuli (Reynolds & Roth, 2018). It has been proposed that visually guided behavior in the newborn infant is largely controlled by subcortical structures such as the superior colliculus and pulvinar (Johnson, 2005); however, the influence of this system declines during the second month of life, perhaps due to inhibition by developing cortical circuits. A second brain system depending upon a degree of cortical maturity and exposure to faces over the first month or two begins to control infant orienting preferences from around two or three months of age.

Evidence shows, however, that cortical structures are deeply involved in early face preference, thus leading to qualify this subcortical-to-cortical model. For example, newborns show a preference for attractive faces that is driven by the internal features of the face and upright orientation (for a review, see Damon, Mottier, Méary, & Pascalis, 2017), which suggests the presence of more advanced face processing skills than what would be expected if face processing was exclusively under the control of a subcortical system.

Note that these early face processing skills are also hard to reconcile with a purely expertise-based account of brain development because newborns seem to already have a facial representation that is quite elaborate (Slater et al., 2010). Furthermore, both neuroimaging and behavioral studies have highlighted the involvement of cortical visual pathways during face processing in two-month-old infants. A positron-emission tomography (PET) study in two-month-olds revealed the activation of a distributed network of cortical areas that largely overlapped the adult face-processing network (with even some additional regions active) when infants viewed faces as compared to a moving dot array (Tzourio-Mazoyer et al., 2002). Additional support for the possibility that cortical structures are involved in early face processing comes from finding that two-month-olds display a preference for upright over inverted faces despite the use of stimuli that activated only the short-wave sensitivity cones in the retina (S cone), which are exclusively conveyed through the geniculocortical visual pathway (Nakano & Nakatani, 2014). Thus, in the first weeks of life, the face-specialized neural network probably develops through interactions between the subcortical and cortical visual pathways (Nakano & Nakatani, 2014).

Face processing becomes increasingly more specialized across the first year of life (Guy, Zieber, & Richards, 2016), but its development is protracted during childhood and early adulthood (Golarai, Liberman, Yoon, & Grill-Spector, 2010). Adult imaging methods such as functional magnetic resonance imaging are not easily applicable to healthy infants, for both technical and ethical reasons. Thus, developmental researchers have relied on electroencephalography and more particularly on measures of changes in event-related potentials (ERPs) during the perception of social stimuli such as faces.

In adults, human faces elicit an ERP component called the "N170"—a negative-going deflection that occurs after approximately 170 milliseconds recorded over occipitotemporal regions of the scalp (for review, see Rossion &

Jacques, 2011). The N170 is typically of larger amplitude for faces than for nonface objects and, rather surprisingly, for inverted than for upright faces. Although the exact underlying neural generators of the N170 are currently still debated (Rossion & Jacques, 2011), source localization studies support the idea that it reflects activity of the core cortical network described earlier, and that the specificity of response of this component can be taken as an index of the degree of specialization of cortical processing for human upright faces. For this reason, a series of studies was undertaken to examine the development of the N170 over the first weeks and months of postnatal life.

A challenge for researchers is that the latency and topography of ERP components thought to be functionally equivalent change across development. For instance, an infant ERP component that has many of the properties associated with the adult N170 but with a slightly longer latency was identified and termed the "N290"—that is, a component peaking at approximately 290 to 350 milliseconds after stimulus onset (de Haan, Pascalis, & Johnson, 2002; Halit, Csibra, Volein, & Johnson, 2004; Halit, de Haan, & Johnson, 2003). Among the cortical sources of the N290 are the fusiform gyrus, lateral occipital area, and superior temporal sulcus, which is fairly similar to what was reported for the N170 in adults (Hoehl & Peykarjou, 2012). The examination of the response properties of this component at three, six, and twelve months of age revealed that the component is present from at least three months of age, but it undergoes gradual changes during the first year of life: for instance, the onset of the N290 is about 350 milliseconds in three-month-olds, and about 290 milliseconds in twelve-month-olds (Halit et al., 2003).

This component becomes more specifically tuned to human upright faces with increasing age. For example, the N290 begins to show differential responses to upright and inverted faces between six and twelve months of age (de Haan, Pascalis, & Johnson, 2002; Halit et al., 2003). In particular, twelve-month-olds and adults showed different ERP responses to upright and inverted faces but three- and six-month-olds did not (de Haan, Pascalis, & Johnson, 2002; Halit et al., 2003).

Furthermore, in infants more than one component is necessary to account for the different dimensions of the faces the adult N170 is sensitive to, which supports the idea that initial cortical processing of faces is broad, poorly tuned, and supported by different regions at different ages. More specifically, in six-month-olds the N290 responds to the species of faces but not to their orientation; the P400 (i.e., a positive-going deflection peaking at about 390–450 milliseconds after stimulus onset) shows sensitivity to the orientation of faces but not their species (Halit et al., 2003). Moreover, the scalp distribution of the P400 changes with age; it is prominent over medial recording sites in three- and six-month-olds but stronger in lateral recording sites in twelve-month-olds.

These two components might reflect processes that spread out in different brain regions or might reflect different processing stages that later become integrated to produce the adult N170 (de Haan, Pascalis, & Johnson, 2002), although the exact mechanisms producing the N290 and P400 across development remain unclear (Hoehl & Peykarjou, 2012). It has been argued that N290 amplitude becomes more strongly associated with fusiform gyrus activation with age (Guy et al., 2016). Overall, the developmental trajectory of these face-sensitive ERP components suggests that the brain-function mapping is dynamic over development, and that the same behavior can be supported by different neural substrate at different ages.

The lines of evidence we have reviewed are inconsistent with a strictly maturational view that posits the addition of new components of processing with development. Rather, the evidence available to date suggests "shrinkage" of the cortical activity associated with face processing during development. Although increased specialization and localization are specifically predicted by the interactive specialization view, these changes are not necessarily inconsistent with the expertise-based view. However, the available evidence from newborns weakens the expertise-based hypothesis, while the evidence on the neurodevelopment of face processing over the first months and years of life is consistent with the kinds of dynamic changes in processing expected from the interactive specialization, not the maturational, approach.

Summary and Conclusion

Human infants begin the journey of social cognition armed with a strong bias for faces, and more generally they present a sensitivity for social stimuli (Reynolds & Roth, 2018). These early biases serve to direct subsequent experience, shaping the face-space to optimize the recognition of the individuals present in the infant's environment, thus progressively adjusting infants to their social group (Pascalis et al., 2014). This attunement to the social group translates at first into visual preferences for adult female own-race faces, and then into better recognition accuracy with own-race and own-species faces, and finally in the development of an early in-group–out-group face categorization.

These infant face-processing abilities represent the core basis for the social cognition skills that will develop throughout life. Infants' attentional bias to faces has been seen as an early marker of social development linked with the development of affective empathy (Peltola et al., 2018), while early perturbations in social stimuli sensitivity are frequently associated with developmental disorders such as ASD (Klin et al., 2015). Furthermore, infant social-cognitive abilities can be related to preschoolers' understanding of others' mental states and beliefs ("theory of mind") by shared domain-specific elements (Yamaguchi, Kuhlmeier, Wynn, & Vanmarle, 2009). In other words, there is a form of continuity in social cognitive development such that capacities specific to the social domain in infancy are linked to later theory of mind skills. Building a social brain needs an early start.

References

Balas, B. J. (2012). Bayesian face recognition and perceptual narrowing in face-space. *Developmental Science, 15,* 579–588. doi: 10.1111/j.1467-7687.2012.01154.x

Bernstein, M., & Yovel, G. (2015). Two neural pathways of face processing: A critical evaluation of current models. *Neuroscience and Biobehavioral Reviews, 55,* 536–546. doi: 10.1016/j.neubiorev.2015.06.010

Bushnell, I. W. R. (2001). Mother's face recognition in newborn infants: Learning and memory. *Infant and Child Development, 10,* 67–74. doi: 10.1002/icd.248

Damon, F., Méary, D., Quinn, P. C., Lee, K., Simpson, E. A., Paukner, A., … Pascalis, O. (2017). Preference for facial averageness: Evidence for a common mechanism in human and macaque infants. *Scientific Reports, 7,* 46303. doi: 10.1038/srep46303

Damon, F., Mottier, H., Méary, D., & Pascalis, O. (2017). A review of attractiveness preferences in infancy: From faces to objects. *Adaptive Human Behavior and Physiology, 3,* 321–336. doi: 10.1007/s40750-017-0071-2

De Haan, M., Humphreys, K., & Johnson, M. H. (2002). Developing a brain specialized for face perception: A converging methods approach. *Developmental Psychobiology, 40,* 200–212. doi: 10.1002/dev.10027

De Haan, M., Pascalis, O., & Johnson, M. H. (2002). Specialization of neural mechanisms underlying face recognition in human infants. *Journal of Cognitive Neuroscience, 14,* 199–209. doi: 10.1162/089892902317236849

Duchaine, B. C., & Yovel, G. (2015). A revised neural framework for face processing. *Annual Review of Vision Science, 1,* 393–416. doi: 10.1146/annurev-vision-082114-035518

Elsabbagh, M., Gliga, T., Pickles, A., Hudry, K., Charman, T., & Johnson, M. H. (2013). The development of face orienting mechanisms in infants at-risk for autism. *Behavioural Brain Research, 251,* 147–154. doi: 10.1016/j.bbr.2012.07.030

Gobbini, M. I., & Haxby, J. V. (2007). Neural systems for recognition of familiar faces. *Neuropsychologia, 45,* 32–41. doi: 10.1016/j.neuropsychologia.2006.04.015

Golarai, G., Liberman, A., Yoon, J. M. D., & Grill-Spector, K. (2010). Differential development of the ventral visual cortex extends through adolescence. *Frontiers in Human Neuroscience, 3,* 80. doi: https://doi.org/10.3389/neuro.09.080.2009

Grossmann, T. (2015). The development of social brain functions in infancy. *Psychological Bulletin, 141,* 1266–1287.

Guy, M. W., Zieber, N., & Richards, J. E. (2016). The cortical development of specialized face processing in infancy. *Child Development, 87,* 1581–1600. doi: 10.1111/cdev.12543

Halit, H., Csibra, G., Volein, Á., & Johnson, M. H. (2004). Face-sensitive cortical processing in early infancy. *Journal of Child Psychology and Psychiatry, 45,* 1228–1234. doi: 10.1111/j.1469-7610.2004.00321.x

Halit, H., de Haan, M., & Johnson, M. H. (2003). Cortical specialisation for face processing: Face-sensitive event-related potential components in 3- and 12-month-old infants. *NeuroImage, 19,* 1180–1193. doi: 10.1016/S1053-8119(03)00076-4

Haxby, J. V., Hoffman, E. A., & Gobbini, M. I. (2000). The distributed human neural system for face perception. *Trends in Cognitive Sciences, 4,* 223–233. doi: 10.1016/S1364-6613(00)01482-0

Heron-Delaney, M., Anzures, G., Herbert, J. S., Quinn, P. C., Slater, A. M., Tanaka, J. W., … Pascalis, O. (2011). Perceptual training prevents the emergence of the other-race effect during infancy. *PLoS ONE, 6*(5), e19858. doi: 10.1371/journal.pone.0019858

Hoehl, S., & Peykarjou, S. (2012). The early development of face processing—what makes faces special? *Neuroscience Bulletin, 28,* 765–788. doi: 10.1007/s12264-012-1280-0

Humphreys, K., & Johnson, M. H. (2007). The development of "face-space" in infancy. *Visual Cognition, 15,* 578–598. doi: 10.1080/13506280600943518

Johnson, M. H. (2005). Subcortical face processing. *Nature Reviews Neuroscience, 6,* 766–774. doi: 10.1038/nrn1766

Johnson, M. H. (2011). Interactive specialization: A domain-general framework for human functional brain development? *Developmental Cognitive Neuroscience, 1,* 7–21. doi: 10.1016/j.dcn.2010.07.003

Johnson, M. H., Senju, A., & Tomalski, P. (2015). The two-process theory of face processing: Modifications based on two decades of data from infants and adults. *Neuroscience and Biobehavioral Reviews, 50,* 169–179. doi: 10.1016/j.neubiorev.2014.10.009

Jones, W., & Klin, A. (2013). Attention to eyes is present but in decline in 2–6-month-old infants later diagnosed with autism. *Nature, 504,* 427–31. doi: 10.1038/nature12715

Karmiloff-Smith, A. (2015). An alternative to domain-general or domain-specific frameworks for theorizing about human evolution and ontogenesis. *AIMS Neuroscience, 2,* 91–104. doi: 10.3934/Neuroscience.2015.2.91

Klin, A., Shultz, S., & Jones, W. (2015). Social visual engagement in infants and toddlers with autism: Early developmental transitions and a model of pathogenesis. *Neuroscience and Biobehavioral Reviews, 50,* 189–203. doi: 10.1016/j.neubiorev.2014 .10.006

Lee, K., Anzures, G., Quinn, P. C., Pascalis, O., & Slater, A. M. (2011). Development of face processing expertise. In A. Calder, G. Rhodes, M. H. Johnson, & J. V Haxby (Eds.), *The Oxford handbook of face perception* (pp. 753–778). Oxford: Oxford University Press.

Lord, C., Rutter, M., DiLavore, P. C., & Risi, S. (2008). *Autism diagnostic observation schedule: ADOS manual.* Torrance, CA: Western Psychological Services.

Maurer, D., & Werker, J. F. (2014). Perceptual narrowing during infancy: A comparison of language and faces. *Developmental Psychobiology, 56,* 154–178. doi: 10.1002/ dev.21177

Nakano, T., & Nakatani, K. (2014). Cortical networks for face perception in two-month-old infants. *Proceedings of the Royal Society B: Biological Sciences, 281,* 20141468. doi: 10.1098/rspb.2014.1468

Pascalis, O., Loevenbruck, H., Quinn, P. C., Kandel, S., Tanaka, J. W., & Lee, K. (2014). On the links among face processing, language processing, and narrowing during development. *Child Development Perspectives, 8,* 65–70. doi: 10.1111/ cdep.12064

Pascalis, O., Scott, L. S., Kelly, D. J., Shannon, R. W., Nicholson, E., Coleman, M., & Nelson, C. A. (2005). Plasticity of face processing in infancy. *Proceedings of the National Academy of Sciences of the United States of America, 102,* 5297–5300. doi: 10.1073/pnas.0406627102

Peltola, M. J., Yrttiaho, S., & Leppänen, J. M. (2018). Infants' attention bias to faces as an early marker of social development. *Developmental Science, 21*(6), 1–14. doi: 10.1111/desc.12687

Quinn, P. C., Lee, K., & Pascalis, O. (2019). Face processing in infancy and beyond: The case of social categories. *Annual Review of Psychology, 70,* 165–189. doi: https://doi.org/10.1146/annurev-psych-010418-102753

Rennels, J. L., & Davis, R. E. (2008). Facial experience during the first year. *Infant Behavior and Development, 31,* 665–678. doi: 10.1016/j.infbeh.2008.04.009

Reynolds, G. D., & Roth, K. C. (2018). The development of attentional biases for faces in infancy: a developmental systems perspective. *Frontiers in Psychology, 9,* 1–16. doi: 10.3389/fpsyg.2018.00222

Rossion, B., & Jacques, C. (2011). The N170: understanding the time-course of face perception in the human brain. In Luck S. J. & Kappenman E. S. (Eds.), *The Oxford handbook of event-related potential components* (pp. 115–142). Oxford: Oxford University Press. doi: 10.1093/oxfordhb/9780195374148.013.0064

Rossion, B., & Michel, C. (2012). An experience-based holistic account of the other-race face effect. In G. Rhodes, A. Calder, M. H. Johnson, & J. V. Haxby (Eds.), *Oxford handbook of face perception* (pp. 1–40). Oxford: Oxford University Press. doi: 10.1093/oxfordhb/9780199559053.013.0012

Scott, L. S., & Monesson, A. (2009). The origin of biases in face perception. *Psychological Science, 20,* 676–680. doi: 10.1111/j.1467-9280.2009.02348.x

Simion, F., & Di Giorgio, E. (2015). Face perception and processing in early infancy: Inborn predispositions and developmental changes. *Frontiers in Psychology, 6,* 969. doi: 10.3389/fpsyg.2015.00969

Slater, A. M., Quinn, P. C., Kelly, D. J., Lee, K., Longmore, C. A., McDonald, P. R., & Pascalis, O. (2010). The shaping of the face space in early infancy: Becoming a native face processor. *Child Development Perspectives, 4,* 205–211. doi: 10.1111/j.1750-8606.2010.00147.x

Sugden, N. A., & Marquis, A. R. (2017). Meta-analytic review of the development of face discrimination in infancy: Face race, face gender, infant age, and methodology moderate face discrimination. *Psychological Bulletin, 143,* 1201–1244. doi: 10.1037/bul0000116

Sugden, N. A., & Moulson, M. C. (2018). These are the people in your neighbourhood: Consistency and persistence in infants' exposure to caregivers', relatives', and strangers' faces across contexts. *Vision Research, 157,* 230–241. doi: 10.1016/j.visres.2018.09.005

Todorov, A. (2017). *Face value: The irresistible influence of first impressions.* Princeton, NJ: Princeton University Press. doi: 10.1080/00220380412331322771

Tzourio-Mazoyer, N., De Schonen, S., Crivello, F., Reutter, B., Aujard, Y., & Mazoyer, B. (2002). Neural correlates of woman face processing by 2-month-old infants. *NeuroImage, 15*(2), 454–461. doi: 10.1006/nimg.2001.0979

Valentine, T., Lewis, M. B., & Hills, P. J. (2016). Face-space: A unifying concept in face recognition research. *Quarterly Journal of Experimental Psychology, 69,* 1996–2019. doi: 10.1080/17470218.2014.990392

Werker, J. F., & Hensch, T. K. (2015). Critical periods in speech perception: New directions. *Annual Review of Psychology, 66,* 173–196. doi: 10.1146/annurev-psych -010814-015104

Wilkinson, N. N., Paikan, A., Gredebäck, G., Rea, F., & Metta, G. (2014). Staring us in the face? An embodied theory of innate face preference. *Developmental Science, 17,* 809–825. doi: 10.1111/desc.12159

Xiao, N. G., Quinn, P. C., Liu, S., Ge, L., Pascalis, O., & Lee, K. (2018). Older but not younger infants associate own-race faces with happy music and other-race faces with sad music. *Developmental Science, 21,* 1–10. doi: 10.1111/desc.12537

Xiao, N. G., Wu, R., Quinn, P. C., Liu, S., Tummeltshammer, K. S., Kirkham, N. Z.,…Lee, K. (2018). Infants rely more on gaze cues from own-race than other-race adults for learning under uncertainty. *Child Development, 89,* e229–e244. doi: 10.1111/cdev.12798

Yamaguchi, M., Kuhlmeier, V. A., Wynn, K., & Vanmarle, K. (2009). Continuity in social cognition from infancy to childhood. *Developmental Science, 12,* 746–752. doi: 10.1111/j.1467-7687.2008.00813.x

3 Principles and Concepts in Early Moral Cognition

Fransisca Ting, Melody Buyukozer Dawkins, Maayan Stavans, and Renée Baillargeon

Overview

According to Graham and colleagues (2013), a "first draft" of moral cognition emerges early and universally in development and is then gradually revised by experience and culture. In this chapter, we explore some of the moral principles and concepts that might be included in this initial draft of morality. First, we review evidence that at least four moral principles guide infants' reasoning about how individuals should act toward others: fairness, harm avoidance, in-group support, and authority. These principles regulate actions at different levels of the social landscape and interact in various ways. Next, we report recent results concerning four moral concepts that support infants' principle-based expectations: moral obligation, moral status, moral circle, and moral character. Together, these findings indicate that early human moral cognition is remarkably sophisticated and provides a rich foundation for infants' adaptation to their social worlds.

Introduction

Over the past two decades, a wealth of research has examined infants' evaluations of social actions, to uncover both how these evaluations are formed and how they affect infants' responses to others. Initially, it appeared as though early sociomoral evaluations reflected mainly infants' ability to distinguish between positive and negative actions (Bloom & Wynn, 2016; Hamlin, 2013b). In a series of experiments, for example, Hamlin and colleagues familiarized infants aged three to twenty-one months to scenarios depicting interactions among nonhuman individuals (e.g., blocks with eyes, nonverbal puppets; Hamlin, 2013a, 2014, 2015; Hamlin & Wynn,

2011; Hamlin, Wynn, & Bloom, 2007, 2010; Hamlin, Wynn, Bloom, & Mahajan, 2011). Each scenario involved a positive event, in which one character acted positively toward a protagonist (e.g., helped it reach the top of a steep hill, returned a ball it had dropped), and a negative event, in which another character acted negatively toward the same protagonist (e.g., knocked it down to the bottom of the hill, stole its ball). Across ages and scenarios, infants tended to look equally at the positive and negative events, suggesting that they did not expect the characters either to help the protagonist achieve its goal or to refrain from harming it.

These negative results did not stem from infants' inability to understand the events presented. After watching these events, three- to eleven-month-olds tested with a social-preference task chose the positive over the negative character (Hamlin, 2015; Hamlin et al., 2007, 2010; Hamlin & Wynn, 2011; for a meta-analysis, see Margoni & Surian, 2018); eight-month-olds did so even if the characters' positive and negative actions were unsuccessful, indicating that infants attended primarily to the characters' intentions in evaluating their actions (Hamlin, 2013a); ten-month-olds tested with a violation-of-expectation task expected the protagonist to also prefer the positive character and detected a violation (as indexed by longer looking times) when the protagonist approached the negative character instead (Hamlin et al., 2007); and twenty-one-month-olds chose the positive character when asked to give away a treat, but chose the negative character when asked to take away a treat (Hamlin et al., 2011).

Together, these results suggested that early sociomoral sensitivities included no particular expectations about how individuals would act toward others. Nevertheless, infants possessed notions of welfare and harm that enabled them to evaluate actions as positive (beneficial to the targets of the actions), neutral, or negative (detrimental to the targets of the actions). These evaluations, in turn, drove affiliative attitudes: Infants preferred, and expected others to prefer, individuals who produced positive actions over individuals who produced negative actions.

In time, however, it became clear that infants possess richer moral sensitivities than was initially thought. One of the turning points in this research came from new evidence that (a) infants do hold expectations about how individuals will act toward others, but (b) these expectations are often context-sensitive and arise only when specific preconditions are met. For example, thirteen- to seventeen-month-olds viewed providing help as obligatory when individuals belonged to the same social group but as optional when individuals belonged to different groups or when group memberships were unspecified (Jin & Baillargeon, 2017; Ting, He, & Baillargeon, 2019a). Similarly, seventeen-month-olds viewed directly intervening in a transgression against an in-group victim as obligatory for a group leader but as optional for a nonleader equal in rank to the victim (Stavans & Baillargeon, 2019).

These and other findings (reviewed in the next sections) have led us to propose a new characterization of infant moral cognition that rests on five assumptions. First, as is the case for early physical and psychological reasoning (Baillargeon, Scott, & Bian, 2016; Lin, Stavans, & Baillargeon, in press), early sociomoral reasoning is guided by a skeletal framework of principles and concepts that emerges early and universally in development (Baillargeon et al., 2015; Buyukozer Dawkins, Ting, Stavans, & Baillargeon, in press).

Second, early sociomoral principles include *fairness, harm avoidance, in-group support*, and *authority*—all of which have long been the focus, explicitly or implicitly, of research across the social sciences (Baumard, André, & Sperber, 2013; Brewer, 1999; Dupoux & Jacob, 2007; Graham et al., 2013; Rai & Fiske, 2011; Shweder, Much, Mahapatra, & Park, 1997). Each principle is normative in nature and specifies what is obligatory and forbidden in its context of application.

Third, each principle is tied to a particular social distinction or structure. Thus, fairness, the broadest of the principles, applies to interactions among any individuals (i.e., entities with moral status); harm avoidance applies to interactions among individuals who are identified as members of the same moral circle (e.g., humans); in-group support applies to interactions among individuals in a moral circle who are identified as members of the same social group (e.g., sports team); and authority applies to interactions among individuals in a social group who are identified as leaders and followers (e.g., coach and players). Thus, with each successive social differentiation, additional expectations come into play that help regulate interactions at this new level.

Fourth, when two or more principles apply to the same situation, they interact in various ways. In some cases, for example, one principle may intensify expectations set by another principle (e.g., even less harm is tolerated when individuals belong not only to the same moral circle but also to the same group; Ting, He & Baillargeon, 2019b). In other cases, two principles may suggest different courses of action and must then be rank ordered (e.g., fairness may dictate that a resource be divided equally between in-group and out-group recipients, but in-group support may dictate that it be reserved for in-group recipients, particularly when it is scarce or otherwise valuable; Bian, Sloane & Baillargeon, 2018).

Finally, the last assumption is that different cultures implement, stress, and rank order the principles differently, resulting in the diverse moral landscape that exists in the world today.

Graham et al. (2013) described their work on moral foundations as "a theory about the universal first draft of the moral mind and about how that draft gets revised in variable ways across cultures" (p. 65). From this perspective, efforts to specify the principles and concepts that shape early sociomoral reasoning thus help shed light on "the universal first draft" of human moral cognition. In this chapter, we first review some of the evidence that principles of fairness, harm avoidance, in-group support, and authority guide early sociomoral expectations. Next, we discuss four key concepts implicated in these expectations: moral obligation, moral status, moral circle, and moral character.

Moral Principles

Fairness.   When watching interactions among individuals, or entities with moral status, infants bring to bear an equity-based principle of fairness: All other things being equal, individuals are expected to give others their just deserts—that is, to treat them as they deserve to be treated in the situation at hand. Thus, in situations involving the distribution of *windfall resources,* four- to fifteen-month-olds expected a distributor to divide items equally between two similar potential recipients (Buyukozer Dawkins, Sloane, & Baillargeon, 2019; Meristo, Strid, & Surian, 2016; Schmidt & Sommerville, 2011); ten- to fifteen-month-olds found it unexpected when an unfair (but not a fair) distributor was rewarded or praised (DesChamps, Eason, & Sommerville, 2015; Meristo & Surian, 2013); thirteen- to seventeen-month-olds preferred a fair over an unfair distributor when asked to choose between them (Burns & Sommerville, 2014; Geraci & Surian, 2011; Lucca, Pospisil, & Sommerville, 2018); twenty- to thirty-month-olds chose to help a fair as opposed to an unfair distributor (Surian & Franchin, 2017); and twenty-one-month-olds took into account recipients' preexisting resources and expected resource-poor recipients to receive a larger share than resource-rich recipients (Buyukozer Dawkins & Baillargeon, 2019).

In situations involving the dispensation of *rewards* for efforts, ten-month-olds found it unexpected when an experimenter gave equal praise to an assistant who had done a puzzle and an assistant who had not (Buyukozer Dawkins, Sloane, & Baillargeon, 2017); seventeen-month-olds expected a resource acquired by two workers to be shared according to the amount of effort each had exerted (Wang & Henderson, 2018); and twenty-one-month-olds found it unexpected when an experimenter gave the same reward to a worker who had done an assigned chore and a slacker who had done no work (Sloane, Baillargeon, & Premack, 2012). Similarly, in a situation involving the meting out of *punishments,* twenty-one-month-olds found it unexpected when two assistants were both punished even though only one of them had disobeyed an instruction (Buyukozer Dawkins et al., 2017).

Together, these results provide converging evidence that, from a young age, infants possess an equity-based expectation of fairness: Individuals are expected to receive the treatment they deserve in each situation, be it an equitable share of windfall resources, a reward commensurate with their efforts, or a punishment befitting their misdeeds.

Harm avoidance. When individuals belong to the same moral circle, infants bring to bear an abstract principle of harm avoidance that sets broad limits on the amount of unprovoked harm the individuals can inflict on one another. Thus, when watching interactions between two individuals, A and B, who belonged to the same moral circle but gave no indication of belonging to the same social group (e.g., two humans who belonged to different groups, two humans whose group memberships were unspecified, or two puppets with the power of speech—henceforth verbal puppets—who belonged to different animal kinds[1]), thirteen- to thirty-three-month-olds did not find it

---

1. Infants appear to view puppets with the power of speech as members of the human moral circle or "honorary humans" (e.g., Big Bird, Elmo). Moreover, just as infants assign humans to different social groups, based on various cues (e.g., Burns & Sommerville, 2014; Jin & Baillargeon, 2017; Liberman, Woodward, Sullivan, & Kinzler, 2016; Ting et al., 2019a), they assign verbal puppets from different animal kinds (e.g., English-speaking monkeys, giraffes, rabbits, and dogs) to different social groups (Bian et al., 2018; Ting & Baillargeon, 2018a).

unexpected when A directed a relatively *mild* negative action toward B: for example, when A ignored B's need for instrumental assistance (Jin & Baillargeon, 2017), threw an object B needed on the floor (Ting et al., 2019a), crumpled a drawing done by B (Ting & Baillargeon, 2018a), or knocked down one block from one of three towers built by B (Ting et al., 2019b). However, infants did detect a violation when A's negative actions were *more intense:* for example, when A directed several mild negative actions toward B (Ting & Baillargeon, 2018a) or knocked down one of three towers built by B (Ting et al., 2019b). These last results contrast sharply with those obtained when A and B gave no cue of belonging to the same moral circle (e.g., two different nonverbal puppets or geometric figures), as the amount of unprovoked harm deemed acceptable was then much greater. Thus, six- to fifteen-month-olds detected no violation when A repeatedly hit B (Kanakogi et al., 2017; Premack & Premack, 1997), growled at and fought with B over the possession of a toy (Rhodes, Hetherington, Brink & Wellman, 2015), or knocked B down a steep hill, causing it to roll end-over-end to the bottom of the hill (Hamlin, 2015). Together, these results suggest that from a young age, infants expect harm avoidance to apply within but not between moral circles.

In-group support. When individuals in a moral circle belong to the same social group, infants bring to bear an abstract principle of in-group support. This principle carries numerous expectations that can be roughly divided into two sets related to in-group care and in-group loyalty.

With respect to *in-group care,* four- to twelve-month-olds expected a woman alone with a crying baby (who presumably belonged to the same group as the woman) to attempt to comfort the baby, and they found it unexpected when she ignored the baby instead (Jin, Houston, Baillargeon, Groh, & Roisman, 2018). Similarly, thirteen- to twenty-nine-month-olds expected a woman to provide help to another woman in need of instrumental assistance when the two belonged to the same group, but they held no expectation about the provision of help when the two women belonged to different groups or when their group memberships were unspecified (Jin & Baillargeon, 2017; Ting et al., 2019a). Finally, thirteen- to twenty-nine-month-olds expected an individual who had observed a transgression against an in-group victim to later refrain from helping the wrongdoer (in a form of indirect third-party punishment intended to deter future transgressions), but they held no such expectation when the transgression involved an out-group victim (Rhodes et al., 2015; Ting et al., 2019a).

In-group care also interacts with harm avoidance to set limits on how much unprovoked and provoked harm can be inflicted on in-group as opposed to out-group members. Thus, thirteen- to thirty-three-month-olds expected individuals to refrain from *any* unprovoked harm toward in-group members, and they therefore detected a violation when mild negative actions that were deemed permissible against out-group members (e.g., throwing a needed object on the floor, crumpling a drawing, or knocking down part of a tower) were directed at in-group members (Jin & Baillargeon, 2017; Ting & Baillargeon, 2018a; Ting et al., 2019a, 2019b). In the same vein, three-year-olds who heard stories about two novel social groups predicted that a harmful action (e.g., stealing a block) would be more likely to be directed at an out-group as opposed to an in-group victim (Rhodes, 2012). Finally, in the case of provoked harm, eighteen-month-olds expected less retaliation (or second-party punishment) toward an in-group as opposed to an out-group wrongdoer for the same transgression (Ting et al., 2019b).

Turning to *in-group loyalty*, six- to ten-month-olds preferred a native speaker of their language over a foreign speaker (Kinzler, Dupoux, & Spelke, 2007); twelve-month-olds preferred toys or snacks endorsed by a native speaker over those endorsed by a foreign speaker (Kinzler, Dupoux, & Spelke, 2012; Shutts, Kinzler, McKee, & Spelke, 2009); and fourteen- to nineteen-month-olds were more likely to imitate a novel conventional action modeled by a native as opposed to a foreign speaker (Buttelmann, Zmyj, Daum, & Carpenter, 2013; Howard, Henderson, Carrazza, & Woodward, 2015). Similarly, after watching two groups of nonhuman characters perform distinct novel conventional actions, seven- to twelve-month-olds detected a violation if a member of one group chose to imitate the other group's conventional action (Powell & Spelke, 2013); twelve-month-olds expected an individual to choose an in-group over an out-group member as a play partner (Bian & Baillargeon, 2019); and after watching adult characters soothe baby characters (Spokes & Spelke, 2017), sixteen-month-olds detected a violation if one baby chose as a play partner a baby who had been soothed by a different adult (and hence presumably belonged to a different group) over a baby who had been soothed by the same adult (and hence presumably belonged to the same group).

Finally, in resource-distribution situations invoking both fairness and in-group loyalty, nineteen-month-olds to three-year-olds expected fairness to prevail when there were sufficient resources for all in-group and out-group individuals present, but they expected in-group loyalty to trump fairness when there were only enough resources for the in-group individuals (Bian et al., 2018; Lee, Esposito, & Setoh, 2018; Olson & Spelke, 2008; Renno & Shutts, 2015).

Together, these results provide converging evidence that from a young age, an abstract principle of in-group support guides infants' reasoning about interactions within social groups and carries rich expectations of in-group care and in-group loyalty.

Authority. According to Rai and Fiske (2011), the principle of authority carries moral obligations for both leaders and followers: Leaders are expected "to lead, guide, direct, and protect" their followers, whereas followers are expected "to respect, obey, and pay deference" to their leaders (p. 63). Building on prior evidence that young infants detect power asymmetries (Pun, Birch, & Baron, 2016; Thomsen, Frankenhuis, Ingold-Smith, & Carey, 2011) and expect them to be stable over time and to extend across situations (Enright, Gweon, & Sommerville, 2017; Gazes, Hampton, & Lourenco, 2017; Mascaro & Csibra, 2012), recent research has begun to examine whether infants share some of the same authority-based expectations as adults.

In experiments focusing on expectations about *leaders* (Stavans & Baillargeon, 2019), seventeen-month-olds watched a group of three bear puppets who served as the protagonist, wrongdoer, and victim. The protagonist brought in two toys for the other bears to share, but the wrongdoer unfairly seized both toys, leaving none for the victim. The protagonist then either took one toy away from the wrongdoer and gave it to the victim (*intervention* event) or approached each bear in turn without redistributing a toy (*nonintervention* event). Infants expected an intervention when the protagonist was portrayed as a leader, but they held no particular expectation for intervention when the protagonist was portrayed as a nonleader equal in rank to the other bears.

In other experiments focusing on expectations about *followers* (Margoni, Baillargeon, & Surian, 2018), twenty-one-month-olds saw computer-animated events in which three characters were playing in a field next to a house when a protagonist arrived and ordered them to go to bed; the characters then filed into the house and could be seen through its front window. Next, the protagonist left the scene, and the characters either returned to the field (*disobedience* event) or remained in the house and went to sleep (*obedience* event). When the protagonist was portrayed as a leader, infants expected the characters to continue to obey her order after she left the scene. When the protagonist was portrayed as a bully, however, infants held no expectation about whether the characters would continue to obey her after she left. Consistent results have also been obtained with social-preference tasks (Thomas, Thomsen, Lukowski, Abramyan, & Sarnecka, 2018). When character A deferred to character B by yielding the right-of-way on a narrow platform, twenty-five-month-olds preferred B over A, suggesting that they viewed B as a leader or other high-power figure. When B knocked A out of the way, however, toddlers then viewed B as a bully and preferred A over B.

Together, these results suggest that by the second year of life, an abstract principle of authority guides infants' expectations about interactions between leaders and followers within social groups.

Moral Concepts

Embedded in our descriptions of early principle-based expectations are several key moral concepts. In this section, we discuss four such concepts: We consider whether infants understand *moral obligations,* on what entities they confer *moral status,* how they identify *moral circles,* and how they evaluate individuals' *moral characters.*

Moral obligations. We have suggested that infants conceptualize the various principle-based expectations reviewed in the preceding section as *obligations* that specify how individuals should act toward others (e.g., individuals should act fairly). However, another possibility is that infants construe these expectations non-normatively as *behavioral regularities* that capture how individuals in their social environments typically act toward others (e.g., individuals typically act fairly). How can we decide between these two possibilities? This is a challenging question to address experimentally.

One approach has been to examine whether infants exhibit a negative attitude and expect others to exhibit a negative attitude toward individuals who do not behave as expected. In the case of fairness, for example, there is evidence that infants not only expect distributors to divide resources fairly between similar potential recipients (Buyukozer Dawkins et al., 2019; Meristo et al., 2016), but also (a) prefer fair over unfair distributors (Geraci & Surian, 2011; Lucca et al., 2018), (b) are more likely to help fair as opposed to unfair distributors (Surian & Franchin, 2017), (c) detect a violation when individuals choose to reward unfair as opposed to fair distributors (Meristo & Surian, 2013), and (d) associate praise with fair distributors and admonishment with unfair distributors (DesChamps et al., 2015). Together, these results suggest that infants who observe an unfair distribution do not simply represent it as a deviation from a behavioral regularity; rather, they view it as a norm violation, with consequences for their and others' attitude toward the wrongdoer.

To provide converging evidence that infants understand moral obligations, we have adopted a different approach focused on *virtuous* actions—namely, positive actions that are not obligatory in the situation and hence go beyond what is morally expected. In a typical experiment, two individuals, A and B, produce the same positive action toward the same protagonist, one at a time. For A, this action is obligatory (e.g., is dictated by the principle of in-group support); for B, this action is optional. Next, infants are presented with both individuals in a visual social-preference task (adapted from Hamlin et al., 2010). The rationale is that if infants look preferentially at B over A, it indicates that they understand moral obligations: (a) they recognize that A's positive action is obligatory but B's is not; (b) they infer that B is acting virtuously (e.g., out of kindness) and going beyond what is morally required in the situation; and hence (c) they evaluate B more favorably than A.

To illustrate our approach, in a series of experiments (Ting & Baillargeon, 2018b), eighteen-month-olds received two orientation trials, two familiarization trials, and one preferential-looking trial. In the first orientation trial, infants were introduced to two groups of puppets; there were two owls (O1 and O2) and two frogs (F1 and F2), who all spoke English and wore different accessories to help distinguish them. In the second orientation trial, one puppet (e.g., O1) built a tower by stacking five discs of decreasing sizes, while the other puppets watched. In the two familiarization trials, O1 now needed help to complete her tower. In the *in-group-helps* trial, O1 was initially alone building her tower and was joined midway through by O2. O1 was unable to reach the last disc, which lay across the apparatus floor from her but within O2's reach (Warneken & Tomasello, 2006). O2 helped by bringing the disc closer to O1, who then completed her tower. The *out-group-helps* trial was identical except that F2 joined and helped O1. Finally, in the preferential-looking trial, infants saw O2 and F2 standing on either side of the apparatus; infants' looking at each puppet was coded frame by frame offline.

Confirming prior results, infants looked equally at the in-group-helps and out-group-helps familiarization trials. During the preferential-looking trial, however, infants looked significantly longer at the out-group helper than at the in-group helper. Infants thus implicitly preferred the puppet who had helped out of kindness (i.e., F2 had no obligation to help O1, so her actions provided evidence that she was kind) over the puppet who might have helped solely out of duty (i.e., O2 was obligated to help O1, so her actions provided no evidence about whether she was kind or not). This positive result was replicated in additional experiments but was eliminated when no helper came in the familiarization trials so O1 was unable to complete her tower; when O1 did not need help and simply exchanged friendly greetings with O2 and F2; and when O1 did not need help and O2 and F2 brought closer an extraneous object. Together, these results suggest two conclusions. First, infants' sociomoral expectations capture moral obligations rather than mere behavioral regularities: Here, in particular, infants viewed individuals as having an obligation to help in-group members in need of assistance. Second, infants already appreciate virtuous actions that go beyond what is morally required.

The preceding results mirror recent findings with adults reported by McManus, Kleiman-Weiner, and Young (in press). In a series of experiments, adults were presented with vignettes about individuals who did or did not provide help to kin or strangers. Consistent with the principle of in-group support, adults judged individuals who failed to help kin less favorably than they did individuals who failed to help strangers. Critically, and in line with our infant results above, adults also judged individuals who helped strangers more favorably than they did individuals who helped kin, presumably because they realized that whereas the latter individuals were morally obligated to help, the former individuals were not and hence went beyond what was morally expected when they showed kindness to strangers.

Moral status.   How do infants identify novel individuals, endowed with the moral obligations and rights we have been discussing, such as the obligation to treat others fairly and the right to be treated fairly by others? To put it another way, on what novel entities are infants likely to confer moral status? The early sociomoral literature suggests that infants confer moral status on any novel entities they identify as *animate*. According to research on early biological reasoning (Baillargeon et al., 2016; Setoh, Wu, Baillargeon, and Gelman, 2013), a novel entity is animate if it gives evidence of having both a functioning body (e.g., it is capable of self-propulsion, which is a sign of internal energy) and a functioning mind (e.g., it interacts with its environment in an agentive manner, which is a sign of internal control). Consistent with this analysis, studies on sensitivity to fairness in resource-distribution tasks indicate that infants expect a fair treatment for novel entities that are both self-propelled and agentive (and hence are identified as animate), but they hold no particular expectation about the fair treatment of novel entities that are only self-propelled, only agentive, or neither self-propelled nor agentive (Buyukozer Dawkins et al., 2019; Meristo et al., 2016; Sloane et al., 2012; Ting & Baillargeon, 2018c; Ziv & Sommerville, 2017).

For example, in recent experiments (Ting & Baillargeon, 2018c), fourteen-month-olds were first introduced to two colorful boxes that were devoid of any morphological similarities to humans and other animals. While an experimenter watched, the boxes gave evidence of being both self-propelled and agentive (they moved on their own and beeped contingently to each other, as though taking turns in a conversation; self-propelled/agentive condition), only self-propelled (only one box moved and beeped in each trial; self-propelled/non-agentive condition), or only agentive (the two boxes beeped contingently to each other but never moved; agentive/non-self-propelled condition). Next, the experimenter divided two identical toys between the two boxes, either fairly (*equal* event) or unfairly (*unequal* event). Infants in the self-propelled/agentive condition looked significantly longer at the unequal than at the equal event, whereas infants in the other two conditions looked equally at the events. The two boxes were thus granted moral status—and, with it, the right to be treated fairly—only when they were shown to have both a functioning body and a functioning mind and were therefore perceived as animates.

The preceding results dovetail well with evidence by Weisman, Dweck, and Markman (2017) that adults' intuitive conception of mental life has three fundamental components: capacities related to having a body (e.g., initiating one's motion, experiencing hunger), capacities related to having a mind (e.g., detecting one's environment, pursuing goals), and capacities related to having a heart, in its metaphorical sense (e.g., telling right from wrong, exercising self-restraint). When infants confer moral status on a novel entity and endow it with moral obligations and rights, they presumably also endow it with the suite of mental capacities necessary to apply these obligations and rights (e.g., to determine, in a reward-dispensation situation, what is the equitable treatment for each potential recipient). If we think of granting these capacities as akin to granting a heart, it suggests that (a) infants' conception of individuals includes a functioning heart as well as a functioning body and a functioning mind, and (b) the tripartite ontology of mental life described by Weisman and colleagues emerges early in life, is highly abstract, and supports the identification of novel, unfamiliar individuals such as faceless animate boxes.

Moral circle.   We have suggested that when infants assign individuals to the same moral circle, they bring to bear a principle of harm avoidance that sets broad limits on the amount of harm the individuals can inflict on each other. But how do infants identify moral circles? At present, little is known about this issue. One possibility might be that infants initially posit a single moral circle, peopled by humans and "honorary humans" such as verbal puppets. In this view, infants would expect harm avoidance in interactions between humans, broadly defined, but would hold no particular expectation about harm avoidance in interactions (a) between humans and nonhumans or (b) between nonhumans. Another possibility might be that infants tend to assign different animate kinds to different moral circles. From an evolutionary perspective, such a tendency could be rooted in the never-ending battle for survival between predators and prey: Tigers and gazelles cannot belong to the same moral circle, nor can foxes and chickens, birds and worms, hawks and mice, and humans and most other animals. (The world of fiction is, of course, replete with clashes between humans and fictitious moral circles, such as the xenomorphic creatures in the *Alien* movie franchise). In this view, infants would expect harm avoidance *within* each animate kind, but would hold no expectation for harm avoidance *between* animate kinds.

Evidence for this second possibility comes from recent experiments (Jin & Baillargeon, 2019) in which twelve-month-olds watched interactions between two novel animate characters with eyes and stick arms. In the same-kind condition, the characters differed only in size (e.g., a large and a small blue square); in the different-kind condition, they differed in size, color, and shape (e.g., a large green circle and a small blue square). In each condition, infants saw two events: a *harm event* in which the first character picked up a stick and hit the second character three times, and a *no-harm* event in which the first character picked up a stick and jumped next to the second character three times. Infants in the same-kind condition found the harm event unexpected, but infants in the different-kind condition did not (even though in each case the same character was hit in the same way). These results thus provide evidence that infants expect harm avoidance between two novel animates from the same kind (e.g., two squares), but have no expectation for harm avoidance between two novel animates from different kinds (e.g., a circle and a square).

Could it be that infants always expect some degree of harm avoidance but simply have a much higher threshold for what constitutes unacceptable harm in interactions between members of different moral circles? Evidence against this suggestion comes from recent experiments on early biological reasoning (Ting, Setoh, Gelman, & Baillargeon, 2019). These experiments built on prior evidence that young infants expect animate entities to have insides (Li, Carey, & Kominsky, 2019; Setoh et al., 2013). To start, eight-month-olds were introduced to two novel animate entities, a closed cube and a closed cylinder, and were shown that each was both self-propelled (i.e., initiated and changed its course of motion) and agentive (i.e., conversed with an experimenter using either beeps or quacks). Next, infants saw two events in which the experimenter manipulated the two entities. In the *no-harm* event, she picked up one of the entities (e.g., the cube), rotated it to show infants its closed bottom, and then tilted it from side to side. In the *harm* event, she selected the other entity (e.g., the cylinder), cut out its bottom with a knife, removed its biological-like insides in two large handfuls, rotated it toward the infants to show it was now hollow, and finally tilted it from side to side. Infants looked about equally at the two events, supporting the suggestion that even grievous harm is deemed acceptable when directed at an entity from a different moral circle.

This negative result was not due to infants' inability to understand the events presented. In additional experiments, infants detected a violation when the gutted entity moved on its own again, suggesting that they expected it to no longer be able to function once its insides were removed. This effect was eliminated if the entities were initially shown to be self-propelled but not agentive (i.e., not animate). Thus, although young infants understand that a novel animate entity's insides support its function, they detect no violation when a member of a different moral circle (in this case a human experimenter) removes these insides, causing loss of function or life. More generally, our discussions of moral status and moral circles both highlight the important role that biological animacy plays in morality, at least in the first years of life. Infants confer moral status on animate entities, and they assign different kinds of animate entities to different moral circles.

The suggestion that infants expect harm avoidance within but not across moral circles fits well with the large body of research with adults on the adverse consequences of *dehumanization,* or perceiving others as less than fully human. When individuals or groups are pushed to the edges of the human moral circle, or even excluded from it, they lose some or all of their human right to be protected from harm. This means that more harm is deemed acceptable against dehumanized individuals or groups (Bastian, Denson, & Haslam, 2013; Goff, Jackson, Di Leone, Culotta, & DiTomasso, 2014; Haslam & Loughnan, 2014; Kteily, Bruneau, Waytz, & Cotterill, 2015; Smith, 2011). To illustrate, Kteily and colleagues (2015) asked adults to locate various social groups, including their own, on a scale depicting human evolution from ape to modern human. Participants endorsed harsher treatments (e.g., less support for immigration, fewer relief donations) for groups they perceived to be less evolved or less human, and they also showed less compassionate responses to harms (e.g., acts of discrimination) suffered by these groups.

The evidence that dehumanization leads to less protection from harm has been found for a wide range of human targets, including individuals with reduced or impaired mental capacities (Capozza, di Bernardo, Falvo, Vianello, & Caio, 2016; Khamitov, Rotman, & Piazza, 2016; Martinez, Piff, Mendoza-Denton, & Hinshaw, 2011; Waytz, Gray, Epley, & Wegner, 2010). According to Waytz and colleagues (2010), for example, if someone "is seen as relatively mindless, then he or she receives diminished moral standing, and might be treated like an animal or an object … Those denied competence, civility, and agency come to be seen as subservient or animalistic, licensing people to contain them against their will and to rob them of human rights" (p. 386).

In the early psychological-reasoning literature, there is substantial evidence that infants negatively evaluate individuals who behave irrationally. For example, infants aged fourteen to eighteen months were less likely to learn new information from, or to direct inquiries to, individuals who mislabeled or misused familiar objects, or who enthused over empty containers (Begus & Southgate, 2012; Koenig & Woodward, 2010; Poulin-Dubois, Brooker, & Polonia, 2011; Zmyj, Buttelmann, Carpenter, & Daum, 2010). Similarly, sixteen-month-olds found it unexpected when an individual chose to affiliate with, as opposed to disengage from, someone who produced inefficient, irrational actions (Liberman, Kinzler, & Woodward, 2018). Building on these results, one could ask whether infants would hold weaker expectations about irrational individuals' moral rights, such as the right to be protected from harm. Such evidence would suggest that a disposition to dehumanize different others is already present in infancy, thus opening new avenues for intervention. (Conversely, but in the same spirit, it would also be interesting to explore what *humanization* interventions might bring infants to view harming nonhuman entities as unacceptable.)

Moral character. The evidence reviewed in this chapter suggests that infants evaluate individuals' actions along a broad spectrum that includes (a) *forbidden* actions, which are negative actions that violate moral principles; (b) *dubious* or questionable actions, which also have a negative valence but do not unambiguously violate moral principles; (c) *neutral* actions; (d) *obligatory* actions, which are positive actions that are required by one or more moral principles; and (e) *virtuous* actions, which are positive actions that go beyond what is morally required and reveal virtues such as kindness and courage. This spectrum maps well onto the results from social-preference tasks: For example, infants have been shown to prefer individuals who produce virtuous actions over those who produce obligatory (Ting & Baillargeon, 2018b), neutral (Kanakogi et al., 2017), or dubious (Dunfield & Kuhlmeier, 2010; Hamlin et al., 2007) actions; to prefer individuals who produce obligatory actions over those who produce forbidden actions (Buon et al., 2014; Geraci & Surian, 2011); and to prefer individuals who produce neutral actions over those who produce dubious actions (Hamlin et al., 2007).

These results raise important questions about the consequences of these evaluations, particularly with respect to issues of *moral character.* While moral principles specify how individuals *should* act, moral characters reflect how they *do* act. After all, an individual may know right from wrong but still choose to act selfishly, with little concern for others. From this perspective, an individual with a good moral character is thus one who acts morally rather than selfishly, whereas an individual with a bad moral character is one who tends to do the reverse (Tooby & Cosmides, 2010).

Do infants take individuals' actions to reveal something pervasive and stable about their moral characters, which allows predictions about how they are likely to act in new contexts? To address this question, we asked in two recent experiments whether two-year-old toddlers who saw an individual violate a moral principle (a) would not find it unexpected if the individual then violated a different moral principle, but (b) would find it unexpected if the individual then acted generously toward others (Ting & Baillargeon, 2018a). In the first experiment, toddlers were first introduced to rabbit and dog puppets who spoke in female voices and wore distinguishing accessories (because the puppets were all capable of speech, we expected toddlers to see them as "honorary humans" and hence as members of the same moral circle). In the familiarization scenario, toddlers saw one puppet (e.g., a rabbit, R1) harm, either once or three times, a puppet from her own group (R2) or a puppet from the other group (D2). Next, in the test scenario, R1 divided two toys either fairly (*equal* event) or unfairly (*unequal* event) between two other puppets (in-group R3 and R4; similar results were obtained when R1 divided the toys between out-group D3 and D4). When R1 harmed out-group D2 only once (e.g., destroyed her drawing), infants still expected R1 to act fairly, suggesting that they did not draw broad negative inferences about her moral character (consistent with prior findings that a mild negative action toward out-group individuals in the same moral circle is not viewed as forbidden). However, when R1 harmed out-group D2 three times (e.g., destroyed her drawing, puzzle, and tower) or when she harmed in-group R2 either once or three times, infants looked equally at the two events (see also Surian, Ueno, Itakura, & Meristo, 2018). Together, these results suggest that when R1 violated harm avoidance and/or in-group support, toddlers inferred that there were broad and enduring deficiencies in her moral character, so they did not find it unexpected when she acted unfairly in a new context, demonstrating once again her lack of concern for others.

In the second experiment, toddlers first saw R1 harm R2 three times. In the test scenario, six toys were introduced for R1 and another in-group puppet, R3, to share. Infants detected a violation when R1 divided the toys generously (taking only one for herself and giving the other five to R3), but not when R3 did so (generous or lavish sharing is deemed acceptable among in-group members; Jin, Bian, & Baillargeon, 2017). These and control results make clear that toddlers did not simply conclude that no predictions could be made about R1's behavior because she did not follow moral rules. Rather, they concluded that R1 had a deficient moral character and was unlikely to act generously toward others.

Together, these results suggest four conclusions about toddlers' evaluations of moral character. First, toddlers perceive the different moral principles as deeply interrelated and pertaining to the same domain of morality;

thus, a violation of one principle affects predictions about other principles. Second, toddlers understand that the moral principles help keep self-interest in check in interactions with others; thus, an individual with a poor moral character is unlikely to act generously toward others. Third, infants expect individuals' moral characters to be stable over time and to extend across situations. Finally, in some cases at least, a single negative action (e.g., like R1 harming her in-group R2 once) may be sufficient to lead infants to draw negative inferences about an individual's moral character.

These results fit well with evidence that adults place a great deal of importance on individuals' moral characters and view them as stable and predictive of future actions (Bollich et al., 2016; Brambilla & Leach, 2014; Goodwin, 2015; Heiphetz, Strohminger, & Young, 2017; Strohminger & Nichols, 2014; Uhlmann, Pizarro, & Diermeier, 2015). Research with adults also shows that a single negative action may at times be sufficient to tarnish an individual's character, and that a bad impression, once formed, may require multiple positive actions to change that impression (Baumeister, Bratslavsky, Finkenauer, &Vohs, 2001; Riskey & Birnbaum, 1974; Rozin & Royzman, 2001; Skowronski & Carlston, 1992). An interesting direction for future developmental research will thus be to examine how many positive actions may be required to reverse toddlers' attribution of a deficient moral character to an individual.

Concluding Remarks

The evidence reviewed in this chapter suggests that at least four moral principles (fairness, harm avoidance, in-group support, and authority) and four moral concepts (moral obligation, moral status, moral circle, and moral character) shape early sociomoral reasoning. Although a great deal of research is still needed to support these conclusions, the "first draft" of moral cognition appears to be remarkably sophisticated and to provide a rich foundation for infants' adaptation to their social worlds, both within the confines of their families and beyond.

Acknowledgments

Preparation of this chapter was supported by a grant from the John Templeton Foundation (to R.B.). We thank Elizabeth Enright for helpful comments and suggestions.

References

Baillargeon, R., Scott, R. M., & Bian, L. (2016). Psychological reasoning in infancy. *Annual Review of Psychology, 67,* 159–186.

Baillargeon, R., Scott, R. M., He, Z., Sloane, S., Setoh, P., Jin, K., & Bian, L. (2015). Psychological and sociomoral reasoning in infancy. In M. Mikulincer & P. R. Shaver (Eds.), *APA handbook of personality and social psychology: Vol. 1. Attitudes and social cognition* (pp. 79–150). Washington, DC: American Psychological Association.

Bastian, B., Denson, T. F., & Haslam, N. (2013). The roles of dehumanization and moral outrage in retributive justice. *PloS ONE, 8*(4), e61842.

Baumard, N., André, J. B., & Sperber, D. (2013). A mutualistic approach to morality: The evolution of fairness by partner choice. *Behavioral and Brain Sciences, 36*(1), 59–78.

Baumeister, R. F., Bratslavsky, E., Finkenauer, C., & Vohs, K. D. (2001). Bad is stronger than good. *Review of General Psychology, 5*(4), 323–370.

Begus, K., & Southgate, V. (2012). Infant pointing serves an interrogative function. *Developmental Science, 15*(5), 611–617.

Bian, L., & Baillargeon, R. (2019, March). *Toddlers and infants hold an abstract expectation of ingroup loyalty.* Paper presented at the Biennial Meeting of the Society for Research in Child Development, Baltimore, MD.

Bian, L., Sloane, S., & Baillargeon, R. (2018). Infants expect ingroup support to override fairness when resources are limited. *Proceedings of the National Academy of Sciences of the United States of America, 115*(11), 2705–2710.

Bloom, P., & Wynn, K. (2016). What develops in moral development? In D. Barner & A. S. Baron (Eds.), *Core knowledge and conceptual change* (pp. 347–364). New York: Oxford University Press.

Bollich, K. L., Doris, J. M., Vazire, S., Raison, C. L., Jackson, J. J., & Mehl, M. R. (2016). Eavesdropping on character: Assessing everyday moral behaviors. *Journal of Research in Personality, 61,* 15–21.

Brambilla, M., & Leach, C. W. (2014). On the importance of being moral: The distinctive role of morality in social judgment. *Social Cognition, 32*(4), 397–408.

Brewer, M. B. (1999). The psychology of prejudice: Ingroup love or outgroup hate? *Journal of Social Issues, 55*(3), 429–444.

Buon, M., Jacob, P., Margules, S., Brunet, I., Dutat, M., Cabrol, D., Dupoux, E. (2014). Friend or foe? Early social evaluation of human interactions. *PloS ONE 9*(2):e88612.

Burns, M. P., & Sommerville, J. A. (2014). "I pick you": The impact of fairness and race on infants' selection of social partners. *Frontiers in Psychology, 5,* 93.

Buttelmann, D., Zmyj, N., Daum, M., & Carpenter, M. (2013). Selective imitation of in-group over out-group members in 14-month-old infants. *Child Development, 84,* 422–428.

Buyukozer Dawkins, M., & Baillargeon, R. (2019, May). *Infants' conflicting expectations and attitudes toward the wealthy.* Paper presented at the Annual Meeting of the Human Behavior and Evolution Society, Boston, MA.

Buyukozer Dawkins, M., Sloane, S., & Baillargeon, R. (2017, August). *Evidence for an equity-based sense of fairness in infancy.* Paper presented at the Dartmouth Workshop on Action Understanding, Hanover, NH.

Buyukozer Dawkins, M., Sloane, S., & Baillargeon, R. (2019). Do infants in the first year of life expect equal resource allocations? *Frontiers in Psychology, 10,* 116.

Buyukozer Dawkins, M., Ting, F., Stavans, M., & Baillargeon, R. (in press). Early moral cognition: A principle-based approach. In D. Poeppel, G. R. Mangun, & M. S. Gazzaniga (Eds.), *The cognitive neurosciences VI.* Cambridge, MA: MIT Press.

Capozza, D., Di Bernardo, G. A., Falvo, R., Vianello, R., & Caio, L (2016). Individuals with intellectual and developmental disabilities: Do educators assign them a fully human status? *Journal of Applied Social Psychology, 46*(9), 497–509.

DesChamps, T. D., Eason, A. E., & Sommerville, J. A. (2016). Infants associate praise and admonishment with fair and unfair individuals. *Infancy, 21*(4), 478–504.

Dunfield, K., & Kuhlmeier, V. A. (2010). Intention-mediated selective helping in infancy. *Psychological Science, 21,* 523–527.

Dupoux, E., & Jacob, P. (2007). Universal moral grammar: A critical appraisal. *Trends in Cognitive Sciences, 11*(9), 373–378.

Enright, E. A., Gweon, H., Sommerville, J. A. (2017). "To the victor go the spoils": Infants expect resources to align with dominance structures. Cognition, 164, 8–21.

Gazes, R. P., Hampton, R. R., & Lourenco, S. F. (2017). Transitive inference of social dominance by human infants. *Developmental Science, 20*(2), e12367.

Geraci, A., & Surian, L. (2011). The developmental roots of fairness: Infants' reactions to equal and unequal distributions of resources. *Developmental Science, 14*(5), 1012–1020.

Goff, P. A., Jackson, M. C., Di Leone, B. A. L., Culotta, C. M., & DiTomasso, N. A. (2014). The essence of innocence: Consequences of dehumanizing Black children. *Journal of Personality and Social Psychology, 106*(4), 526–545.

Goodwin, G. P. (2015). Moral character in person perception. *Current Directions in Psychological Science, 24*(1), 38–44.

Graham, J., Haidt, J., Koleva, S., Motyl, M., Iyer, R., Wojcik, S. P., & Ditto, P. H. (2013). Moral foundations theory: The pragmatic validity of moral pluralism. *Advances in Experimental Social Psychology, 47,* 55–130.

Hamlin, J. K. (2013a). Failed attempts to help and harm: Intention versus outcome n preverbal infants' social evaluations. *Cognition, 18,* 451–474.

Hamlin, J. K. (2013b). Moral judgment and action in preverbal infants and toddlers evidence for an innate moral core. *Current Directions in Psychological Science, 22,* 186–193.

Hamlin, J. K. (2014). Context-dependent social evaluation in 4.5-month-old human infants: The role of domain-general versus domain-specific processes in the development of social evaluation. *Frontiers in Psychology, 5,* 614.

Hamlin, J. K. (2015). The case for social evaluation in preverbal infants: Gazing toward one's goal drives infants' preferences for Helpers over Hinderers in the hill paradigm. *Frontiers in Psychology, 5,* 1563.

Hamlin, J. K., & Wynn, K. (2011). Young infants prefer prosocial to antisocial others. *Cognitive Development, 26,* 30–39.

Hamlin, J. K., Wynn, K., & Bloom, P. (2007). Social evaluation by preverbal infants. *Nature, 450,* 557–559.

Hamlin, J. K., Wynn, K., & Bloom, P. (2010). Three-month-olds show a negativity bias in their social evaluations. *Developmental Science, 13,* 923–929.

Hamlin, J. K., Wynn, K., Bloom, P., & Mahajan, N. (2011). How infants and toddlers react to antisocial others. *Proceedings of the National Academy of Sciences of the United States of America, 108,* 19931–19936.

Haslam, N., & Loughnan, S. (2014). Dehumanization and infrahumanization. *Annual Review of Psychology, 65,* 399–423.

Heiphetz, L., Strohminger, N., & Young, L. L. (2017). The role of moral beliefs, memories, and preferences in representations of identity. *Cognitive Science, 41*(3), 744–767.

Howard, L. H., Henderson, A. M., Carrazza, C., & Woodward, A. L. (2015). Infants' and young children's imitation of linguistic in-group and out-group informants. *Child Development, 86*(1), 259–275.

Jin, K., & Baillargeon, R. (2017). Infants possess an abstract expectation of ingroup support. *Proceedings of the National Academy of Sciences of the United States of America, 114,* 8199–8204.

Jin, K., & Baillargeon, R. (2019). Infants' expectations about harm avoidance within and between moral circles. Manuscript in preparation.

Jin, K., Bian, L., & Baillargeon, R. (2017, April). *What behaviors lead toddlers to infer an ingroup relation? The case of sharing.* Paper presented at the Biennial Meeting of the Society for Research in Child Development, Austin, TX.

Jin, K. S., Houston, J. L., Baillargeon, R., Groh, A. M., & Roisman, G. I. (2018). Young infants expect an unfamiliar adult to comfort a crying baby: Evidence from a standard violation-of-expectation task and a novel infant-triggered-video task. *Cognitive Psychology, 102,* 1–20.

Kanakogi, Y., Inoue, Y., Matsuda, G., Butler, D., Hiraki, K., & Myowa-Yamakoshi, M. (2017). Preverbal infants affirm third-party interventions that protect victims from aggressors. *Nature Human Behaviour, 1*(2), 0037.

Khamitov, M., Rotman, J. D., & Piazza, J. (2016). Perceiving the agency of harmful agents: A test of dehumanization versus moral typecasting accounts. *Cognition, 146,* 33–47.

Kinzler, K. D., Dupoux, E., & Spelke, E. S. (2007). The native language of social cognition. *Proceedings of the National Academy of Sciences of the United States of America, 104,* 12577–12580.

Kinzler, K. D., Dupoux, E., & Spelke, E. S. (2012). 'Native' objects and collaborators: Infants' object choices and acts of giving reflect favor for native over foreign speakers. *Journal of Cognition and Development, 13*(1), 67–81.

Koenig, M. A., & Woodward, A. L. (2010). Sensitivity of 24-month-olds to the prior inaccuracy of the source: possible mechanisms. *Developmental Psychology, 46*(4), 815.

Kteily, N., Bruneau, E., Waytz, A., & Cotterill, S. (2015). The ascent of man: Theoretical and empirical evidence for blatant dehumanization. *Journal of Personality and Social Psychology, 109*(5), 901–931.

Lee, K. J. J., Esposito, G., & Setoh, P. (2018). Preschoolers favor their ingroup when resources are limited. *Frontiers in Psychology, 9,* 1752.

Li, Y., Carey, S., & Kominsky, J. (2019, September). Infants' inferences about insides. Paper presented at the Biennial Meeting of the Cognitive Development Society, Louisville, KY.

Liberman, Z., Kinzler, K. D., & Woodward, A. L. (2018). The early social significance of shared ritual actions. *Cognition, 171,* 42–51.

Liberman, Z., Woodward, A. L., Sullivan, K. R., Kinzler, K. D. (2016). Early emerging system for reasoning about the social nature of food. *Proceedings of the National Academy of Sciences of the United States of America, 113,* 9480–9485.

Lin, Y., Stavans, M., & Baillargeon, R. (in press). Infants' physical reasoning and the cognitive architecture that supports it. In O. Houdé & G. Borst (Eds.), *Cambridge handbook of cognitive development.* Cambridge, England: Cambridge University Press.

Lucca, K., Pospisil, J., & Sommerville, J. A. (2018). Fairness informs social decision making in infancy. *PloS ONE, 13*(2), e0192848.

Margoni, F., Baillargeon, R., & Surian, L. (2018). Infants distinguish between leaders and bullies. *Proceedings of the National Academy of Sciences of the United States of America, 115*(38), E8835–E8843.

Margoni, F., & Surian, L. (2018). Infants' evaluation of prosocial and antisocial agents: A meta-analysis. *Developmental Psychology, 54*(8), 1445–1455.

Martinez, A. G., Piff, P. K., Mendoza-Denton, R., & Hinshaw, S. P. (2011). The power of a label: Mental illness diagnoses, ascribed humanity, and social rejection. *Journal of Social and Clinical Psychology, 30*(1), 1–23.

Mascaro, O., & Csibra, G. (2012). Representation of stable social dominance relations by human infants. *Proceedings of the National Academy of Sciences of the United States of America, 109*(18), 6862–6867.

McManus, R., Kleiman-Weiner, M., & Young, L. (in press). What we owe to family: The impact of social obligations on moral judgment. *Psychological Science.*

Meristo, M., Strid, K., & Surian, L. (2016). Preverbal infants' ability to encode the outcome of distributive actions. *Infancy, 21*(3), 353–372.

Meristo, M., & Surian, L. (2013). Do infants detect indirect reciprocity? *Cognition, 129*(1), 102–113.

Olson, K. R., & Spelke, E. S. (2008). Foundations of cooperation in young children. *Cognition, 108*(1), 222–231.

Poulin-Dubois, D., Brooker, I., & Polonia, A. (2011). Infants prefer to imitate a reliable person. *Infant Behavior and Development, 34*(2), 303–309.

Powell, L. J., & Spelke, E. S. (2013). Preverbal infants expect members of social groups to act alike. *Proceedings of the National Academy of Sciences of the United States of America, 110,* 3965–3972.

Premack, D., & Premack, A. J. (1997). Infants attribute value± to the goal-directed actions of self-propelled objects. *Journal of Cognitive Neuroscience, 9*(6), 848–856.

Pun, A., Birch, S. A., & Baron, A. S. (2016). Infants use relative numerical group size to infer social dominance. *Proceedings of the National Academy of Sciences of the United States of America, 113*(9), 2376–2381.

Rai, T. S., & Fiske, A. P. (2011). Moral psychology is relationship regulation: Moral motives for unity, hierarchy, equality, and proportionality. *Psychological Review, 118,* 57–75.

Renno, M. P., & Shutts, K. (2015). Children's social category-based giving and its correlates: Expectations and preferences. *Developmental Psychology, 51*(4), 533.

Rhodes, M. (2012). Naïve theories of social groups. *Child Development, 83*(6), 1900–1916.

Rhodes, M., Hetherington, C., Brink, K., & Wellman, H. M. (2015). Infants' use of social partnerships to predict behavior. *Developmental Science, 18*(6), 909–916.

Riskey, D. R., & Birnbaum, M. H. (1974). Compensatory effects in moral judgment: Two rights don't make up for a wrong. *Journal of Experimental Psychology, 103,* 171–173.

Rozin, P., & Royzman, E. B. (2001). Negativity bias, negativity dominance, and contagion. *Personality and Social Psychology Review, 5*(4), 296–320.

Schmidt, M. F., & Sommerville, J. A. (2011). Fairness expectations and altruistic sharing in 15-month-old human infants. *PloS ONE, 6*(10), e23223.

Setoh, P., Wu, D., Baillargeon, R., & Gelman, R. (2013). Young infants have biological expectations about animals. *Proceedings of the National Academy of Sciences of the United States of America, 110*(40), 15937–15942.

Shutts, K., Kinzler, K. D., McKee, C. B., & Spelke, E. S. (2009). Social information guides infants' selection of foods. *Journal of Cognition and Development, 10*, 1–17.

Shweder, R. A., Much, N. C., Mahapatra, M. & Park, L. (1997). The "big three" of morality (autonomy, community, divinity) and the "big three" explanations of suffering. In A. M. Brandt & P. Rozin (Eds.), *Morality and health* (pp. 119–169). New York: Routledge.

Skowronski, J. J., & Carlston, D. E. (1992). Caught in the act: When impressions based on highly diagnostic behaviours are resistant to contradiction. *European Journal of Social Psychology, 22*(5), 435–452.

Sloane, S., Baillargeon, R., & Premack, D. (2012). Do infants have a sense of fairness? *Psychological Science, 23*, 196–204.

Smith, D.L. (2011). *Less than human: Why we demean, enslave, and exterminate others.* New York: St. Martin's Press.

Spokes, A. C., & Spelke, E. S. (2017). The cradle of social knowledge: Infants' reasoning about caregiving and affiliation. *Cognition, 159*, 102–116.

Stavans, M., & Baillargeon, R. (2019). Infants expect leaders to right wrongs. *Proceedings of the National Academy of Sciences of the United States of America, 116*(33), 16292–16301.

Strohminger, N., & Nichols, S. (2014). The essential moral self. *Cognition, 131*, 159–171.

Surian, L., & Franchin, L. (2017). Toddlers selectively help fair agents. *Frontiers in Psychology, 8*, 944.

Surian, L., Ueno, M., Itakura, S., & Meristo, M. (2018). Do infants attribute moral traits? Fourteen-month-olds' expectations of fairness are affected by agents' antisocial actions. *Frontiers in Psychology, 9*, 1649.

Thomas, A. J., Thomsen, L., Lukowski, A. F., Abramyan, M., & Sarnecka, B. W. (2018). Toddlers prefer those who win but not when they win by force. *Nature Human Behaviour, 2*(9), 662–669.

Thomsen, L., Frankenhuis, W., Ingold-Smith, M., & Carey, S. (2011). Big and mighty: Preverbal infants mentally represent social dominance. *Science, 331*, 477–480.

Ting, F., & Baillargeon, R. (2018a, July). *2-year-old toddlers make broad inferences about moral characters*. Paper presented at the Biennial International Congress on Infant Studies, Philadelphia, PA.

Ting, F., & Baillargeon, R. (2018b, July). *Do infants understand moral obligations?* Paper presented at the Biennial International Congress on Infant Studies, Philadelphia, PA.

Ting, F., & Baillargeon, R. (2018c, July). *14-month-olds use animacy cues to determine who has moral rights*. Paper presented at the Annual Meeting of the Society for Philosophy and Psychology, Ann Arbor, MI.

Ting, F., He, Z., & Baillargeon, R. (2019a). Toddlers and infants expect individuals to refrain from helping an ingroup victim's aggressor. *Proceedings of the National Academy of Sciences of the United States of America, 116*(13), 6025–6034.

Ting, F., He, Z., & Baillargeon, R. (2019b, March). *Group membership modulates early expectations about retaliatory harm*. Paper presented at the Biennial Meeting of the Society for Research in Child Development, Baltimore, MD.

Ting, F., Setoh, P., Gelman, R., & Baillargeon, R. (2019, September). *Young infants expect an animate's insides to drive its function*. Paper presented at the Biennial Meeting of the Cognitive Development Society, Louisville, KY.

Tooby, J., and Cosmides, L. (2010). Groups in mind: The coalitional roots of war and morality. In H. Høgh-Olesen (Ed.), *Human morality and sociality: Evolutionary and comparative perspectives* (pp. 191–234). New York: Palgrave MacMillan.

Uhlmann, E. L., Pizarro, D. A., & Diermeier, D. (2015). A person-centered approach to moral judgment. *Perspectives on Psychological Science, 10*(1), 72–81.

Wang, Y., & Henderson, A. M. (2018). Just rewards: 17-month-old infants expect agents to take resources according to the principles of distributive justice. *Journal of Experimental Child Psychology, 172*, 25–40.

Warneken, F., & Tomasello, M. (2006). Altruistic helping in human infants and young chimpanzees. *Science, 311*(5765), 1301–1303.

Waytz, A., Gray, K., Epley, N., & Wegner, D. M. (2010). Causes and consequences of mind perception. *Trends in Cognitive Sciences, 14*(8), 383–388.

Weisman, K., Dweck, C. S., & Markman, E. M. (2017). Rethinking people's conceptions of mental life. *Proceedings of the National Academy of Sciences of the United States of America, 114*, 11374–11379.

Ziv, T., & Sommerville, J. A. (2017). Developmental differences in infants' fairness expectations from 6 to 15 months of age. *Child Development, 88*(6), 1930–1951.

Zmyj, N., Buttelmann, D., Carpenter, M., & Daum, M. M. (2010). The reliability of a model influences 14-month-olds' imitation. *Journal of Experimental Child Psychology, 106*(4), 208–220.

4   Early Social Cognition: Exploring the Role of the Medial Prefrontal Cortex

Tobias Grossmann

Overview

Among the brain areas implicated in the adult social brain, processes in the prefrontal cortex (PFC), particularly medial prefrontal cortex (mPFC), play an important role for social cognition. However, it has long been thought that the PFC is functionally silent during infancy, and until recently little research has been dedicated to examining the role of mPFC in early social cognition. In this chapter I discuss an emerging body of neuroimaging studies with infants, providing evidence that mPFC assumes a functional role in social information processing from early in human ontogeny, and I argue that the mPFC involvement in infancy reflects self-relevance detection and might play an important role in learning during social encounters. This look at early social cognition through the lenses of neuroscience closes a gap between work with adults, identifying the mPFC as a key region involved in social cognition, and behavioral work attesting sophisticated social-cognitive abilities in infants.

Background

In the course of evolution, humans have developed numerous higher cognitive skills such as language, reasoning, planning, and complex social behavior. The prefrontal cortex (PFC) can be seen as the neural substrate that underpins much of this higher cognition (Wood & Grafman, 2003). The PFC refers to the regions of the cerebral cortex that are anterior to premotor cortex and the supplementary motor area (Zelazo & Müller, 2002). In humans the PFC makes up approximately a quarter to a third of the cortex (Fuster, 2008). Although the PFC in humans may not be disproportionately

enlarged with respect to other brain regions when compared to our clos-
est living relatives the great apes (Semendeferi, Lu, Schenker, & Damasio,
2002), there have been suggestions that certain parts of the PFC possess a
number of human-specific structural and functional properties that may
underpin human-unique social and cognitive abilities (see Saxe, 2006).

Based on its neuroanatomical connections, the PFC can be broadly
divided into two sections: (a) the medial PFC (mPFC) and (b) the lateral
PFC (lPFC) (Fuster, 2008; Wood & Grafman, 2003). The mPFC includes the
medial portions of Brodmann areas (BA) 9–12, and BA 25, and has reciprocal
connections with brain regions that are implicated in emotional processing (amygdala), memory (hippocampus), and higher-order sensory regions
(within temporal cortex) (for more detailed information, see Fuster, 2008;
Wood & Grafman, 2003). The lPFC includes the lateral portions of BA 9–12,
BA 44 and 45, and BA 46, and has reciprocal connections with brain regions
that are implicated in motor control (basal ganglia, premotor cortex, supplementary motor area), performance monitoring (cingulate cortex), and
higher-order sensory processing (within temporal and parietal cortex) (for
more detailed information, see Fuster, 2008; Wood & Grafman, 2003). Furthermore, the lPFC and mPFC are reciprocally connected, allowing for information exchange and integration across these two broad sections of the PFC.

Critically, the distinction between the lPFC and mPFC in neuroanatomical terms maps onto general differences in brain function. Namely, while
the mPFC is thought to be mainly involved in processing, representing, and
integrating social and affective information, the lPFC is thought to support
cognitive control processes (Fuster, 2008; Wood & Grafman, 2003). That it
is important to consider this distinction when thinking about PFC function
is also reflected in evolutionary models, according to which the lPFC developed much later than the mPFC; the lPFC is considered to have evolved
from the motor regions of the brain (Fuster, 2008). Despite this distinction, it should be noted that the mPFC and lPFC are parts of a coordinated
system, and they normally work together in the service of guiding human
behavior and decision-making.

From a developmental perspective, it has long been thought that the
PFC, as the seat of most higher brain functions, is functionally silent during most of infancy (for a review, see Zelazo & Müller, 2002). Contrary to
this notion, much of the work concerned with the precise mapping (localization) of brain activation in human infants has provided compelling

evidence that the PFC exhibits functional activation much earlier than previously thought. In fact, a systematic review of the activation patterns across neuroimaging studies mainly based on functional near-infrared spectroscopy (fNIRS) reveals that PFC function can be broadly divided into two distinct anatomical clusters with different functional properties (Grossmann, 2013a). One cluster of activations falls within the region of the mPFC and is mainly involved in social and affective processes; another cluster is located in the lPFC and shows sensitivity to cognitive processes such as memory and attention. This functional distinction between the mPFC and lPFC observed in human infant brains is in line with adult data and evolutionary models and may thus represent a developmentally continuous organization principle of PFC function.

The mPFC has been shown to play a fundamental role in a wide range of social cognitive abilities such as self-reflection, person perception, social engagement, and theory of mind in adults (Amodio & Frith, 2006). This involvement of the mPFC in social cognition and interaction has led to the notion that the mPFC serves as a key region in understanding self and others (Frith & Frith, 2006). Although this is not the focus of this chapter, it should be noted that apart from its recruitment during social cognitive functions in adults, the mPFC has also been shown to be more generally involved in a number of processes related to decision-making in adults (e.g., Heekeren, Marrett, & Ungerleider, 2008). In particular, a unifying model has been proposed that views the mPFC as a region concerned with learning and predicting the likely outcomes of actions (Alexander & Brown, 2011).

Only little is known about the role of the mPFC in the development of social cognition and interaction. This is particularly true for infancy, specifically the earliest steps of postnatal development (the first year of life). Addressing the question of whether the mPFC plays a role in infant social cognition and, if it does, to theorize about what role this might be is the goal of this chapter. Such a look at early social cognition and interaction during infancy through the lens of social neuroscience is critical because it allows us (a) to understand the nature and developmental origins of mPFC function and (b) to close a gap between the extensive behavioral work showing rather sophisticated infant social cognitive skills (Baillargeon, Scott, & He, 2010; Spelke & Kinzler, 2007; Woodward, 2009) and the social neuroscience work with adults examining mature mPFC functions (Amodio & Frith, 2006; Lieberman, 2006).

Review of Studies Implicating the mPFC in Social Information Processing during Infancy

Lesion Research

That the mPFC plays an important role in the development of social cognition is evident in the work that has examined mPFC lesions. For example, a study comparing early onset (during infancy) and adult onset lesions to the mPFC showed that, despite typical basic cognitive abilities, patients with mPFC lesions had severely impaired social behavior (Anderson, Bechara, Damasio, Tranel, & Damasio, 1999). More specifically, regardless of when the mPFC lesion had occurred, symptoms were shared across patients with mPFC damage, including an insensitivity to future consequences of actions, defective autonomic responses to punishment contingencies, and failure to respond to interventions that would change behavior (Anderson et al., 1999). Critically, this study revealed that over and above the shared symptomatology, acquired damage to the mPFC during infancy had a much more severe impact on social functioning, as signified by striking defects in social and moral reasoning leading to a syndrome that closely resembled psychopathy.

In this study, it was found that early onset damage to the mPFC was related to antisocial behaviors such as stealing, violence against persons and property, and severe impairment of social-moral reasoning and in verbal generation of responses to social situations. Specifically, in adults with early onset lesions to the mPFC, moral reasoning was conducted at a much lower level than expected by their age, such that moral dilemmas were mainly approached from an egocentric perspective characterized by avoiding punishment. Furthermore, early onset damage of the mPFC was related to a limited consideration of the emotional implications of one's own behavior toward others and far fewer responses generated to resolve interpersonal conflict. This suggests that the mPFC plays a critical role in the acquisition of social and moral behaviors early during ontogeny. It also suggests that, in contrast with many other brain regions where damage and especially damage early in ontogeny can be compensated (Thomas & Johnson, 2008), the mPFC is less plastic and more vulnerable. This in turn indicates that there might be a sensitive period in development during which the mPFC is required so that individuals develop and learn socially and morally appropriate behaviors.

Even though the study of patients with lesions to the mPFC is of great importance in illuminating mPFC function, patients with circumscribed mPFC lesions acquired during infancy, as reported by Anderson and colleagues (1999), are extremely rare and hence can provide only limited insight into these early stages of developing mPFC function. It is therefore crucial to employ functional neuroimaging to shed light on the development of mPFC function during infancy if we strive to better characterize its role in early social cognition.

Neuroimaging Research

Recent advances in applying functional imaging technology to infants—specifically, the advent of using functional near-infrared spectroscopy (fNIRS)—has made it possible to study the infant brain at work. This optical imaging method measures hemodynamic responses from cortical regions, permitting the localization of brain activation (Lloyd-Fox, Blasi, & Elwell, 2010).

Other neuroimaging techniques that are well established in adults are limited in their use with infants because of methodological concerns. For example, functional magnetic resonance imaging (fMRI) requires the participant to remain very still and exposes them to a noisy environment. Although fMRI has been used with infants (Dehaene-Lambertz, Dehaene, & Hertz-Pannier, 2002; Dehaene-Lambertz et al., 2006), most of the work is restricted to the study of auditory processes in sleeping, sedated, or very young infants (but see Deen et al., 2017, for recent fMRI research on face processing in awake infants).

The method of fNIRS seems better suited for infant research because it can accommodate a good degree of movement from the infants, enabling them to sit upright on their parent's lap and behave relatively freely while watching or listening to certain stimuli. In addition, unlike fMRI, fNIRS systems are portable. Finally, despite its inferior spatial resolution also in terms of obtaining responses from deeper (subcortical) brain structures, fNIRS (like fMRI) measures localized patterns of hemodynamic responses in the cortical regions, thus allowing for a comparison of infant fNIRS data with adult fMRI data.

In the last decade, there has been a surge of fNIRS studies with infants, including a number of studies that have looked at PFC activation during a wide range of experimental tasks (for a review, see Grossmann, 2013a). In the following sections, I shall review the available experimental evidence that implicates the mPFC in infant social cognition and provide an overview of the range of social contexts during which infants employ the mPFC. The review of the empirical work is organized according to the two main sensory modalities—visual (face) and auditory (voice)—in which social stimuli were presented to infants. After presenting the experimental evidence, I will discuss a number of issues that arise from these studies. Finally, based on these findings, I will outline an account of what role mPFC may play in the early development of social cognition during infancy.

Before I begin with the review of the infant neuroimaging research, it needs to be emphasized that any account of the emergence of social information processing needs to acknowledge and incorporate the fact that, contrary to what was often assumed in the past (James, 1890/2007; Piaget, 1952), human infants enter the world tuned to their social environment and readily prepared for social interaction. At birth, infants exhibit a number of biases that preferentially orient them to socially relevant stimuli. In particular, it has been shown that newborns prefer faces over other kinds of visual stimuli (Johnson, Dziurawiec, Ellis, & Morton, 1991; Johnson & Morton, 1991), voices over other kinds of auditory stimuli (DeCasper & Fifer, 1980; Vouloumanos, Hauser, Werker, & Martin, 2010), and biological motion over other kinds of motion (Simion, Regolin, & Bulf, 2008). While these biases may provide an important foundation for the emergence of social cognitive abilities, biases observed in newborns are thought to be only broadly tuned to social stimuli and are assumed to mainly be mediated by subcortical mechanisms (Johnson, 2005a, 2005b). It is thus critical to understand how the cortical systems develop, allowing the human infant to make sense of the social world. These are the processes that this chapter is most concerned with and aims to conceptualize.

Voices.   The human voice, apart from having obvious functions in linguistic communication, also carries a wealth of socially relevant information such as age, gender, and emotional state (Belin, Fecteau, & Bedard, 2004). Newborns have been shown to display significantly increased responses in the mPFC to their own mother's voice reading a story in infant-directed speech (IDS) compared to their mothers reading the same story in adult-directed speech (ADS) (Saito, Aoyama, et al., 2007). This indicates that newborn infants discriminate between these two forms of speech and dedicate increased mPFC processing resources to IDS, which is of high social and emotional relevance to the infant. In another study, Saito, Kondo, and colleagues (2007) also showed that mPFC activation can be obtained in response to non-maternal emotional speech. This finding suggests that it is the emotional tone of voice that characterizes the positive affect in speech that drives this effect on the mPFC in newborns.

Older infants (four to thirteen months of age) were presented with IDS and ADS sentences spoken by their own mother or a female stranger, and prefrontal and temporal cortex responses were measured using fNIRS (Naoi et al., 2012). This study showed that while the infants' temporal cortex discriminated between IDS and ADS regardless of speaker, the prefrontal cortex (including the mPFC in the left hemisphere) was engaged only when the mother spoke with IDS. Together with the data from newborns already presented, this suggests that mPFC responses undergo change during infancy and become more finely tuned to the primary caregiver's voice. Indeed, in agreement with behavioral work showing that by seven to nine months of age infants have the strongest preference for their primary caregivers, prefrontal responses change during infancy such that at seven to nine months of age an infant's mPFC is most sensitive to the mother's IDS.

Faces.   Another important area of investigation is the work on the perception of visual social stimuli. The human face provides an infant with a wealth of socially and affectively relevant information such as age, gender, and attentional and emotional state. From birth, human infants preferentially attend to faces (Johnson & Morton, 1991). For example, Tzourio-Mazoyer and colleagues (2002) presented a small sample of six two-month-old infants with a face or a control stimulus, while measuring brain activity using positron emission tomography (PET). (Note that PET is not commonly used with infants because it exposes them to small amounts of radiation. The infants scanned in this study were tested in an intensive care unit as part of a clinical follow-up evaluation for a neonatal syndrome known as hypoxic-ischemic encephalopathy.) In this study, when they viewed faces compared to nonsocial visual control stimuli (diodes), infants not only activated regions in the temporal cortex but also showed significant activation within the mPFC in the right hemisphere.

More recently, the finding that the infant mPFC is preferentially engaged when processing facial stimuli has been replicated and extended using fMRI by Deen and colleagues (2017). In this study, four- to six-month-old infants showed clear activation of the mPFC when viewing dynamic faces as compared with nonsocial control stimuli (dynamic movies of natural environments). This suggests that already at this young age infants recruit parts of the so-called extended face processing network that are considered to be crucial in assigning social and affective significance to faces (Haxby, Hoffman, & Gobbini, 2000).

In fact, on the basis of the research by Deen and colleagues (2017), it has even been argued that activity in the mPFC during face processing/social interaction may, directly or indirectly, shape and guide the specialization of face regions in the temporal cortex. This proposal is based on the idea that infants look at faces to engage in positive and reciprocal social interaction, and that this is reflected in mPFC involvement. This activity in the mPFC may influence or even potentiate responses to faces in the temporal cortex (Powell, Kosakowski, & Saxe, 2018).

An important communicative signal conveyed by faces is eye gaze. The monitoring of eye gaze direction is essential for effective social learning and communication among humans (Csibra & Gergely, 2009), with eye contact being one of the most powerful modes of establishing a communicative link between humans (Kampe, Frith, & Frith, 2003). In an fNIRS study, four-month-old infants watched two kinds of dynamic scenarios in which a face either established eye contact or averted its gaze followed by a smile (Grossmann et al., 2008). The results revealed that, similar to what is known from adults (Kampe et al., 2003; Pelphrey, Viola, & McCarthy, 2004), processing eye contact activates not only the superior temporal cortex, implicated in processing information from biological motion cues, but also the mPFC, important for social and affective communication. Moreover, in the same study, measuring electrical brain responses over the prefrontal cortex in another group of four-month-old infants showed that only a smile that was preceded by eye contact evoked increased prefrontal cortex responses in four-month-old infants (Grossmann et al., 2008), supporting the notion that already in infancy the mPFC plays a role in interpreting social and affective information directed at the self. The finding that infants' mPFC is involved during eye contact (mutual gaze) and positive engagement has been replicated in a live fNIRS paradigm in which the infants directly interacted with an experimenter rather than a virtual agent on a screen in a peek-a-boo game (Urakawa, Takamoto, Ishikawa, Ono, & Nishijo, 2015).

That smiling at an infant while making eye contact is a powerful cue for triggering mPFC activation has also been demonstrated in another fNIRS study (Minagawa-Kawai et al., 2009), in which nine- to twelve-month-old infants were presented with videos of either their own mother or a female stranger smiling at them or looking neutrally at them. Smiling at the infants evoked greater activity in the mPFC regardless of the familiarity with the face, suggesting that the mPFC is flexibly employed during positive social interactions. Nevertheless, mPFC activity was significantly greater in response to the infant's own mother smiling when compared to the female stranger smiling. This suggests that an infant's mPFC responses are particularly sensitive to affective cues from the primary caregiver. Interestingly, in this study it was shown that mothers exhibited a very similar mPFC response when looking at their own infant smiling, thus pointing to a shared neural mechanism engaged during social interaction between caregivers and infants.

Eye gaze also plays an important role in coordinating attention during triadic interactions between self, other, and the environment. During a typical triadic interaction, a person may establish eye contact with another person and then direct that person's gaze to an object or event. Functional NIRS has been used to localize infant prefrontal brain responses during triadic social interactions with a virtual agent presented on a screen (Grossmann & Johnson, 2010). Specifically, Grossmann and Johnson (2010) used fNIRS to examine brain responses in the PFC of five-month-old infants during triadic social interactions. To investigate whether young infants engage specialized prefrontal brain processes while engaged in joint attention, infants were presented with scenarios in which a social partner (a virtual agent presented on a screen) (a) engaged in joint attention by gaze cueing the infant's attention to an object after establishing eye contact (joint attention condition); (b) gaze cued the infant's attention to an empty location (no referent condition); or (c) looked at an object without prior eye contact with the infant (no eye contact condition). Only in the joint attention condition did infants recruit a specific brain region within the mPFC, demonstrating that five-month-old infants are sensitive to triadic interactions. Moreover, this study showed that five-month-old infants recruited a similar region within the mPFC as adults when engaged in triadic social interaction (Schilbach et al., 2010), suggesting that young infants engage adult-like brain processes during these kinds of interactions.

These studies provided first evidence that the mPFC is implicated in joint engagement in infancy, but an important outstanding question was whether infants are also sensitive to when a social partner follows their gaze. This is a particularly critical question because it can inform theories that posit that brain processes are shared and flexibly engaged by self- and other-initiated actions and interactions (Meltzoff, 2007; Schilbach et al., 2013). Recently, Schilbach and colleagues (2010) showed that in adults there are key brain regions, such as the left medial dorsal PFC, involved in both responding to joint attention and to initiating joint attention. This suggests that adults flexibly engage specific brain processes that are shared between self- and other-initiated gaze interactions.

Grossmann, Lloyd-Fox, and Johnson (2013) examined five-month-olds' sensitivity to when a social partner follows their gaze by measuring infant brain responses using fNIRS during scenarios in which a social partner either followed the infants' gaze to an object that they had previously looked at (congruent condition) or a social partner shifted attention to look at a different object (incongruent condition). The fNIRS data revealed that a region in the left mPFC showed an increased response when compared to baseline during the congruent condition but not during the incongruent condition, suggesting that infants are sensitive to when someone follows their gaze.

From a developmental perspective, this finding is in line with theories emphasizing the importance of the early emergence of social cognitive abilities required to engage in joint action (Csibra & Gergely, 2009; Tomasello, Carpenter, Call, Behne, & Moll, 2005) and supports theories positing a link between brain processes implicated in actions performed by self and by others (Meltzoff, 2007). From a neuroscience perspective, this finding further strengthens accounts that—in contrast to the commonly held notion of a late maturation of PFC function—assign a pivotal functional role to the PFC in infant cognition in general (Grossmann, 2013a) and the mPFC in infant social cognition in particular (Grossmann, 2013b).

The finding that specific parts of the mPFC play a role in triadic interactions receives more support from work examining the perception of human action (Lloyd-Fox, Blasi, Everdell, Elwell, & Johnson, 2011). In this study, when five-month-olds were presented with actions (hand movements) while being addressed through eye contact and thereby creating a triadic interaction, they showed increased activation within the mPFC. The same regions of the mPFC were not active when infants were observing human actions that were purely dyadic in nature such as mouth movement or eye gaze shifts.

Face and Voice

In adults, initiating a social interaction by eye contact and calling a person's name results in overlapping activity in the mPFC (Kampe, Frith, & Frith,

2003), suggesting that, regardless of modality, the intention to make contact is detected by the same brain region. In a fNIRS study to examine the neural basis of detecting social interactive signals across modalities, five-month-old infants watched faces that either signaled eye contact or directed their gaze away from the infant, and they also listened to voices that addressed them with their own name or another name (Grossmann, Parise, & Friederici, 2010). The results of this study revealed that infants recruit adjacent regions in the mPFC when they process eye contact and their own name. Moreover, five-month-old infants who responded sensitively to eye contact in the one mPFC region were also more likely to respond sensitively to their own name in the adjacent mPFC region as revealed in a correlation analysis, suggesting that responding to communicative signals in these two regions is functionally related. These fNIRS results suggest that infants at the age of five months selectively process and flexibly attend to social interactive signals across modalities.

Integration and Discussion

This review presented an overview of the current experimental evidence on infants' mPFC involvement during the processing of vocal and facial social information. We have seen that infants employ the mPFC during a range of contexts, including the perception of emotional and IDS cues in the auditory domain and the perception of faces and eye gaze cues in the visual domain. These findings support the central thesis that the mPFC is important for social information processing from early in ontogeny and likely plays a vital role in the development of social cognitive abilities during infancy. This notion stands in contrast to the prevailing idea, primarily based on research with adolescents, that the mPFC is a brain region that only matures later in ontogeny when a more explicit understanding of the social world is achieved (Blakemore, 2008; Singer, 2006).

On the basis of the evidence presented here, I could even argue that the mPFC is more important earlier in development than later in development because it is critically involved in the acquisition of social cognitive abilities from birth and becomes somewhat less important once social cognitive and interactive abilities have been robustly acquired. That this might be the case is evident in the mPFC lesion work presented earlier where it was shown that early onset compared to adult onset lesions in the mPFC resulted in more severe outcomes in terms of social and moral impairments (Anderson et al., 1999). More support for this view of the mPFC playing a greater role earlier in development comes from neuroimaging work on social cognition with adolescents, which has shown that the engagement of posterior regions of cortex increases with age, but mPFC involvement in social cognitive tasks decreases with age during adolescence (for a review, see Johnson, Grossmann, & Cohen Kadosh, 2009). This can be seen as evidence for a reduction of the involvement of the mPFC in social cognition during development, which concurs with another line of work demonstrating that prefrontal regions play a greater role during the acquisition of a new skill as shown in perceptual learning tasks (Gilbert & Sigman, 2007).

Therefore, one implication of the work presented here is that the mPFC plays a role in the acquisition of social cognitive skills from early in ontogeny. In general, this notion is in line with views that conceive of infants as competent and actively engaged social agents and learners, who enter the world readily prepared for social interaction and social cognition (Csibra & Gergely, 2009; Meltzoff, 2007; Spelke & Kinzler, 2007).

But what is the functional role that the mPFC takes in the early development of social cognition during infancy? I would like to propose that mPFC involvement in infancy (and beyond) is likely important for the detection of self-relevant information. This proposal is based on (a) the observed pattern of mPFC involvement in the studies we have just reviewed, and (b) an extensive body of evidence from prior work with adults that implicates the mPFC in the assessment and representation of information with reference to the self (for a review, see Amodio & Frith, 2006). This proposal can thus be seen as a developmental extension of prior accounts of adult mPFC function into infancy.

More specifically, as shown earlier, the mPFC is involved in infants' responding to social interactive cues, which index information that is relevant to the self—such as when listening to IDS or their own name, perceiving eye contact, or experiencing a triadic interaction. This increased sensitivity to self-relevant information might serve a critical learning function because it highlights potentially useful information that others present to the infant (Csibra & Gergely, 2009; Sperber & Wilson, 1995). In support of this view, it has been shown that infants' learning is influenced and improved when they are addressed by IDS and eye contact (Senju & Csibra, 2008; Singh, Morgan, & White, 2004; Yoon, Johnson, & Csibra, 2008). The mPFC might thus be involved in learning from others by detecting the relevance of others' actions with reference to the self.

Obviously, this sensitivity to self-relevant information in infancy does not imply that infants have an explicit (conceptual) understanding of the self (Rochat, 2003, 2011). However, one may argue that the sensitivity to self-relevant information serves as a powerful foundation for developing a sense of self because it provides infants with the opportunity to experience when the self is being addressed in an interaction. In fact, it has been argued that early social interactions during infancy and the experiences gained therein can be considered the cradle of self-development (Reddy, 2003).

One intriguing implication of this proposal is that by measuring mPFC involvement in a given context, one might be able to examine the extent to which an infant perceives information as self-relevant. For example, on a trial-by-trial basis one could look at infants' mPFC response to eye contact and then see whether they are more likely to show gaze-following in response to an eye gaze shift of a social partner. The prediction based on the proposal I presented earlier is that in trials when infants show mPFC involvement on seeing eye contact, they should be more likely to gaze follow. In behavioral work, it has already been shown that infants are more likely to gaze follow when they had previously been presented with eye contact or heard IDS (Senju & Csibra, 2008), although it is unclear what underlying neural processes correlate with this behavioral phenomenon.

Moreover, this proposed approach might also be useful in assessing interindividual differences in the perception of relevance to the self in response to identical stimuli. In such a scenario, we might be able to identify infants who tend to show little sensitivity to perceptual social signals indicating self-relevance but also infants who are overly sensitive to social information even if it is not directed at them. The potential existence of extreme biases in either direction in early development might have serious detrimental effects on social development in the long term. For example, a strongly reduced sensitivity to self-relevant information might be linked to a neurodevelopmental disorder such as autism, where it has been shown that lacking behavioral sensitivity to self-relevant signals such as eye contact and name cues is one of the earliest detectable warning signs for the later development of autism (Elsabbagh & Johnson, 2007; Elsabbagh et al., 2012; Zwaigenbaum et al., 2005). The development of biomarkers such as brain-based measures to guide an early identification of developmental disorders is still in its infancy but has been shown to be of great promise, especially when relying on measures that assess infants' responses to eye contact (Elsabbagh et al., 2012).

Another potential implication that arises from the proposal that the mPFC processes self-relevant information is that activating the mPFC may result in top-down regulation of activity in functionally connected parts of the brain, especially within the face and voice-sensitive regions in the temporal cortex. In other words, mPFC engagement may be elicited in a bottom-up manner through signals indexing self-relevance such as eye contact or IDS and then, as a result, impact face and voice processing in the posterior parts of the brain in a top-down manner. Critically, behavioral work exists that suggests that top-down modulation instantiated through the mPFC, which has the potential to affect learning and behavior, may occur.

More specifically, in behavioral experiments, facial and vocal cues that activate the mPFC in neuroimaging experiments have been shown to affect infants' learning and behavior (Csibra & Gergely, 2009). For example, Farroni, Massaccesi, Menon, and Johnson (2007) have shown that eye contact, when compared with averted gaze, enables four-month-old infants' face recognition in a face-identity learning paradigm. This behavioral finding is also in line with the recently proposed notion that mPFC involvement during face processing/social interaction may, directly or indirectly, shape and guide the specialization of face regions in temporal cortex (Powell, Kosakowski, & Saxe, 2018). Despite its plausibility and theoretical appeal, to date no direct neuroimaging evidence exists to show that the mPFC plays a functional role in top-down regulation of processes in the posterior cortex in human infants.

Limitations and Suggestions for Future Research

Despite the progress that has been made in elucidating the role of the mPFC in early development, to gain a better and more complete picture of mPFC function in infancy it is vital to address a number of remaining issues. First, more work is needed to precisely map and compare activation within the mPFC across social tasks during infancy. Specifically, as far as the infant fNIRS data presented in this review are concerned, no standardized anatomical mapping of the functional activation in PFC has been employed that would allow us to compare and integrate the information about the mPFC across studies and tasks in a meta-analysis.

This issue becomes particularly important when one considers the fact that in adults there appear to be considerable functional divisions within the mPFC (Amodio & Frith, 2006; Bzdok et al., 2013). The first steps have been taken toward standardizing the analysis of infant fNIRS data, which promises to provide a better basis for carrying out such comparisons (Cristia et al., 2013). Nonetheless, a remaining issue is the limited depth resolution of fNIRS, as commonly used in infant studies, which obtains most of its signal from superficial cortical structures but is virtually blind to deeper cortical sources (Lloyd-Fox et al., 2010).

Second, very little work has been done to compare mPFC activation across ages during infancy. Most work to date has focused on one particular age group. The few studies that have looked at various age groups during infancy revealed intriguing insights into how mPFC function changes and becomes more finely tuned to social signals from the caregiver (Naoi et al., 2012). A systematic examination of mPFC function across infancy will provide important information concerning the functional specialization of this brain region.

Third, another important aspect to consider is that although we have observed activation of individual mPFC regions during infancy, we do not know whether the activity of the mPFC and other brain regions is coordinated into functional networks, as seen in adults. Work using resting-state fMRI with infants has indicated that some of the functional connections between certain parts of the mPFC and posterior cortical regions known in adults are not yet developed in infants (Fransson et al., 2007). Furthermore, resting-state studies testing infants across various ages have shown that this long-range integration of cortical activity emerges throughout the first few years of life (Fransson, Aden, Blennow, & Lagercrantz, 2011; Gao et al., 2009; Homae et al., 2010). The relevance that these changes in resting-state activity have for infants' brain function while they are actively involved in one of the experimental tasks reviewed here is unclear, and this requires attention in future work.

Conclusions

Taken together, the findings from the studies presented here provide evidence that the mPFC plays an important role in social cognition from early in development. Based on the reviewed experimental data, I put forward the proposal that mPFC involvement in social information processing in infancy is related to the detection of self-relevant information and may thereby impact social learning and behavior. This look at early social cognition through the lens of social neuroscience allows us to better understand the nature and developmental origins of mPFC function by closing a gap between the extensive behavioral work showing sophisticated social cognitive skills in infants and work with adults concerning the pertinent role of the mPFC in social cognition. It is my hope that this chapter will help stimulate work to illuminate the neural basis of social cognition in infancy and foster the cross-talk between developmental psychologists and social neuroscientists.

References

Alexander, W. H., & Brown, J. W. (2011). Medial prefrontal cortex as an action-outcome predictor. *Nature Neuroscience, 14,* 1338–1344.

Amodio, D. M., & Frith, C. D. (2006). Meeting of minds: The medial frontal cortex and social cognition. *Nature Reviews Neuroscience, 7*(4), 268–277.

Anderson, S. W., Bechara, A., Damasio, H., Tranel, D., & Damasio, A. R. (1999). Impairment of social and moral behavior related to early damage in human prefrontal cortex. *Nature Neuroscience, 2,* 469–479.

Baillargeon, R., Scott, R. M., & He, Z. (2010). False-belief understanding in infants. *Trends in Cognitive Sciences, 14*(3), 110–118.

Belin, P., Fecteau, S., & Bedard, C. (2004). Thinking the voice: Neural correlates of voice perception. *Trends in Cognitive Sciences, 8,* 129–135.

Blakemore, S. J. (2008). The social brain in adolescence. *Nature Reviews Neuroscience, 9,* 267–277.

Bzdok, D., Langner, R., Schilbach, L., Engemann, D.A., Laird, A.R., Fox, P.T., ... Eickhoff, S.B. (2013). Segregation of the human medial prefrontal cortex in social cognition. *Frontiers in Human Neuroscience, 7,* 232. doi:10.3389/fnhum.2013.00232

Cristia, A., Dupoux, E., Hakuno, Y., Lloyd-Fox, S., Schuetze, M., Kivits, J., ... Minagawa-Kawai, Y. (2013). An online database of infant functional near infrared spectroscopy studies: A community-augmented systematic review. *PLoS One, 8*(3), e58906. doi: 10.1371/journal.pone.0058906

Csibra, G., & Gergely, G. (2009). Natural pedagogy. *Trends in Cognitive Science, 13,* 148–153. doi: 10.1016/j.tics.2009.01.005

DeCasper, A. J., & Fifer, W. P. (1980). Of human bonding: Newborns prefer their mothers' voices. *Science, 280,* 1174–1176.

Deen, B., Richardson, H., Dilks, D. D., Takahashi, A., Keil, B., Wald, L. L.,...Saxe, R. (2017). Organization of high-level visual cortex in human infants. *Nature Communications, 8,* 13995. doi: 10.1038/ncomms13995

Dehaene-Lambertz, G., Dehaene, S., & Hertz-Pannier, L. (2002). Functional neuroimaging of speech perception in infants. *Science, 298,* 2013–2015.

Dehaene-Lambertz, G., Hertz-Pannier, L., Dubois, J., Meriaux, S., Roche, A., Sigman, M., & Dehaene, S. (2006). Functional organization of perisylvian activation during presentation of sentences in preverbal infants. *Proceedings of the National Academy of Sciences of the United States of America, 103,* 14240–14245.

Elsabbagh, M., & Johnson, M. H. (2007). Infancy and autism: Progress, prospects, and challenges. *Progress in Brain Research, 164,* 355–383.

Elsabbagh, M., Mercure, E., Hudry, K., Chandler, S., Pasco, G., Charman, T.,...Johnson, M. H. (2012). Infant neural sensitivity to dynamic eye gaze is associated with later emerging autism. *Current Biology, 22*(4), 338–342.

Farroni, T., Massaccesi, S., Menon, E., & Johnson, M. H. (2007). Direct gaze modulates face recognition in young infants. *Cognition, 102,* 396–404.

Fransson, P., Aden, U., Blennow, M., & Lagercrantz, H. (2011). The functional architecture of the infant brain as revealed by resting-state fMRI. *Cerebral Cortex, 21,* 145–154.

Fransson, P., Skiöld, B., Horsch, S., Nordell, A., Blennow, M., Lagercrantz, H., & Aden, U. (2007). Resting-state networks in the infant brain. *Proceedings of National Academy of Sciences of the United States of America, 104,* 15531–15536.

Frith, C. D., & Frith, U. (2006). The neural basis of mentalizing. *Neuron, 50,* 531–534.

Fuster, J. M. (2008). *The prefrontal cortex* (4th ed.). London: Elsevier.

Gao, W., Zhu, H., Giovanello, K. S., Shen, D., Smith, J. K., Shen, D.,...Lin, W. (2009). Evidence on the emergence of the brain's default network from 2-week-old to 2-year-old healthy pediatric subjects. *Proceedings of the National Academy of Sciences of the United States of America, 106,* 6790–6795.

Gilbert, C. D., & Sigman, M. (2007). Brain states: Top-down influences in sensory processing. *Neuron, 54,* 677–696.

Grossmann, T. (2013a). Mapping prefrontal cortex functions in human infancy. *Infancy, 18*(3), 303–324. doi: 10.1111/infa.12016

Grossmann, T. (2013b). The role of medial prefrontal cortex in early social cognition. *Frontiers in Human Neuroscience, 7,* 340. doi: 10.3389/fnhum.2013.00340

Grossmann, T., & Johnson, M. H. (2010). Selective prefrontal cortex responses to joint attention in early infancy. *Biology Letters, 6*(4), 540–543. doi: 10.1098/rsbl.2009.1069

Grossmann, T., Johnson, M. H., Lloyd-Fox, S., Blasi, A., Deligianni, F., Elwell, C., & Csibra, G. (2008). Early cortical specialization for face-to-face communication in human infants. *Proceedings of the Royal Society of London, Series B: Biological Sciences, 275*, 2803–2811.

Grossmann, T., Lloyd-Fox, S., & Johnson, M. H. (2013). Brain responses reveal young infants are sensitive to when a social partner follows their gaze. *Developmental Cognitive Neuroscience, 6*, 155–161.

Grossmann, T., Parise, E., & Friederici, A. D. (2010). The detection of communicative signals directed at the self in infant prefrontal cortex. *Frontiers in Human Neuroscience, 4*, 201. doi: 10.3389/fnhum.2010.00201

Haxby, J. V., Hoffman, E. A., & Gobbini, M. I. (2000). The distributed human neural system for face perception. *Trends in Cognitive Science, 4*, 223–233. doi: 10.1016/S1364-6613(00)01482-0

Heekeren H. R, Marrett S, & Ungerleider L. G. (2008). The neural systems that mediate human perceptual decision making. *Nature Reviews Neuroscience, 9(6)*, 467–479.

Homae, F., Watanabe, H., Otobe, T., Nakano, T., Go, T., Konishi, Y., & Taga, G. (2010). Development of global cortical networks in early infancy. *Journal of Neuroscience, 53*, 4877–4882.

James, W. (2007). *The principles of psychology* (Vol. 1). New York: Cosimo. (Original work published in 1890)

Johnson, M. H. (2005a). *Developmental Cognitive Neuroscience* (2nd ed.). Oxford: Blackwell.

Johnson, M. H. (2005b). Subcortical face processing. *Nature Reviews Neuroscience, 6(10)*, 766–774.

Johnson, M. H., Dziurawiec, S., Ellis, H. D., & Morton, J. (1991). Newborns' preferential tracking of face-like stimuli and its subsequent decline. *Cognition, 40*, 1–19. doi: 10.1016/0010-0277(91)90045-6

Johnson, M. H., Grossmann, T., & Cohen Kadosh, K. (2009). Mapping functional brain development: Building a social brain through interactive specialization. *Developmental Psychology, 45(1)*, 151–159. doi:10.1037/a0014548

Johnson, M. H., & Morton, J. (1991). *Biology and cognitive development: The case for face recognition.* Oxford: Blackwell.

Kampe, K. K. W., Frith, C. D., & Frith, U. (2003). "Hey John": Signals conveying communicative intention toward the self activate brain regions associated with "mentalizing," regardless of modality. *Journal of Neuroscience, 23(12)*, 5258–5263.

Lieberman, M. D. (2006). Social cognitive neuroscience: A review of core processes. *Annual Review of Psychology, 58*, 18.11–18.31.

Lloyd-Fox, S., Blasi, A., & Elwell, C. E. (2010). Illuminating the developing brain: The past, present and future of functional near infrared spectroscopy. *Neuroscience and Biobehavioral Reviews, 34*(3), 269–284.

Lloyd-Fox, S., Blasi, A., Everdell, N., Elwell, C. E., & Johnson, M. H. (2011). Selective cortical mapping of biological motion processing in young infants. *Journal of Cognitive Neuroscience, 23,* 2521–2532.

Meltzoff, A. N. (2007). 'Like me': A foundation for social cognition. *Developmental Science, 10*(1), 126–134. doi: 10.1111/j.1467-7687.2007.00574.x

Minagawa-Kawai, Y., Matsuoka, S., Dan, I., Naoi, N., Nakamura, K., & Kojima, S. (2009). Prefrontal activation associated with social attachment: Facial-emotion recognition in mothers and infants. *Cerebral Cortex, 19*(2), 284–292.

Naoi, N., Minagawa-Kawai, Y., Kobayashi, A., Takeuchi, K., Nakamura, K., Yamamoto, J., & Kojima, S. (2012). Cerebral responses to infant-directed speech and the effect of talker familiarity. *NeuroImage, 59*(2), 1735–1744.

Pelphrey, K. A., Viola, R. J., & McCarthy, G. (2004). When strangers pass: Processing of mutual and averted gaze in the superior temporal sulcus. *Psychological Science, 15,* 598–603.

Piaget, J. (1952). *The origins of intelligence in children.* New York: International Universities Press.

Powell, L. J., Kosakowski, H. L., & Saxe, R. (2018). Social origins of cortical face areas. *Trends in Cognitive Sciences, 22*(9), 752–763.

Reddy, V. (2003). On being the object of attention: Implications for self other consciousness. *Trends in Cognitive Sciences, 7*(9), 397–402.

Rochat, P. (2003). Five levels of self-awareness as they unfold early in life. *Consciousness and Cognition, 12*(4), 717–731. doi: 10.1016/S1053-8100(03)00081-3

Rochat, P. (2011). The self as phenotype. *Consciousness and Cognition, 20*(1), 109–119. doi: http://dx.doi.org/10.1016/j.concog.2010.09.012

Saito, Y., Aoyama, S., Kondo, T., Fukumoto, R., Konishi, N., Nakamura, A., ... Toshima, T. (2007). Frontal cerebral blood flow change associated with infant-directed speech. *Archives of Disease in Childhood, 92,* F113–F116.

Saito, Y., Kondo, T., Aoyama, S., Fukumoto, R., Konishi, N., Nakamura, K., ... Toshima, T. (2007). The function of the frontal lobe in neonates for response to a prosodic voice. *Early Human Development, 83,* 225–230.

Saxe, R. (2006). Uniquely human social cognition. *Current Opinion Neurobiology, 16,* 235–239.

Schilbach, L., Timmermans, B., Reddy, V., Costall, A., Bente, G., Schlicht, T., & Vogeley, K. (2013). Toward a second-person neuroscience. *Behavioral and Brain Sciences, 36*(4), 393–414. doi: 10.1017/S0140525X12000660

Schilbach, L., Wilms, M., Eickhoff, S. B., Romanzetti, S., Tepest, R., Bente, G.,... Vogeley, K. (2010). Minds made for sharing: Initiating joint attention recruits reward-related neurocircuitry. *Journal of Cognitive Neuroscience, 22*(12), 2702–2715.

Semendeferi, K., Lu, A., Schenker, N., & Damasio, H. (2002). Humans and great apes share a large frontal cortex. *Nature Neuroscience, 5,* 272–276.

Senju, A., & Csibra, G. (2008). Gaze following in human infants depends on communicative signals. *Current Biology, 18,* 668–671. doi: 10.1016/j.cub.2008.03.059

Simion, F., Regolin, L., & Bulf, H. (2008). A predisposition for biological motion in the newborn baby. *Proceedings of the National Academy of Sciences of the United States of America, 15,* 809–813.

Singer, T. (2006). The neuronal basis and ontogeny of empathy and mind reading: Review of literature and implications for future research. *Neuroscience and Biobehavioral Reviews, 30*(6), 855–863. doi: 10.1016/j.neubiorev.2006.06.011

Singh, L., Morgan, J. L., & White, K. S. (2004). Preference and processing: The role of speech affect in early spoken word recognition. *Journal of Memory and Language, 51,* 173–189.

Spelke, E. S., & Kinzler, K. D. (2007). Core knowledge. *Developmental Science, 10*(1), 89–96. doi: 10.1111/j.1467-7687.2007.00569.x

Sperber, D., & Wilson, D. (1995). *Relevance: Communication and cognition* (2nd ed.). Oxford: Blackwell.

Thomas, M. S. C., & Johnson, M. H. (2008). New advances in understanding sensitive periods in brain development. *Current Directions in Psychological Science, 17*(1), 1–5. doi: 10.1111/j.1467-8721.2008.00537.x

Tomasello, M., Carpenter, M., Call, J., Behne, T., & Moll, H. (2005). Understanding and sharing intentions: The origins of cultural cognition. *Behavioral and Brain Sciences, 28,* 675–691.

Tzourio-Mazoyer, N., De Schonen, S., Crivello, F., Reutter, B., Aujard, Y., & Mazoyer, B. (2002). Neural correlates of woman face processing by 2-month-old infants. *NeuroImage, 15,* 454–461.

Urakawa, S., Takamoto, K., Ishikawa, A., Ono, T., & Nishijo, H. (2015). Selective medial prefrontal cortex responses during live mutual gaze interactions in human infants: An fNIRS study. *Brain Topography, 28*(5), 691–701.

Vouloumanos, A., Hauser, M. D., Werker, J. F., & Martin, A. (2010). The tuning of human neonates' preference for speech. *Child Development, 81,* 517–527.

Wood, J. N., & Grafman, J. (2003). Human prefrontal cortex: Processing and representational perspectives. *Nature Reviews Neuroscience, 4,* 139–147.

Woodward, A. (2009). Infants' grasp of others' intentions. *Current Directions in Psychological Science, 18,* 53–57.

Yoon, J. M. D., Johnson, M. H., & Csibra, G. (2008). Communication-induced memory biases in preverbal infants. *Proceedings of the National Academy of Sciences of the United States of America, 105*(36), 13690–13695. doi: 10.1073/pnas.0804388105

Zelazo, P. D., & Müller, U. (2002). Executive functions in typical and atypical development. In U. Goswami (Ed.), *Handbook of childhood cognitive development* (pp. 445–469). Oxford: Blackwell.

Zwaigenbaum, L., Bryson, S., Rogers, T., Roberts, W., Brian, J., & Szatmari, P. (2005). Behavioral manifestation of autism in the first year of life. *International Journal of Developmental Neuroscience, 23,* 143–152. doi: 10.1016/j.ijdevneu.2004.05.001

5 Foundations of Imitation

Virginia Slaughter

Overview

This chapter challenges the idea that human infants possess an innate mirror neuron system that enables them to imitate other people's actions from birth. New evidence from large-scale behavioral research and electromyography experiments with infants has found no evidence of imitation in the first few months of life. This accords with an in-depth analysis of tongue and throat development that suggests that coordination of these organs for breathing and feeding takes precedence in the newborn period, making imitation unlikely. An alternative model is that infants learn to imitate in a rich social environment that engenders associations between the body movements they produce and those that they observe. This model lines up with evidence that brain mechanisms underlying imitation are malleable and depend on experience.

Introduction

Humans are prolific imitators, and imitation takes many forms. Simply put, imitation involves reproducing the body movements that you observe in another. This simple definition belies the tremendous complexity of this behavior, which can be classified along several overlapping dimensions. One dimension is the motivation to imitate, which we do to learn new skills and also to connect with others. Familiarity is another dimension: sometimes we imitate actions we have already practiced, and other times we imitate something completely new. Complexity is yet another dimension; we can imitate simple gestures such as sticking out the tongue, all the way up to complex action sequences such as line dancing. Imitation of the latter may involve reproducing a goal-directed action, and when imitating to achieve a goal we may imitate with perfect fidelity, or we might do something similar to get the same outcome. Finally, imitation can be deliberate or inadvertent. When imitation is simple and inadvertent, such as when we move our jaws while spooning food into a baby's mouth, it is sometimes called automatic mimicry or mirroring. Within the developmental literature going back to Piaget (1951), all instances of reproducing another's body movements are termed *imitation,* so in this chapter I will use that term for all imitative acts, even those as simple as copying a model's tongue protrusion.

The various forms of imitation are likely to draw on different neuro-cognitive resources, but it is thought that an underlying mechanism that links the actions we observe with those that we produce is at the heart of all imitative behaviors (Heyes, 2011; Ferrari, Bonini, & Fogassi, 2009). That mechanism may seem straightforward, but in fact it must solve the "correspondence problem" (Heyes, 2011) of translating a visual image of someone else's body movements into a matching motor pattern in one's own body. The correspondence problem is especially acute when considering imitation of "opaque" gestures such as facial expressions because without reference to a mirror it is not possible to view simultaneously one's own and another's face to make an explicit match.

A widespread view is that imitation is an inborn capacity, underpinned by a dedicated mirror neuron mechanism that "directly maps a pictorial or kinematic description of the observed action onto an internal motor representation of the same action" (Iacoboni et al., 1999, p. 2526; see also Meltzoff, 2010; Simpson, Murray, Paukner, & Ferrari, 2014). This chapter outlines the origins of this view, reviews evidence that challenges it, and presents an alternative model of the foundations of imitation in humans.

Development of Imitation

The first imitative behavior seen in human infants takes the form of copying simple body gestures. In his classic observational studies of his own three infants, Piaget (1951) documented the emergence of imitation in the first half-year of life. In line with his constructivist perspective on cognitive development, Piaget assumed imitation came about as infants played around with moving their bodies and linked those movements with environmental events. As such, Piaget claimed that imitation emerged gradually over the first year of life, first as repetition of familiar actions and sounds, followed by imitation of novel acts when the infant could see both her own body movement and that of the model. It was only in Sensorimotor Stage 6 (ages eighteen months to twenty-four months) that the acquisition of mental representational capacities make it possible for infants to imitate opaque behaviors because these require a cognitive bridge between action observation and execution.

Building on Piaget's work, Uzgiris and Hunt (1975) created a standardized developmental scale for imitation, which presents first familiar and then novel gestures (some involving acting on objects) that the infant can see herself perform. The most advanced test of the scale is imitation of opaque gestures, such as tapping the head.

In a now-classic paper published in *Science*, Meltzoff and Moore (1977) reported that infants aged just twelve to twenty-eight days imitated the gestures of tongue protrusion, mouth opening, lip pursing, and finger waving as performed by an adult model. Besides challenging the Piagetian view of how imitation develops, this research pioneered a rigorous methodology for testing imitation. This involved an adult modeling more than one "target" gesture to newborn infants in a tightly controlled procedure. The "cross-target" analysis involved comparing the infants' production of target gestures in response to different models—for example, comparing infants' tongue protrusions when the adult was modeling tongue protrusion versus mouth opening. This procedure was designed to rule out alternative explanations for imitation, such as newborns showing an overall increase in activity, or an increase in one specific behavioral response, when engaged with the adult model.

Meltzoff and Moore's (1977) data suggested that human infants have an inborn mechanism that solves the correspondence problem, enabling newborns to match up their vision of an adult's body movements with their own body's production of those same movements. Soon afterward, an experiment by Field, Woodson, Greenberg, and Cohen (1982) extended the scope of neonatal imitation with data suggesting that newborns could also imitate distinct facial expressions. Building on this and other replications of neonatal imitation, Meltzoff and Moore (1997) proposed that humans possess an innate active intermodal matching (AIM) mechanism, which maps our own body movements to those of social partners. This proposal gained momentum with the discovery of mirror neurons.

First documented in the inferior premotor cortex of macaques, the distinctive quality of mirror neurons is that they fire both when the animal makes a particular movement, such as grasping an object, and when the animal observes another individual making the very same movement (Gallese, Fadiga, Fogassi, & Rizzolatti, 1996). Subsequent studies with humans using neuroimaging techniques suggested that similar functionality is present in the human brain (Molenberghs, Cunnington, & Mattingley, 2012). Thus mirror neurons appeared to be a neural substrate that fit the theoretical description of the AIM mechanism, by representing both one's own body movement and the same movement executed by another. Reports that chimpanzees and macaques also imitate from birth have bolstered the view that imitation is an inborn behavioral capacity in primates, underpinned by mirror neurons (Ferrari et al., 2006; Myowa-Yamakoshi, Tomonaga, Tanaka, & Matsuzawa, 2004).

This research gave rise to a compelling evolutionary and developmental theory. Imitation is a fundamentally social behavior by which we share and communicate our experiences and knowledge. Furthermore, there is ample evidence that we like those who imitate us, and we imitate those we like (Chartrand & Lakin, 2013). An innate mirroring mechanism, enabling newborns to imitate adult models, could have evolved to facilitate interaction between infants and their caregivers at a time when the infant is still relatively helpless (Ferrari et al., 2009). In their earliest form of communication, newborns and caregivers signal via imitation that "you are like me" and "we like each other" (Meltzoff, 2007). This mutual attraction prompts caregivers to invest in newborns, thereby increasing the chances that infants will survive the first months of life when they cannot explicitly communicate their needs but are at their most vulnerable. And this initial "like me" connection is a foundation for later-developing social-cognitive skills including empathy, mindreading and social learning.

Imitation appears to drop out of the newborn behavioral repertoire at around two months of age (Abranvel & Sigafoos, 1984). The explanation for this is that development of the motor cortex enables more mature, deliberate behaviors that suppress newborns' automatic tendency to imitate (Ferrari et al., 2009; Meltzoff & Moore, 1992). At that stage, infants begin to communicate with caregivers via social smiling and vocalizations that, according to the theory, replace the simple imitative signal. Then in the latter half of the first year of life, more complex and controlled forms of imitation begin to emerge. One proposal is that a subcortical "direct mirror pathway" that links premotor and posterior parietal regions to the primary motor cortex is responsible for neonatal imitation and automatic imitation in older children and adults (Ferrari et al., 2009), while an "indirect mirror pathway" is responsible for voluntary forms of imitation (e.g., selective, delayed) via connections with the prefrontal cortex.

With the exception of one experimental study of adult patients being treated for epilepsy (Mukamel, Ekstrom, Kaplan, Iacoboni, & Fried, 2010), all of the research on human mirror neurons has relied on neurocognitive techniques that identify brain regions that register activity both when participants are executing motor acts and when they are observing them. Functional imaging studies commonly find such correlated activity in the inferior frontal gyrus, the premotor cortex, the inferior and superior parietal lobule, and the superior temporal sulcus in the adult brain. These regions are often referred to as the human "mirror neuron system" although there are also indications of mirror activity outside of these brain regions (Molenberghs et al., 2012; Mukamel et al., 2010). In human infants, who are too young to lie still for functional imaging, mirror activity is investigated with electroencephalography. The most frequently used technique measures mu rhythm activity over the motor cortex, with desynchronization of the rhythm suggesting activation in the brain region. Mu desynchronization during observation of hand actions has been recorded in infants as young as eight months (Marshall & Meltzoff, 2014), suggesting that movement production and movement perception draw on the same areas in the infant brain.

Is Imitation Innate?

The status of newborn imitation is an enduring controversy in developmental psychology. In the four decades since Meltzoff and Moore's (1977) initial discovery, there have been over thirty published experiments on newborn imitation. Across the board, the findings are mixed: some have reported replications of Meltzoff and Moore (1977) and extensions of the range of gestures imitated, others have reported imitation of just a subset of gestures, and others have found no evidence that newborns copy adults' gestures. Even the systematic literature reviews have drawn different conclusions (e.g., Ray & Heyes, 2011; Simpson et al., 2014).

Within the mixed results, tongue protrusion emerged as the most reliable gesture imitated (Anisfeld, 1996; Ray & Heyes, 2011), leading to alternative views of this behavior. One view is that if tongue protrusion is the only gesture that newborns copy, then such "imitation" is better characterized as a simple reflex, as opposed to a product of a mirror neuron mechanism. However, it is not unreasonable to assume that the newborn's mirror neuron system might be limited to mapping just one gesture as a starting state. Another proposal is that tongue protrusion "imitation" is pure coincidence. The argument here is that when newborns get excited, they poke out their tongues as a primitive exploratory response, and an adult face with a protruding tongue—by far the most commonly presented model in the neonatal imitation studies—is particularly exciting (Jones, 2009).

It is difficult to untangle the competing perspectives, partly because the vast majority of neonatal imitation studies assess the behavior at only one time point. A longitudinal perspective is crucial because repeated testing enables assessment of the reliability of the behavior. If newborn imitation is driven by an inborn mirror mechanism, which automatically generates motor responses matching the gestures of another person, then it should operate consistently over the first two months of life before being overtaken by cortically mediated behaviors. Longitudinal testing also assists researchers to evaluate the possibility that newborns differ in their tendency to imitate, and those individual differences mask a population-level effect (Simpson et al., 2014).

The only large-scale longitudinal tests of neonatal imitation to date produced null results. Oostenbroek and colleagues (2016) tested over one-hundred infants on four occasions during the first two months of life. Infants were presented with a female adult modeling a range of facial, vocal, and hand gestures. Longitudinal analyses using the "cross-target" methodology indicated that newborns failed to reliably imitate any of the gestures. For example, across the four testing points, infants showed no more tongue protrusion in response to the adult poking out her tongue than they did in response to a happy face, a sad face, or a mouth opening. Furthermore, there was no intraindividual consistency in infants' responses, undermining the idea that some newborns are better imitators than others.

As with many of the previous failures to replicate the neonatal imitation effect, the null findings of Oostenbroek and colleagues (2016) prompted critiques of their methodology. Most of the critiques revolve around neonates' limitations in state regulation, eyesight, and motor control. For instance, it is argued that the adult model should be unfamiliar to the newborn to maintain the infant's interest during testing, that the model's face should be brightly lit to ensure the newborn can see the target gestures, that testing should be done with the newborn in a padded seat for postural stability, and that modeling and response periods must be sufficiently long to account for newborns' slow motor organization (see Meltzoff, et al., 2018). These points may be valid for eliciting infants' best performances in the experimental context (or not—for discussion, see Oostenbroek et al., 2018). However the methodological critiques appear to run counter to the theory that an innate mirror neuron system has been shaped by evolution to produce imitation facilitating newborn-caregiver emotional bonds. If imitation in neonates is so fragile that it can only be elicited under strict seating, lighting, and timing conditions, then it is not likely to manifest in the everyday life of a newborn infant. And if an unfamiliar face is essential, that undermines the idea that imitation creates an enduring connection between infants and their caregivers during the first weeks of life.

Newborns' physiological immaturity certainly challenges researchers to devise valid tests of their imitative abilities. But newborn physiological limitations pose a more serious problem for the claim that newborns imitate. A close analysis of human aerodigestive development suggests that human neonates may be physically incapable of imitating oral gestures (Keven & Akins, 2017). This is because, in the weeks after birth, newborns must develop oral motor schemes that enable coordinated breathing, sucking, and swallowing. Although these movements first appear in utero, taking in air and liquids postnatally requires adaptation of the basic motor patterns. This is essential for survival. It seems that this developmental process generates rhythmic oral stereotypies, including poking the tongue in and out and opening and closing the mouth, which newborns practice as they acquire coordinated jaw, tongue, and throat movements. These oral gestures are stimulated by subcortically mediated sensorimotor connections, making it unlikely that infants could produce them in response to a visual model. As with earlier accounts of tongue protrusion, Keven and Akins (2017) have proposed that neonates are aroused in face-to-face interactions with adults and this stimulates oral stereotypies that coincidentally match tongue-protrusion and mouth-opening models. These mouth gestures are not social responses, and their apparent disappearance at two months of age is a consequence of oral coordination and control rather than the emergence of new social skills.

The observation that newborn imitation disappears after two months in humans (which has not gone unchallenged; see Ray & Heyes, 2011) is problematic for the innate mirror neuron account. A prominent proposal is that newborn imitation is mediated by a "direct mirror pathway" that automatically translates visual input to motor output. This pathway is also supposed to produce automatic imitation effects in adults (Ferrari et al., 2009). But recent studies of automatic imitation in infants, using the same electromyography (EMG) techniques that capture the phenomenon in adults, suggest that it is not fully evident until at least six months of age. This leaves a four-month gap between neonatal imitation and automatic imitation, which makes little sense if they are both generated by an innate direct mirror pathway.

Using EMG, automatic imitation is measured via activation of facial and/ or hand muscles while participants are watching other people moving their bodies. Electrodes are attached to the relevant body part to detect potentiation in the underlying muscles. Imitation that is not necessarily visible in overt behavior is operationally defined with EMG as selective increases in activation of the muscles corresponding to those used by the model demonstrating the target gesture. For instance, imitation of a smile gesture is evident if the participant's cheek but not eyebrow muscles activate during observation of a smile. This technique is well established with adults (Heyes, 2011) and recently has been applied in infant research.

Kaiser, Crespo-Llado, Turati, and Geanu (2017) assessed four- and seven-month-olds' responses to adult faces expressing happiness, anger, and fear. They put electrodes over the infants' zygomaticus (on the cheek, used for smiling), corrugator (on the forehead, used for frowning), and frontalis (on the forehead, used for fearful eyebrow raising). This experiment revealed no selective activation of facial muscles in the four-month-olds. The older infants produced congruent facial muscle activation to the happy and fearful face models, but there was no matching response for the angry face. In a similar study, Isomura and Nakano (2016) tested five-month-olds' facial muscle potentiation in response to happy and sad face models. They observed congruent facial activation in the infants, but only if an emotionally matching vocal track was presented along with the face (e.g., laughter accompanying the smile and crying accompanying the frown). Taken together, these two studies undermine the idea that imitation of facial expressions is innate because there was no evidence of such imitation at four months of age; it appeared at five months, but only with voice accompaniment; and at seven months, it was evident for just two of three facial emotions.

Another study by de Klerk, Hamilton, and Southgate (2018) tested four-month-olds for automatic imitation of facial and manual gestures, and used near-infrared spectroscopy to investigate the brain regions that were active when infants watched the model's movements. These researchers also manipulated whether or not the model was gazing at the infant, based on the notion that imitation is a social signal and therefore more likely to occur when accompanied by eye contact. They found no indication of imitation in response to manual gestures (hand opening/closing, finger actions) in either gaze condition. Of the facial gestures they presented—eyebrow raising, frowning, tongue protrusion, and mouth opening—automatic imitation was only detected in response to the eyebrow-raising model and only in the direct gaze condition.

Now, the fact that this form of automatic imitation was only observed with the model's direct gaze does not itself undermine the theory of a "direct mirror pathway" linking neonatal and automatic imitation. It could be that eye contact is crucial for eliciting imitation via this pathway; this has never been tested in newborns. However, none of the other facial gesture models, in either gaze condition, elicited a significant pattern of corresponding muscle activation in the infants. As with the failure to find consistent evidence of facial expression imitation via EMG, the lack of automatic imitation effects at four months challenges the idea that such imitation is innately generated by a direct mirror pathway in the brain.

While the infants watched the models in the experiment of de Klerk and colleagues (2018), there was increased brain activity over the left posterior superior temporal sulcus (STS) during the direct gaze conditions. Mirror neuron activity in the STS appears to play a role in imitation (Molenberghs et al., 2009), which might suggest that mirror neurons were active while the infants viewed the model's gestures. However, the STS is generally responsive to social stimuli, so it is also possible that perceiving faces with direct gaze, irrespective of the gestures being modeled, stimulated this brain activity.

Automatic imitation can also be tested behaviorally. The typical paradigm involves pitting action execution against action observation (Heyes, 2010). Participants view a modeled gesture at the same time they are instructed to perform either a matching or mismatching gesture. According to the direct mirror pathway model, visual input of a gesture automatically stimulates the matching motor pattern. So when participants are instructed to produce the matching gesture, performance is facilitated, whereas production of a mismatching gesture is slowed by interference. Automatic imitation is operationalized as the reaction time difference between these two conditions.

McKyton, Ben-Zion, and Zohary (2018) used this paradigm in a study of newly sighted children who had been blind for at least the first seven years of life due to cataracts. The researchers reasoned that the children's automatic imitation should be intact if it is underpinned by an inborn mirror neuron system. In the experiment, children were instructed to tap a table with their right or left hand in response to the color of a video prompt. Besides the color, the videos depicted hands tapping. Children were told to ignore the hands and focus on the color, yet an automatic imitation effect was evident in slower tapping when the video hands mismatched the child's correct response. Compared to a typically developing control group, the newly sighted children showed a significantly reduced automatic imitation effect, meaning that they were less influenced by the sight of mismatching hands. This suggests that early blindness weakened the tendency to automatically imitate.

Interestingly, the newly sighted children were fully capable of producing deliberate imitation by moving the same body part as the model when asked to do so. These findings are hard to reconcile with the direct and indirect mirror pathway model. Why would the direct mirror pathway be impaired as a result of early blindness, yet the indirect pathway responsible for deliberate imitation remain intact? It is possible that the newly sighted children were also slower to produce deliberate imitation; this was not reported by the study. If so, that would suggest the impairment affected both mirror pathways. Based on their data, McKyton and colleagues (2018) concluded that visual input is essential for development of a fully functional mirror neuron system.

Mirror Learning

The study by McKyton and colleagues (2018) suggests that (lack of) visual input affects automatic imitation over the long term. These results are in line with numerous experiments demonstrating short-term learning effects for adults' automatic imitation (see Catmur, Walsh, & Heyes, 2009 for a review). For example, when adult participants repeatedly practiced hand opening while watching a hand close and vice versa, automatic imitation effects for these two gestures were abolished. On the flip side, practicing hand opening and closing actions in conjunction with a robot hand performing matching gestures significantly increased adults' automatic imitation in response to the robot hand. Thus, changing the visual input associated with specific gestures can modify automatic imitation. Although some researchers question the role of mirror neurons in imitation (for discussion, see Hickock, 2014), these short-term imitation learning effects are typically associated with mirror neuron areas in the adult brain (Catmur et al., 2009).

The fact that imitation is modified by experience does not necessarily challenge the proposal that imitation is an innate capacity; inborn abilities can be open to learning. But these effects fit well with associative learning models of imitation, which consider that the correspondence problem is solved through repeated, coincident experience of producing and perceiving specific body gestures (Cook, Bird, Catmur, Press, & Heyes, 2014; Del Guidice, Manera, & Keysers, 2009; Quadrelli & Turati, 2016). So how might these associations come about in development? One route would be manual self-observation. Starting in the first weeks of life, infants watch their own hand gestures—finger movements, grasping movements, waving—which means that they experience proprioceptive and visual input for those gestures at the same time. This perfect contingency of action observation and execution could produce mirror mechanisms (if not individual mirror neurons) that respond to both. Then down the track in development, the visual input of, for instance, a hand grasp would activate the associated matching motor program facilitating imitation. Note that this model does not distinguish direct and indirect mirror pathways; rather, it assumes that mirror mechanisms vary in associative strength according to the learning history, and allows that other correlated inputs, such as sounds, can be incorporated into the associations that stimulate imitation.

Action observation-execution correspondences can also be created when caregivers imitate their infants (Heyes, 2010; Ray & Heyes, 2011). Here the proposed learning mechanism is the same: The infant has a motor experience that is contingent with a matching sensory input. So for instance, an infant makes a kissing sound with her lips, and then her parent does the same thing; as this experience is repeated, the sight of pursed lips and the kissing sound become associated with the matching motor program in the infant's brain. Although some researchers have questioned whether there is sufficient correlated input during natural caregiver–infant interactions to make this model plausible (Ferrari, Tramacere, Simpson, & Iriki, 2013), the available evidence suggests that caregivers imitate their infants around once per two to four minutes and that up to 20 percent of infant vocalizations and gestures are imitated (Malatesta & Haviland, 1982; Kokkinaki & Vitalaki, 2013).

Caregiver imitation may be particularly powerful for learning sensorimotor associations because it taps into infants' perceptual preferences. In the first months of life, there are clear preferences for familiar stimuli and response-contingent stimulation, as well as an attraction to faces (Del Guidice et al., 2009; Slater, 1998). Thus when a caregiver imitates the infant's kissing sound during face-to-face interaction, it may be especially rewarding because it is familiar (e.g., the infant hears the sound she just heard herself make being reproduced), it is contingent (e.g., it is dependent on the infant's own activity), and it involves a moving face, which infants naturally enjoy. These rewarding features of being imitated could stimulate the infant to repeat her initial action, leading to a synchronous loop of mutually reinforcing imitation. This model has yet to be thoroughly tested, but a correlational study by Rayson, Bonaiuto, Ferrair, & Murray (2017) indicated that maternal imitation of infant mouth opening and smiling at two months of age was positively associated with mu rhythm desynchronization over the motor cortex when the infant observed those same two facial gestures at nine months of age. This finding supports the idea that parental imitation helps to cement associations between perception and production of facial gestures in the infant brain.

An obvious problem with the association learning account of imitation is that infants move all the time without seeing their own movements and without being imitated by caregivers. This could create "junk associations" (Catmur et al., 2009) that would dampen the strength of mirror associations. One proposal is that action observation–execution associations are "canalized" or streamlined by infants' visual abilities and preferences, limiting the number of junk associations that would be repeated often enough to compete with mirror associations. For example, infants spend hours watching their own hands in the first couple of months of life (Del Guidice et al., 2009). This intensive experience could produce such strong foundational associations that they would be effectively permanent (although still open to short-term training effects as per the adult-imitation learning experiments). Another suggestion is that observed and executed actions are only associated when one is predicted by the other, as when the infants' gesture is imitated by the caregiver (Catmur et al., 2009).

A developmental model based on associative learning lines up remarkably well with Piaget's observations of the development of imitation. He noted that infants initially imitate familiar gestures when they can simultaneously see the model and their own matching gesture. At this stage, viewing the model would activate a well-practiced associated motor program, enabling imitation. Later in development, infants imitate novel actions that they can see; here the visual input might stimulate motor programs that the infant modifies during execution to match the model. Indeed, Piaget (1951) observed progressively accurate matching when infants imitated novel models. Finally, according to Piaget, infants acquire the ability to imitate opaque gestures. These forms might be slowest to develop because associations that rely on visual input from sources other than the infant's own body, including caregiver imitation and exposure to one's own face in mirrors, are access limited. The association learning model of imitation also accounts for Isomura and Nakano's (2016) finding that six-month-olds automatically imitated facial expressions only when auditory input was also present. This could be because the underlying associations were still relatively weak, requiring multimodal perceptual input to activate the matching motor response.

The model also fits with findings from an automatic imitation experiment involving typically developing three- to seven-year-old children (O'Sullivan, Bijvoet-van den Berg, & Caldwell, 2018). The children were instructed to perform a specific gesture with both hands immediately after seeing the experimenter execute a matching or mismatched gesture. The paired gestures varied according to their assumed learning histories. Clapping and waving were considered "commonly imitated" gestures because they are naturally executed in tandem with other people, as during applause and signaling hello and goodbye. The other two gestures, pointing and opening/closing the hand, were considered rarely imitated because they are not usually performed in synchrony with other people. The researchers predicted that children would be slowest to generate the commonly imitated gestures when the experimenter modeled something different because they would have to overcome stronger observation–execution associations. This was partially upheld. At all ages, children showed automatic imitation effects for all four gestures, but it was strongest for the clap. So the reaction time difference between clapping after a clap (facilitated by mirror associations) and clapping after a wave (inhibited by the mismatch) was particularly large. This could be because the addition of sound created rich sensorimotor associations, or because clapping along with others is so commonplace that it was difficult for children to resist the tendency to imitate (O'Sullivan et al., 2018).

Considering imitation as a learned rather than an innate capability turns around assumptions about its role in development. Rather than imitation being the developmental foundation of social behavior, as per the "like me" hypothesis, imitation may be a product of sociality. This up-ended proposal has some empirical support from research with macaques—whose brains unequivocally possess mirror neurons. Macaque species living in distinct geographical areas have differently structured social relations: Japanese macaques are characterized as "despotic" whereas those living on Sulawesi are known for being "egalitarian." The egalitarian animals have more numerous relationships with conspecifics within their troops, and those relationships are less constrained by dominance hierarchies. An observational study of these two distinct species, living in different zoological parks in Europe, revealed significant differences in imitation. The egalitarian macaques produced facial imitation during play exchanges whereas the despotic animals did not, even though there was no difference in the number of play face signals generated by the different troops (Scopa & Palagi, 2016). This suggests that the macaque brain is more likely to generate imitative behavior in an open, cooperative social environment.

Conclusion

Although careful to document its early development, Piaget (1951) downplayed the significance of imitation in human development. Meltzoff and Moore's (1977) counterproposal that imitation is innate turned a spotlight on this behavior that stimulated forty years of exciting research and theorizing. Currently there is no consensus on the status of neonatal imitation, but new evidence undermines the proposal that imitation and the brain mechanisms that support it are functional at birth. Association learning is emerging as a viable mechanism to explain the early development of imitation; the challenge will be to understand how such learning leads to the myriad forms of human imitative behavior.

References

Anisfeld, M. (1996). Only tongue protrusion modeling is matched by neonates. *Developmental Review, 16,* 149–161.

Catmur, C., Walsh, V., & Heyes, C. M. (2009). Associative sequence learning: The role of experience in the development of imitation and the mirror system. *Philosophical Transactions of the Royal Society Series B: Biological Sciences, 364,* 2369–2380.

Chartrand, T. L., & Lakin, J. L. (2013). The antecedents and consequences of human behavioural mimicry. *Annual Review of Psychology, 64,* 285–308.

Cook, R., Bird, G., Catmur, C., Press, C., & Heyes, C. (2014). Mirror neurons: From origin to function. *Behavioral and Brain Sciences, 37,* 177–192.

De Klerk, C. C., Hamilton, A. F. C., & Southgate V. (2018). Eye contact modulates facial mimicry in 4-month-old infants: an EMG and fNIRS study. *Cortex,* 106, 93–103. doi: 10.1016/j.cortex.2018.05.002

Del Giudice, M., Manera, V., & Keysers, C. (2009). Programmed to learn? The ontogeny of mirror neurons *Developmental Science, 12,* 350–36.

Ferrari, P. F., Bonini, L., & Fogassi, L. (2009). From monkey mirror neurons to primate behaviours: Possible "direct" and "indirect" pathways. *Philosophical Transactions of the Royal Society B: Biological Sciences, 364,* 2311–2323. doi: 10.1098/rstb.2009.0062

Ferrari, P. F., Tramacere, A., Simpson, E. A., & Iriki, A. (2013). Mirror neurons through the lens of epigenetics. *Trends in Cognitive Sciences, 17,* 450–457. doi: 10.1016/j.tics.2013.07.003

Ferrari P. F., Visalberghi E., Paukner A., Fogassi L., Ruggiero A., & Suomi S. J. (2006). Neonatal imitation in rhesus macaques. *PLoS Biology, 4,* e302. doi: 10.1371/journal.pbio.0040302

Field, T., Woodson, R., Greenberg, R., & Cohen, D. (1982). Discrimination and imitation of facial expressions by neonates. *Science, 218*(4568), 179–181.

Gallese V., Fadiga L., Fogassi L., & Rizzolatti G. (1996). Action recognition in the premotor cortex. *Brain, 119,* 593–609.

Heyes, C. (2010). Where do mirror neurons come from? *Neuroscience and Biobehavioral Reviews, 34,* 575–583. doi: 10.1016/j.neubiorev.2009.11.007

Heyes, C. (2011). Automatic imitation. *Psychological Bulletin, 137,* 463–483. doi: 10.1037/a0022288

Hickok, G. (2014). *The myth of mirror neurons: The real neuroscience of communication and cognition.* New York: W. W. Norton.

Iacoboni, M., Woods, R. P., Brass, M., Bekkering, H., Mazziotta, J. C., & Rizzolatti, G. (1999). Cortical mechanisms of human imitation. *Science, 286,* 2526–2528.

Isomura, T., & Nakano, T. (2016). Automatic facial mimicry in response to dynamic emotional stimuli in five-month-old infants. *Proceedings of the Royal Society of London B: Biological Sciences, 283*(1844), 1e8.

Jones, S. S. (2009). The development of imitation in infancy. *Philosophical Transactions of the Royal Society B: Biological Sciences, 364*(1528), 2325–2335.

Kaiser, J., Crespo-Llado, M. M., Turati, C., & Geangu, E. (2017). The development of spontaneous facial responses to others' emotions in infancy: An EMG study. *Scientific Reports, 7*(1), 17500. doi: 10.1038/s41598-017-17556-y

Keven, N., & Akins, K. (2017). Neonatal imitation in context: Sensory-motor development in the perinatal period. *Behavioral and Brain Sciences, 40,* e381.

Kokkinaki, T., & Vitalaki, E. (2013). Exploring spontaneous imitation in infancy: A three generation inter-familial study. *Europe's Journal of Psychology, 9*(2), 259–275. doi: 10.5964/ejop.v9i2.506

Malatesta, C. Z., & Haviland, J. M. (1982). Learning display rules: The socialization of emotion expression in infancy. *Child Development, 53,* 991–1003.

Marshall, P. J., & Meltzoff, A. N. (2014). Neural mirroring mechanisms and imitation in human infants. *Philosophical Transactions of the Royal Society B: Biological Sciences, 369*(1644), 20130620. doi: 10.1098/rstb.2013.0620

McKyton, A., Ben-Zion, I., & Zohary, E. (2018). Lack of automatic imitation in newly sighted individuals. *Psychological Science, 29*(2), 304–310.

Meltzoff, A. N. (2007). 'Like me': A foundation for social cognition. *Developmental Science, 10,* 126–134.

Meltzoff, A. N. (2010). Social cognition and the origins of imitation, empathy, and theory of mind. In U. Goswami (Ed.), *The Wiley-Blackwell handbook of childhood cognitive development* (pp. 49–75). Hoboken, NJ: Wiley-Blackwell.

Meltzoff, A. N., & Moore, M. K. (1977) Imitation of facial and manual gestures by human neonates. *Science, 198*(4312):75–78.

Meltzoff, A. N., & Moore, M. K. (1992). Early imitation within a functional framework: The importance of person identity, movement, and development. *Infant Behavior and Development, 15,* 479–505.

Meltzoff, A. N., & Moore, M. K. (1997). Explaining facial imitation: A theoretical model. *Early Development and Parenting, 6*(3–4), 179–192. doi: 10.1002/(SICI)1099-0917(199709/12)6:3/4<179

Meltzoff, A. N., Murray, L., Simpson, E., Heimann, M., Nagy, E., Nadel, J., ... Ferrari, P. F. (2018). Re-examination of Oostenbroek et al. (2016): Evidence for neonatal imitation of tongue protrusion. *Developmental Science, 21,* e12609.

Molenberghs, P., Cunnington, R., & Mattingley, J. B. (2009). Is the mirror neuron system involved in imitation? A short review and meta-analysis. *Neuroscience and Biobehavioral Reviews, 33,* 975–980.

Molenberghs. P., Cunnington, R., & Mattingley, J. B. (2012). Brain regions with mirror properties: A meta-analysis of 125 human fMRI studies. *Neuroscience and Biobehavioral Reviews, 36,* 341–349.

Mukamel, R., Ekstrom, A. D., Kaplan, J., Iacoboni, M., & Fried, I. (2010). Single neuron responses in humans during execution and observation of actions. *Current Biology, 20*(8), 750–756. doi: 10.1016/j.cub.2010.02.045

Myowa-Yamakoshi M., Tomonaga M., Tanaka M., & Matsuzawa T. (2004). Imitation in neonatal chimpanzees *(Pan troglodytes)*. *Developmental Science, 7,* 437–442. doi: 10.1111/j.1467-7687.2004.00364.x

Oostenbroek, J., Redshaw, J., Davis, J., Kennedy-Costantini, S., Nielsen, M., Slaughter, V., & Suddendorf, T. (2018). Re-evaluating the neonatal imitation hypothesis. *Developmental Science, 22*(2), e12720. https://doi.org/10.1111/desc.12720

Oostenbroek, J., Suddendorf, T., Nielsen, M., Redshaw, J., Kennedy-Costantini, S., Davis, J., ... Slaughter, V. (2016) Comprehensive longitudinal study challenges the existence of neonatal imitation in humans. *Current Biology, 26,* 1334–1338. doi: 10 .1016/j.cub.2016.03.047

O'Sullivan, E. P., Bijvoet-van den Berg, S., & Caldwell, C. A. (2018). Automatic imitation effects are influenced by experience of synchronous action in children. *Journal of Experimental Child Psychology, 171,* 113–130. doi: 10.1016/j.jecp.2018.01.013

Quadrelli, E., & Turati, C. (2016). Origins and development of mirroring mechanisms: A neuroconstructivist framework. *British Journal of Developmental Psychology, 34,* 6–23.

Piaget, J. (1951). *Play, dreams, and imitation in childhood.* London: Routledge & Kegan Paul.

Ray, E., & Heyes, C. (2011). Imitation in infancy: The wealth of the stimulus. *Developmental Science, 14,* 92–105.

Rayson, H., Bonaiuto, J. J., Ferrari, P. F., & Murray, L. (2017). Early maternal mirroring predicts infant motor system activation during facial expression observation. *Scientific Reports, 7,* 11738. doi: 10.1038/s41598-017-12097-w

Scopa, C., & Palagi, E. (2016). Mimic me while playing! Social tolerance and rapid facial mimicry in macaques (*Macaca tonkeana* and *Macaca fuscata*). *Journal of Comparative Psychology, 130,* 153–161. doi: 10.1037/com0000028

Simpson, E. A., Murray, L., Paukner, A., & Ferrari, P. F. (2014). The mirror neuron system as revealed through neonatal imitation: Presence from birth, predictive power and evidence of plasticity. *Philosophical Transactions of the Royal Society B: Biological Sciences, 369,* 20130289. doi: 10.1098/rstb.2013.0289

Slater, A. (1998). *Perceptual development: Visual, auditory, and speech perception in infancy.* East Sussex, United Kingdom: Psychology Press.

Uzgiris, I. C., & Hunt, J. M. (1975). *Assessment in infancy: Ordinal scales of psychological development.* Urbana: University of Illinois Press.

6 The Development of the Social Brain within a Family Context

Diane Goldenberg, Narcis Marshall, Sofia Cardenas, and Darby Saxbe

Overview

The parent–child relationship is often the first and most foundational bond, serving as the groundwork upon which all other social bonds will be built upon. It consists of deeply self-relevant stimuli of such critical evolutionary importance for our survival as a species that it may be deeply embedded within the brain. The current chapter will review research on how parents are represented in their child's brain and how a child is represented in the parent's brain. We will also describe how emerging work is beginning to characterize synchronous neural response in both members of the parent–child dyad in real time, providing insight into potential mechanisms that support the co-construction of these neural representations over time. We will conclude by examining how the parent–child bond may provide a window into the self–other overlap that occurs, to some extent, in all social relationships throughout the life span.

Introduction

Humans, from birth, are wired to connect. Newborns preferentially attend to social information, looking longer at faces than any other objects (Johnson, Dziurawiec, Ellis, & Morton, 1991) and listening longer to human voices than any other sounds (Vouloumanos & Werker, 2007). Even two dots and a line capture an infant's attention when forming the rudimentary configuration of a face (Goren, Sarty, & Wu, 1975). Parental stimuli appear to be of particularly profound importance for infants, likely due to prenatal fetal experiences. For example, the mother's voice becomes familiar to her developing fetus, the vibrations of her vocalizations carried through her bones and reverberating into the amniotic fluid (Moon & Fifer, 2000). In the womb, the boundary between self and other are as permeable as the umbilical cord for the mother and her developing child. In the days and months after birth the infant depends entirely upon parental care for survival. Mothers and fathers provide external regulation to the infant's immature systems, and the line between self and other is further blurred, extending the enmeshment that occurs in the womb into the realm of daily life.

This enmeshment may be reflected through biobehavioral synchrony, in which hormonal, physiological, and behavioral cues are dynamically exchanged between parent and young during social contact (Feldman, 2012). Over time, each member of the dyad becomes sensitized to and influenced by their partner's physiological and behavioral cues, forming the basis for the parent–infant bond. For example, each partners' endocrine reactivity may undergo processes of attunement through the interplay of genetic dispositions, prenatal environment, and postpartum experiences. In this way, the behavior and biology of parent and child becomes entwined, each element shaping the other in a dynamic and bidirectional manner. The parent–child bond may thus provide a unique lens through which to understand the larger psychological phenomenon of the distinction between the self and of the other.

In humans, the question of whether a common representation network between self and other exists (Decety & Sommerville, 2003) has spanned several lines of research, including developmental science, social psychology, and cognitive neuroscience. Similar computational and neural mechanisms are involved during mental representation of one's own action and actions of others. Self–other equivalences exist when the body movements of one person match those of another, such as during the behavioral synchrony that occurs with parents and their infants. As social beings, the self is intimately tied to others, yet is also distinct. Self–other overlap may characterize social relationships, particularly close relationships, throughout the life span. The parent–child relationship is often the first and most foundational bond, serving as the groundwork upon which all other social bonds will be built upon.

From a developmental perspective, the overlap between self and other within the context of the parent–child bond must flexibly adjust to serve the needs of the maturing child. For both parent and child, there may be tight fusion between self and other during infancy. In early childhood, the parent–child bond serves as an anchor for healthy socioemotional development (Morris, Silk, Steinberg, Myers, & Robinson, 2007). The degree of self–other overlap may shrink as the child begins forming bonds with others in contexts outside the home, such as school. Adolescence, in particular, marks a pronounced shift as a period of social reorientation (Blakemore, 2008). A growing need for independence from parents scaffolds the transition into adulthood, and peers develop a new salience. The individual explores aspects of their selves that are distinct from their parents, with the evolutionarily adaptive purpose of forming new relationships outside the family and ultimately forming new families of their own. A cycle begins anew with the once-child now occupying the other side of the parent–child dyad. We thus move through processes of enmeshment and individuation in our relationships with others, rooted within the original parent–child bond. This bond consists of deeply self-relevant stimuli, of such critical evolutionary importance for our survival as a species that it may be deeply embedded within the brain.

Advances in neuroimaging have allowed researchers to investigate the neural underpinnings of the parent–child bond in parents and their children. Although rodent models have provided the groundwork for understanding this research question at a mechanistic level, we will be focusing on humans. Although some studies have examined structural brain changes in parent–child contexts, this chapter focuses on functional neural responses to social stimuli.

Typically, experiments involve participants watching or listening to stimuli related to their parent or child during functional magnetic resonance imaging (fMRI). This methodology provides a proxy for brain activity by indirectly assessing changes in regional blood oxygenation, allowing researchers to evaluate the activity within the brain as a function of various states of mind elicited by cognitive tasks or sensory stimuli. Other studies use electroencephalography (EEG) or magnetoencephalography (MEG), which respectively measure the brain's electrical fields and its magnetic fields via electrodes. Both methods offer complementary strengths: EEG/MEG provides excellent temporal resolution, allowing for in-the-moment testing of behaviorally synchronous interactions. In contrast, fMRI provides optimal spatial resolution, important because many of the structures that have been implicated in the parent–child bond are located deep within the brain and cannot easily be assessed via electrodes. Together, these methodologies provide a more complete view of the parent or child brain.

The current chapter will review research on how parents are represented in their child's brain, and how a child is represented in the parent's brain. We will also describe emerging work that is beginning to characterize synchronous neural response in both members of the dyad in real time, providing insight into potential mechanisms that support the co-construction of these neural representations over time. We will conclude with future directions and closing remarks on how the examination of the parent–child bond may provide a window into the self–other overlap that occurs, to some extent, in all social relationships throughout the life span.

Representation of the Parent within a Child's Brain

Child Neural Response to Parental Stimuli

Parents play a tremendous role in shaping their children's development throughout the life span. Evidence from a growing body of neuroimaging work shows that some of the most important effects parents have on their children may be reflected in the brain. One study using MEG found that children viewing video recordings of their own mother–child interactions (compared with unfamiliar mother–child dyads) activated a greater neural response across distributed brain regions, including the superior temporal sulcus, insula, and fusiform gyrus (Pratt, Goldstein, & Feldman, 2018). These regions are implicated in mentalizing, reaction to motivational salience, and embodiment of others' emotions, respectively, and are collectively posited to be involved in attachment (Feldman, 2015). Furthermore, oscillatory power in response to own mother–child stimuli in the theta and gamma frequency bands was positively linked with the degree of social synchrony between mother and child during coded interactions, providing novel evidence for the link between mother–child synchrony and neural response to attachment cues in children. The unique effects of parental stimuli on child neural response demonstrate the considerable resources devoted to the representation of this relationship, essential for survival, within a child's brain.

Neural Underpinnings of Parental Influence on Child Behavioral Response

Emerging work has uncovered differential patterns of brain activation in response to parent stimuli among important neural regions that regulate reward-response and risk taking. For example, one study implemented a paradigm in which adolescents performed a risky decision-making task during fMRI while aware that their mothers were present in the scan control room and observing their task performance (Guassi-Moreira & Telzer, 2018). Maternal presence had an influence on their adolescents' brains, eliciting greater activation in the ventral striatum, a key reward-related brain region, when making safe decisions and decreased activation when choosing risky decisions. These results indicate an important and unique effect of parental presence as a regulator for decision making and risky behavior at a neural level, which has important implications for their children's behavior.

Another study directly leveraged a parent versus peer contrast in adolescents undergoing fMRI by presenting adolescent participants with brief video clips of their parent or an unfamiliar peer (Saxbe, Del Piero, Immordino-Yang, Kaplan, & Margolin, 2015). When viewing clips of the parent compared with a peer, adolescents showed greater activation in medial prefrontal regions previously identified as more responsive to familiar versus similar others (Krienen, Tu, & Buckner, 2010). In contrast, when viewing a peer compared with parent, adolescents had greater activation in posterior mentalizing structures, such as the precuneus and posterior cingulate cortex. Furthermore, this increased activation in these structures to the peer compared to parent condition was associated with more self-reported risk-taking behavior and affiliations with more risk-taking peers. Results suggest individual differences in the neural correlates that support a shift away from parents and toward peers in adolescents, which may in turn relate to adolescents' real-world risk-taking behaviors.

Another study found that neural activation during social influence was matched in area and magnitude from both peers and parents, suggesting that parental influence remains an important factor for children's neural functioning throughout later development (Welborn et al., 2016). Parents clearly continue to be important social presences for their adolescent children, impacting both their brain and behavior (Guassi-Moreira, Tashjian, Galvan, & Silvers, 2018).

Developmental Trajectories

Parental presence may also help shape the neural maturation necessary for children's successful functioning across domains of development. For example, previously institutionalized youths with a history of maternal deprivation exhibit mature-like neural connectivity patterns that are more representative of adult phenotypes when compared with age-matched controls (Gee et al., 2013). This finding demonstrates that parental presence, or the lack thereof, plays a unique and important role in the maturation of neural circuits. The authors posit that accelerated development of neural networks may be an ontogenetic adaptation in response to maternal absence.

Thus, parental presence may provide scaffolding to support the protracted development of neural connectivity in normative samples. Indeed, in many of the studies that we have reviewed, the presence of parental stimuli was related to more mature-like neural response in their typically developing children, compared to activation elicited during conditions of nonparental stimuli. For example, greater mother–child synchrony in the MEG study (Pratt et al., 2018) was linked to neural oscillations associated with more mature socioemotional brain networks (such as multi-oscillatory patterns of activation, increased gamma-band activation, and greater integration of alpha, beta, and gamma band), supporting the preliminary hypothesis that stronger mother–child attachments may help in the development of key brain networks necessary for healthy social and emotional functioning.

Similarly, in the study examining risky decision making in adolescence (Guassi-Moreira & Telzer, 2018), maternal presence elicited greater functional coupling between the prefrontal cortex and reward-processing regions when making safe decisions, which is more representative of adult phenotypes. Importantly, these effects were only observed for maternal presence and not in the presence of a nonparental adult. Indeed, adolescents showed functional coupling associated with more immature patterns of neural activation and increased risk-taking behavior when in the presence of a nonparent, highlighting the unique importance of parental factors.

Clinical Implications

Children's neural response to parents may be linked with psychological disorders like depression. Research in clinical adolescent samples suggests that symptomology and neural activation may manifest differently in response to parent versus nonparent stimuli. For example, one study found that the degree of adolescent neural response to parent stimuli was related to clinical symptoms of depression (Whittle et al., 2011). Adolescent depressive symptoms were associated with reduced neural activity during exposure to video clips of their own mother's positive behavior, compared with video clips of an unfamiliar mother's positive behavior. Specifically, reduced activation was found in the right anterior cingulate cortex, an area associated with positive emotion and self-referential processing.

Relatedly, children's neural activation may reflect the intergenerational transmission of aggression within families. In a longitudinal study, adolescents' reduced activation in emotion processing regions when rating their parents' emotions mediated the association between parents' past aggression and adolescents' subsequent aggressive behavior toward parents (Saxbe, Del Piero, Immordino-Yang, Kaplan, & Margolis, 2016). These findings suggest that adolescents who show dampened neural responses to their parents' emotions are most at risk for perpetuating cycles of family violence. This work from developmental and social neuroscience underscores the importance of understanding child neural reactivity to parent stimuli as it relates to healthy trajectories as well as points of vulnerability during development. Conversely, parallel and convergent lines of research, largely within the parenting literature, have been conducted to understand how one's own child is represented in the brains of mothers and fathers.

Representation of the Child within a Parent's Brain

Many parents report the birth of their first child as a momentous event in their adult lives, marking the start of a phase in which caring for another becomes their central focus. Given the importance of parenting, more researchers have shifted attention toward understanding how the parental brain changes during the postpartum period to prepare for the parent–infant bond (Feldman, 2015; Swain, 2008). This period may offer a window of neuroplasticity that allows individuals to adjust to the physiological changes and novel experiences of parenthood, adaptively acquiring the sensitive caregiving behaviors upon which their infants depend upon to survive and thrive.

Parent Neural Response to Child Stimuli

Given that parenting is crucial to the survival of our species, we have likely evolved neural and motivational systems that are finely tuned for parenting. Indeed, several fMRI studies have found that human parents respond differentially to cues representing their infant versus other infants (Feldman, 2012). In a seminal study on the parenting brain, first-time mothers who were two to four months postpartum looked at pictures of their own infant versus other, age-matched infants during fMRI. When viewing own versus other infants, mothers had more activation in the orbitofrontal cortex, an area associated with ratings of pleasant mood and positive attachment (Nitschke et al., 2004). This study suggests that mothers have a distinct neural response to their own infant versus another infant, and that this activation relates to feelings supporting the motivation to provide care.

More recent research on the parenting brain has attempted to understand the relationship between hormones, neural activity, and parenting behavior. One study found that fathers with greater striatal activation, implicated in reward, when viewing images of their infants during an fMRI task also had higher testosterone after interacting with their infants in an interaction task (Kuo, Carp, Light, & Grewen, 2012). Typically, lower paternal testosterone across the postpartum period is associated with higher paternal caretaking (Storey, Noseworthy, Delahunty, Halfyard, & McKay, 2011). In many species, testosterone is believed to support the mating effort at the expense of the parenting effort. For example, one study found that plasma testosterone levels and testes volume (aspects of reproductive biology relating to mating effort) were inversely correlated with caregiving (Mascaro, Hackett, & Rilling, 2013). Additionally, when fathers viewed pictures of their own child, activity in a key component of the motivational system (i.e., ventral tegmental area) predicted paternal caregiving and was negatively related to testes volume.

The biology of human males may reflect a trade-off between the mating and parenting efforts. However, researchers theorize that short-term testosterone spikes after infant interaction may encourage protective paternal behaviors (Kuo et al., 2012). Collectively, this research demonstrates that there are neural changes during the transition to parenthood that support the acquisition of new skills and changing motivations required for sensitive caregiving in both mothers and fathers. The parenting brain does not exist in a vacuum, but rather it dynamically interacts with biological processes, such as hormones, as well as parenting behavior.

Neural Underpinnings of Child Influence on Parental Response

Despite recent interest in the parenting brain, little is known about how parental neural response in the scanner relates specifically to real-world parenting behavior. A few studies have attempted to establish the functional significance of activation patterns by linking them with observed behavior outside the scanner. In one study, maternal neural activation while mothers listened to their own infant crying was related to behavior during a free play and cleanup task (Musser, Kaiser-Laurent, & Ablow, 2012). The results demonstrated that greater maternal sensitivity during the behavioral session was associated with increased activation in neural regions involved in regulatory control. The investigators speculated that mothers who exhibit greater regulation of negative response to infant cries at the neural level are better able to demonstrate more sensitive responding at the behavioral level.

Using the same sample, the researchers also found that maternal neural activation during the mothers' own infants' cry related to the attachment behavior of their infants outside the scanner (Laurent & Ablow, 2012). Specifically, mothers with lower activation in regulatory regions of the brain had infants with greater amounts of avoidant behavior during another behavioral session in which the mother and infant were separated and then reunited.

More recently, researchers examined interoceptive sensitivity in the parental brain in infancy as a predictor of child somatization behavior six years later (Abraham, Hendler, Zagoory-Sharon, & Feldman, 2018). Interoception—the representation of key bodily signals—is a key aspect of human caregiving. The results found that greater activation in the anterior insula, a region implicated in interoceptive processes, to own-infant stimuli predicted lower somatic problems in the children six years later. Further, parent sensitivity partially mediated the links between parental neural activation and child somatic symptoms, suggesting a potential pathway for how parental neural response may influence child behavior.

To our knowledge, only one study thus far has examined the neurobiological underpinnings of caregiving behavior by directly measuring neural response while parents perform a behavioral task relevant for caregiving. In this study, mothers performed a parenting decision-making task in an fMRI setting to evaluate neural correlates of child feedback to caregiving decisions (Ho, Konrath, Brown, & Swain, 2014). The task was designed to probe brain circuits underlying goal-directed parenting behaviors in the context of parent–child interactions. The findings demonstrated an interplay between neural function during decision-making, dispositional empathy, and cortisol reactivity to negative child feedback among mothers. These results highlight a need for further research investigating the neural underpinnings of behaviors relevant for sensitive caregiving in parents.

In our own laboratory, we are currently undertaking work in which first-time fathers undergo fMRI while performing a novel task designed to approximate real-world operations for new parents (exerting regulatory control while listening to continuous infant cries). We are using a well-validated task (the go/no-go; Aron & Poldrack, 2006) to evaluate the neural underpinnings of regulatory control, adapted so that new fathers perform the go/no-go during fMRI while listening to recordings of either continuous infant cries or frequency-matched white noise. In this way, we can probe the neural underpinnings of regulatory control exerted during infant cries, compared with distracting, noninfant stimuli. As the field continues to progress, future work can provide further insight into how the parental brain relates to the very caregiving behaviors it is hypothesized to support.

Developmental Trajectories

The transition to parenthood is not often considered a developmental stage, yet, as we have reviewed, it is a period of significant change in multiple domains, including roles, motivations, hormones, and neurobiology. Research has explored the differences between parents and age-matched nonparents. Most human adults have a baseline motivation to care for infants, which likely serves the evolutionarily adaptive purpose of encouraging any individual to step in as an infant caretaker in the absence of a biological parent (Feldman, 2015). However, fundamental differences have been found at the neural level between parents and nonparents in response to infant stimuli. For example, neural activation in response to infant cries or laughter depended upon parental experience, such that parents showed stronger activation in motivational systems from crying and nonparents showed stronger activation from laughing (Seifritz et al., 2003).

Interestingly, these findings were independent of parent sex, suggesting that the brains of mothers and fathers are uniquely tuned to infant stimuli, despite biological differences in their pathways to parenthood. Do mothers and fathers form social attachments to their infants through similar neural mechanisms? The "distinct pathways hypothesis" suggests that the parental brain is shaped by two distinct pathways, a 'maternal pathway' and a 'paternal pathway' (Feldman, 2015). Mothers have a complex cascade of hormonal changes driven by pregnancy and lactation, which have been linked to greater sensitivity to infant cues and maternal caregiving behavior. In contrast, the paternal pathway to attachment is thought to occur through repeated activation of the "mirror network," a group of brain regions associated with social cognition (e.g., inferior frontal cortex and superior parietal lobule; Abraham et al., 2014).

Researchers have attempted to test the distinct pathway hypothesis by comparing the neural response of mothers and fathers to video clips of their infants. The results have suggested that mothers had greater activation in regions associated with emotion salience (e.g., amygdala) and fathers had greater activation in areas associated with cognitive empathy (e.g., tempo-parietal junction) (Atzil, Hendler, Zagoory-Sharon, Winetraub, & Feldman, 2012). These results support the theory that the development of maternal and paternal attachments are facilitated by differential neural pathways.

Clinical Implications

Researchers have begun to address the effect of adversity on the ability of parents to form attachment bonds with their infants. Twenty percent of post-partum women experience anxiety or depressive disorders, which can have detrimental effects on the mother, child, and family (Pawluski, Lonstein, & Fleming, 2017). Mothers with depression, compared with healthy moth-ers, have reduced neural activation in prefrontal areas when encountering infant cues (Kingston, Tough, & Whitfield, 2012). Depressed mothers also appear to have heightened amygdala activation to unfamiliar babies (Wonch et al., 2016), perhaps blunting the neural activation for their own infant. Furthermore, mothers with weaker connectivity between the amygdala and insula reported greater symptoms of depression, while those with stronger connectivity were more responsive to their newborns. Postpartum anxiety and depression may affect a mother's ability to be responsive to her infant's needs, with potentially long-lasting implications for both mother and child.

Synchronous Neural Response in the Parent–Child Dyad

Despite growing evidence suggesting the neural underpinnings of the parent–child bond and its implications for both parent and child well-being, our understanding of the neural processes linking these two minds is only beginning to emerge. Recent theoretical work suggests that more similar neural states allow individuals to connect and be attuned to their environment in a more harmonious way (Wheatley, Kang, Parkinson, & Looser, 2012). Although this concept applies to any two (or more) similar individuals undergoing the same experience, it is especially relevant to the parent–child relationship, in which genotypic and phenotypic similarities are more likely to exist, and experiential similarities are magnified throughout time with the accumulation of shared experiences.

Recent work using novel techniques is unveiling how brain-to-brain synchronization may occur in parent–child dyads, dynamically shaping the brain and behavior of each individual over time. One recent study videorecorded mother–child dyads in their home environments and coded for moments of behavioral synchrony (e.g., simultaneous moments of positive affect) (Levy, Goldstein, & Feldman, 2017). When children and their mothers later observed these videos during MEG, episodes of behavioral synchrony, compared with nonsynchrony, elicited increased neural oscillations in the superior temporal sulcus, a hub of social cognition and mentalizing. Furthermore, this neural pattern was coupled between mother and child and was dyad-specific (i.e., not evident when observing videos of unfamiliar mother–child dyads). This finding suggests that behavior-based processes may drive synchrony between two brains during social interactions.

This neural synchrony is not only evinced during moment-to-moment interactions, it appears to be reflected in the brain's stable and intrinsic architecture, as measured through resting state MRI. Researchers acquired resting state readings in mother–child dyads and daily diary data for two weeks, finding that parents and children with more similar neural architectures also had more similar day-to-day emotional synchrony (Lee, Miernicki, & Telzer, 2017). Importantly, dyadic neural similarity was associated with greater emotional competence in children, suggesting that being neutrally in-tune with parents confers emotional benefits.

This work suggests that neural synchrony in the parent–child dyad exists and is likely built over time, yet the neural mechanisms underlying this synchrony, as it occurs in the moment, are unknown. Two recent studies have used dual-EEG measurement during a real-time interaction in dyads to explore this question. One study found that child behavior predicted maternal neuronal oscillations and maternal behavior predicted the neuronal oscillations of her child, providing novel evidence for bidirectional pathways of influence (Atzaba-Poria, Deater-Deckard, & Bell, 2017). The other study examined adult–infant dyads and found significant mutual neural coupling during a social interaction task (Leong et al., 2017). Furthermore, direct gaze strengthened adult–infant neural connectivity in both directions during communication, suggesting a potential mechanism for how brain-to-brain synchrony occurs.

As the field continues to move forward, studies such as these will illuminate how biobehavioral synchrony is both dynamic and iterative, both members of the dyad shaping each other's biology and behavior over time.

Conclusions and Future Directions

In this chapter, we have described how children respond to parent stimuli and how parents respond to child stimuli, along with preliminary evidence for parent–child neural synchrony. It is clear from the literature that the parent–child relationship is represented in the brain across a broad array of networks and regions, underscoring the importance and evolutionary salience of this bond. In particular, regions involved in mentalizing (e.g., superior temporal sulcus), recognition and reward (e.g., fusiform gyrus; ventral striatum), and embodying the physical states of ourselves and others (e.g., insula) activate in children (to their parents) and in parents (to their children). Collectively, these findings from several lines of research illustrate how the parent–child bond is represented analogously in the brain from both sides of the dyad, drawing upon a larger caregiving network (Feldman, 2015).

Despite the interesting findings reviewed in this chapter, there is a clear need for more research on the neural correlates of the parent–child bond, given its evolutionary significance. Longitudinal research is needed to understand how children view their parents over time. For example, a study that tracked children's responses to parental stimuli over several years (from early to late childhood or from childhood to adolescence) would help to elucidate the changing salience of the parent–child relationship across development. Similarly, parents' responses to own-child stimuli might change over the transition to parenthood and as the child matures.

More prospective research on the functional real-world implications of neural responding to parent–child stimuli is also needed. If children show greater neural activation to their parents compared with strangers does that reflect the strength of their attachment to parents and predict their future socioemotional developmental outcomes? Similarly, although there is some preliminary evidence that parents' responses to infant stimuli are associated with their parenting sensitivity, more research is needed to probe the mechanisms that might underlie this relationship and identify targets for intervention. Both fathers and mothers have participated in several of the studies reviewed in this chapter, which is valuable because fathers play an important role in parenting. However, most studies that have examined child responses to parent stimuli or child–parent neural synchrony have focused exclusively on mothers.

In conclusion, the work reviewed in this chapter has the potential to bridge parenting literature with developmental and social neuroscience in pursuit of a broader understanding of the neural underpinnings of social relations grounded within the evolutionarily conserved primary bond of the parent–child relationship. The parenting brain is an ancient system sensitized by pregnancy and the postpartum period. It underpins parental motivation via dopaminergic reward pathways and supports the ability to resonate with infant states via the anterior insula and cingulate cortex to provide sensitive and attuned caregiving. Similarly, the child's affiliative brain appears to respond in a comparable way to attachment cues.

The partially superimposed neural representations of parent and child may support the degree of overlap between self and other that contributes to each individual's survival-related attunement with the other. In a larger sense, using one's own perspective to simulate the psychological state of another during empathic responding is a process implied to be a "breach of individual separateness" (Singer et al., 2004), as reflected through activations that mirror self-focused responses. Indeed, threat-responsive regions of the brain have been found to represent others in a manner that is very similar to the way they represent the self, but only to the extent that those others are perceived as familiar (Beckes, Coan, & Hasselmo).

These findings suggest that one of the defining features of human social bonding may be increasing levels of overlap between neural representations of self and other. Thus, the self–other overlap that characterizes the exclusive parent–child bond at the neural level may ultimately inform a deeper understanding of the larger neural representations of others and ourselves throughout the life span.

References

Abraham, E., Hendler, T., Shapira-Lichter, I., Kanat-Maymon, Y., Zagoory-Sharon, O., & Feldman, R. (2014). Father's brain is sensitive to childcare experiences.

Proceedings of the National Academy of Sciences of the United States of America, 111(27), 9792–9797.

Abraham, E., Hendler, T., Zagoory-Sharon, O., & Feldman, R. (2018). Interoception sensitivity in the parental brain during the first months of parenting modulates children's somatic symptoms six years later: The role of oxytocin. *International Journal of Psychophysiology, 136,* 39–48.

Aron, A. R., & Poldrack, R. A. (2006). Cortical and subcortical contributions to stop signal response inhibition: Role of the subthalamic nucleus. *Journal of Neuroscience, 26*(9), 2424–2433.

Atzaba-Poria, N., Deater-Deckard, K., & Bell, M. A. (2017). Mother–child interaction: Links between mother and child frontal electroencephalograph asymmetry and negative behavior. *Child Development, 88*(2), 544–554.

Atzil, S., Hendler, T., Zagoory-Sharon, O., Winetraub, Y., & Feldman, R. (2012). Synchrony and specificity in the maternal and the paternal brain: Relations to oxytocin and vasopressin. *Journal of the American Academy of Child and Adolescent Psychiatry, 51*(8), 798–811.

Beckes, L., Coan, J. A., & Hasselmo, K. (2012). Familiarity promotes the blurring of self and other in the neural representation of threat. *Social Cognitive and Affective Neuroscience, 8*(6), 670–677.

Blakemore, S. J. (2008). The social brain in adolescence. *Nature Reviews Neuroscience, 9*(4), 267.

Decety, J., & Sommerville, J. A. (2003). Shared representations between self and other: A social cognitive neuroscience view. *Trends in Cognitive Sciences, 7*(12), 527–533.

Feldman, R. (2012). Parent–infant synchrony: A biobehavioral model of mutual influences in the formation of affiliative bonds. *Monographs of the Society for Research in Child Development, 77*(2), 42–51.

Feldman, R. (2015). The adaptive human parental brain: Implications for children's social development. *Trends in Neurosciences, 38*(6), 387–399.

Gee, D. G., Gabard-Durnam, L. J., Flannery, J., Goff, B., Humphreys, K. L., Telzer, E. H., … Tottenham, N. (2013). Early developmental emergence of human amygdala–prefrontal connectivity after maternal deprivation. *Proceedings of the National Academy of Sciences of the United States of America, 110*(39), 15638–15643.

Goren, C. C., Sarty, M., & Wu, P. Y. (1975). Visual following and pattern discrimination of face-like stimuli by newborn infants. *Pediatrics, 56*(4), 544–549.

Guassi Moreira, J. F., Tashjian, S. M., Galván, A., & Silvers, J. A. (2018). Parents versus peers: Assessing the impact of social agents on decision making in young adults. *Psychological Science, 29*(9), 1526–1539.

Guassi Moreira, J. F., & Telzer, E. H. (2018). Mother still knows best: Maternal influence uniquely modulates adolescent reward sensitivity during risk taking. *Developmental Science, 21*(1), e12484.

Ho, S. S., Konrath, S., Brown, S., & Swain, J. E. (2014). Empathy and stress related neural responses in maternal decision making. *Frontiers in Neuroscience, 8*, 152.

Johnson, M. H., Dziurawiec, S., Ellis, H., & Morton, J. (1991). Newborns' preferential tracking of face-like stimuli and its subsequent decline. *Cognition, 40*(1–2), 1–19.

Kingston, D., Tough, S., & Whitfield, H. (2012). Prenatal and postpartum maternal psychological distress and infant development: A systematic review. *Child Psychiatry and Human Development, 43*(5), 683–714.

Krienen, F. M., Tu, P. C., & Buckner, R. L. (2010). Clan mentality: Evidence that the medial prefrontal cortex responds to close others. *Journal of Neuroscience, 30*, 13906–13915.

Kuo, P. X., Carp, J., Light, K. C., & Grewen, K. M. (2012). Neural responses to infants linked with behavioral interactions and testosterone in fathers. *Biological Psychology, 91*(2), 302–306.

Laurent, H. K., & Ablow, J. C. (2012). The missing link: Mothers' neural response to infant cry related to infant attachment behaviors. *Infant Behavior and Development, 35*(4), 761–772.

Lee, T. H., Miernicki, M. E., & Telzer, E. H. (2017). Families that fire together smile together: Resting state connectome similarity and daily emotional synchrony in parent-child dyads. *Neuroimage, 152*, 31–37.

Leong, V., Byrne, E., Clackson, K., Georgieva, S., Lam, S., & Wass, S. (2017). Speaker gaze increases information coupling between infant and adult brains. *Proceedings of the National Academy of Sciences of the United States of America, 114*(50), 13290–13295.

Levy, J., Goldstein, A., & Feldman, R. (2017). Perception of social synchrony induces mother–child gamma coupling in the social brain. *Social Cognitive and Affective Neuroscience, 12*(7), 1036–1046.

Mascaro, J. S., Hackett, P. D., & Rilling, J. K. (2013). Testicular volume is inversely correlated with nurturing-related brain activity in human fathers. *Proceedings of the National Academy of Sciences of the United States of America, 110*(39), 15746–15751.

Moon, C. M., & Fifer, W. P. (2000). Evidence of transnatal auditory learning. *Journal of Perinatology, 20*(S1), S37.

Morris, A. S., Silk, J. S., Steinberg, L., Myers, S. S., & Robinson, L. R. (2007). The role of the family context in the development of emotion regulation. *Social Development, 16*(2), 361–388.

Musser, E. D., Kaiser-Laurent, H., & Ablow, J. C. (2012). The neural correlates of maternal sensitivity: An fMRI study. *Developmental Cognitive Neuroscience, 2*(4), 428–436.

Nitschke, J. B., Nelson, E. E., Rusch, B. D., Fox, A. S., Oakes, T. R., & Davidson, R. J. (2004). Orbitofrontal cortex tracks positive mood in mothers viewing pictures of their newborn infants. *Neuroimage, 21*(2), 583–592.

Pawluski, J. L., Lonstein, J. S., & Fleming, A. S. (2017). The neurobiology of postpartum anxiety and depression. *Trends in Neurosciences, 40*(2), 106–120.

Pratt, M., Goldstein, A., & Feldman, R. (2018). Child brain exhibits a multi-rhythmic response to attachment cues. *Social Cognitive and Affective Neuroscience, 13*(9), 957–966.

Saxbe, D., Del Piero, L. B., Immordino-Yang, M. H., Kaplan, J. T., & Margolin, G. (2015). Neural correlates of adolescents' viewing of parents' and peers' emotions: Associations with risk-taking behavior and risky peer affiliations. *Social Neuroscience, 10*(6), 592–604.

Saxbe, D., Del Piero, L. B., Immordino-Yang, M. H., Kaplan, J. T., & Margolin, G. (2016). Neural mediators of the intergenerational transmission of family aggression. *Development and Psychopathology, 28*(2), 595–606.

Seifritz, E., Esposito, F., Neuhoff, J. G., Lüthi, A., Mustovic, H., Dammann, G., ... Di Salle, F. (2003). Differential sex-independent amygdala response to infant crying and laughing in parents versus nonparents. *Biological Psychiatry, 54*(12), 1367–1375.

Singer, T., Seymour, B., O'Doherty, J., Kaube, H., Dolan, R. J., & Frith, C. D. (2004). Empathy for pain involves the affective but not sensory components of pain. *Science, 303*(5661), 1157–1162.

Storey, A. E., Noseworthy, D. E., Delahunty, K. M., Halfyard, S. J., & McKay, D. W. (2011). The effects of social context on the hormonal and behavioral responsiveness of human fathers. *Hormones and Behavior, 60*(4), 353–361.

Swain, J. E. (2008). Baby stimuli and the parent brain: Functional neuroimaging of the neural substrates of parent-infant attachment. *Psychiatry (Edgmont), 5*(8), 28–36.

Vouloumanos, A., & Werker, J. F. (2007). Listening to language at birth: Evidence for a bias for speech in neonates. *Developmental Science, 10*(2), 159–164.

Welborn, B. L., Lieberman, M. D., Goldenberg, D., Fuligni, A. J., Galván, A., & Telzer, E. H. (2016). Neural mechanisms of social influence in adolescence. *Social Cognitive and Affective Neuroscience, 11*(1), 100–109.

Wheatley, T., Kang, O., Parkinson, C., & Looser, C. E. (2012). From mind perception to mental connection: Synchrony as a mechanism for social understanding. *Social and Personality Psychology Compass, 6*(8), 589–606.

Whittle, S., Yücel, M., Forbes, E. E., Davey, C. G., Harding, I. H., Sheeber, L.,…Allen, N. B. (2011). Adolescents' depressive symptoms moderate neural responses to their mothers' positive behavior. *Social Cognitive and Affective Neuroscience, 7*(1), 23–34.

Wonch, K. E., de Medeiros, C. B., Barrett, J. A., Dudin, A., Cunningham, W. A., Hall, G. B.,…Fleming, A. S. (2016). Postpartum depression and brain response to infants: Differential amygdala response and connectivity. *Social Neuroscience, 11*(6), 600–617.

II Language and Theory of Mind

7 Infants' Early Competence for Language and Symbols

Ghislaine Dehaene-Lambertz, Ana Fló, and Marcela Peña

Overview

Humans have much more sophisticated communication skills than other species. They are not limited to emotional cries, alarm calls, and soothing demands; they also interpret the inner and outer world in a symbolic way, resulting in a collective intelligence and an accumulation of knowledge called culture. This culture permeates the child and fosters efficient learning, based on the knowledge accumulated through generations. To develop this collective intelligence, it requires (a) a social brain predisposed to learn from conspecifics, (b) awareness of one's mental state and knowledge and those of others, (c) a shared common language of thought, and (d) a communication system for exchanging this information. We insist on the value of symbolic representations as a compressed, necessary format for representing information to ourselves and exchanging information with others. We propose that human cognition has been boosted beyond the cognition of other primates by the multiplicative advantage of codevelopment of social cognition, language but also symbolic thinking that can be observed from the first months of life on.

Introduction

Humans are constantly looking for rules and causal relationships to explain what has happened and predict what will happen. Collaborative thinking in adults allows a significant improvement in prediction accuracy (Bahrami et al., 2010), but collaborating with others requires, on the one hand, having explicit representations of the problem to resolve and, on the other hand, knowing that it is possible to share these representations unambiguously with another mind. This shared cognition implies a set of symbols that efficiently summarize the concepts we want to represent first to ourselves and second to share with others, but also an implicit assumption that the other can understand these symbols in order to capitalize on each other's knowledge. Thus, beyond a theory of mind, this shared cognition requires a pedagogical stance, as proposed by Csibra and Gergely (2009). This pedagogical capacity might have existed since the ancient hominins, if we accept the Oldowan stone tool industry (2.34 million years ago) (d'Errico & Banks, 2015) as one of the oldest testimony of collective elaborated production. But how does it begin in infants?

A Symbolic Brain

In the flow of thoughts, to isolate relevant information that could be shared with others, it is necessary to summarize and discretize sensory information. A first step is to gather different objects sharing common characteristics into a single category, but a more powerful operation would be to further compress this information into a single arbitrary symbolic form. Humans are particularly skilled at creating and using symbolic systems: music notation, traffic signs, equations, even scarification and uniforms are simplified marks that summarize complex information. Language is the first symbolic system acquired by infants and the most productive and versatile. A word can condense the essence of an individual, a category of objects, an action, an abstract concept such as freedom. These symbols can be combined in logical operations such as addition, negation, exclusion, and quantification or even superimposed to create poetic effects. The symbolic power of language is evident in adult exchanges. Infants may also be sensitive to it very early on, when they listen to speech.

No one denies that words are arbitrary labels attached to a semantic concept, but the initial relationship between the label and the concept is disputed. It is conventionally assumed that, because a label is produced associated with an object, infants first learn about the co-occurrence of these two events, the labels being only another characteristic of the object, like the sound it makes when it falls. Gradually, infants understand that the label can be used to refer to the object. Instead, we propose that infants immediately use the label as an internal variable that stands for the object.

We also propose that this variable is explicit, at a high-level node that establishes contact between a global workspace and domain-specific modules.

Because of the location of symbols at a higher-level, infants can explicitly and consciously control their use of labels, notably to share and receive information. They can also use them to combine concepts calculated in underlying modules, such as "to the left of the blue wall," generating new unitary representations (Hermer-Vazquez, Spelke, & Katsnelson, 1999). We support our proposal by examining comparative brain anatomy, a reinterpretation of published studies in infants, and recent studies directly testing the hypothesis of an early symbolic system.

Development of the Frontal Areas in Humans

Symbolic representations and manipulation are assumed to be supported by frontal areas (Nieder, 2009). Indeed, when adult humans and macaques listen to the same tone sequences that vary in either the number of tones or the structure of the sequence, both species detect the changes, but only in humans is the left inferior frontal gyrus, a common region, activated by both changes. This result was interpreted as evidence of a more abstract "change" code in humans compared with the simple response of discrimination in each specific module in macaques (Wang, Uhrig, Jarraya, & Dehaene, 2015).

These cross-species computational differences are supposed to be supported by the expansion of the associative areas of the frontal lobe, the inferior parietal and the posterior part of the superior temporal regions (Chaplin, Yu, Soares, Gattass, & Rosa, 2013), in parallel with the development of large tracts connecting the frontal areas to all other lobes, such as the arcuate fasciculus with the inferior parietal and the superior temporal regions. The major difference in connectivity between macaques and humans is the large connectivity of the inferior frontal regions with the associative auditory cortices in humans (Neubert, Mars, Thomas, Sallet, & Rushworth, 2014).

These particularities are already observed during gestation. The development of the modern human brain differs significantly from monkeys and even from older humans. In particular, its prolonged maturation over many years increases the period of plasticity. Unlike chimpanzees, human fetuses retain rapid brain growth after 22 weeks of gestation, which persists even up to two years (Sakai et al., 2012; DeSilva & Lesnik, 2006; Coqueugniot, Hublin, Veillon, Houet, & Jacob, 2004; Neubauer, Gunz, & Hublin, 2010). During that period, the prefrontal cortex develops faster than the rest of the brain, again unlike chimpanzees (Sakai et al., 2011). Even compared with ancient *Homo sapiens,* the globular shape of the modern human brain is more pronounced, and it develops mainly after birth. Endocasts of *sapiens* and Neanderthal newborns are not very different, while adult shapes differ due to the enlargement of integration cortices (Neubauer, Hublin, & Gunz, 2018; Gunz, Neubauer, Maureille, & Hublin, 2010).

However, because of their slow maturation, associative regions have long been thought to be poorly functional at a young age, and the role of frontal regions in infant cognition has been underestimated. Yet several functional magnetic resonance imaging (fMRI) studies have shown early activation in these areas. Moreover, as any other brain region, the frontal lobe is never activated as a whole but is parceled into regions that show functional similarity with adult responses. For example, when short-term verbal memory is required, inferior frontal regions are involved (Dehaene-Lambertz, Dehaene, & Hertz-Pannier, 2002) whereas attention to long-term memory content depends on more dorsal regions (Dehaene-Lambertz et al., 2006). Similarly to adults, the balance between medial prefrontal and orbitofrontal regions is observed in infants when a familiar rewarding stimulus, such as the maternal voice, and a new and unknown stimulus with a value to evaluate, such as the voice of another mother, are presented (Dehaene-Lambertz et al., 2010).

Frontal areas are not only at the top of a hierarchy of a bottom-up flow of information, they also send feedback information to improve perception and direct learning. The hierarchical organization of brain areas defined by the relative proportion of neurons in supragranular (contributing to feedforward pathways) and infragranular (contributing to feedback pathways) layers is observed since gestation in primates. Feed-forward axons reach the correct target before the end of gestation. By contrast, feedback connectivity is exuberant and progressively pruned after birth (Price et al., 2006). Evidence of top-down activity has been observed with near-infrared spectroscopy in eight-month-old infants who were exposed to pairs of a tone followed by a smiley. From time to time the smiley was absent, but a response in the visual areas was recorded nevertheless, revealing that infants were expecting the image (Emberson, Richards, & Aslin, 2015). Other experiments using electroencephalography also have shown complex expectations from infants at five months (Kabdebon & Dehaene-Lambertz, 2019) and twelve months (Kouider et al., 2015) after they have learned arbitrary sound–image associations.

Frontal neurons through long-distance connectivity participate in a powerful global workspace that offers the possibility of integrating the results of the computations of the many modular brain networks into a common space (Mesulam, 1998; Dehaene & Naccache, 2001). Information in this central space can be maintained for a long time, amplified, and combined with other information but at the cost of slow serial entries. Moreover, these entries can be consciously manipulated—that is, they become explicit for oneself and reportable to others.

The signature of this conscious space is an all-or-none response. If a stimulus dimension is linearly manipulated, such as the duration of presentation of a masked face, the sensory cortices follow the same linear response. In infants, the amplitude and duration of the P400 vary linearly with the duration of the face presentation. By contrast, later responses are only recorded when the stimulus is consciously perceived and not when it remains below the perception threshold displaying a characteristic all-or-none response. In adults, this conscious stage is reached in 300 milliseconds while it is around 1 second in infants (Kouider et al., 2013).

Therefore, the immaturity of the child's brain, which is often apprehended as a contrast between mature low-level regions and immature high-level regions, is more appropriately described as a dynamic competition between parallel circuits whose computational efficiency is controlled by maturation. Differences in the speed of local computations and information transfer can favor one circuit over the other, in particular to enter the global workspace and be amplified and integrated into explicit representations.

In summary, maturation extends over many years in humans, refining connectivity and accelerating local computations and information exchange between distant regions. However, the neural architectural design is fundamentally similar to that of adults, thus leading to identical or similar computational properties, but at a much slower speed. Notably abstract computations, such as the manipulation of symbols and the conscious access to mental representations, can be accessible even to very young children. Do we have evidence of such abilities?

The Power of Words in Infants

The first acquired symbolic system is language, and many studies have pointed out that infants by five to six months of age, if not earlier, might be equipped with a functional referent-label mapping mechanism. As early as six months, they have already noticed a few words and associate them with people ("mon" "dad," the infant's own name; Tincoff & Jusczyk, 1999), body parts (Tincoff & Jusczyk, 2012), and actions ("hug," "eat"; Bergelson & Swingley, 2012). In the laboratory, they easily learn to map arbitrary sounds to objects: at two months, infants can associate one syllable with one familiar object, for example. But very quickly they succeed in more complex tasks (Gogate, 2010; Gogate, Prince, & Matatyaho, 2009). Using event-related potential (ERP), Friedrich and Friederici (2011, 2017) reported that three-month-old and six-month-old infants were able to learn the mapping between eight words and objects. However, only older infants remembered the associations the next day, and sleep seems to be crucial for maintaining learning (Friedrich, Wilhelm, Born, & Friederici, 2015; Friedrich, Wilhelm, Molle, Born, & Friederici, 2017).

What do infants learn? A simple association or more than that? Naming an object helps children in many areas. Ten-month-old infants pay more attention to objects that have previously been named than to those that are silently presented or even pointed at (Baldwin & Markman, 1989). It is not only attention that is amplified but also categorization of objects and their memorization. In a series of behavioral experiments in very young children, Waxman and collaborators studied the influence of language on the formation of conceptual categories. Different objects or images belonging to the same category were successively presented to children of different groups and ages. The presentation was accompanied either by a sentence naming the object with a pseudo-word—"Look at the blicket"—or by musical tones, or in silence. During the test, the children were consistently far more able to distinguish between two new objects—one belonging to the familiar category and the other to a new category—in the naming condition (Waxman & Markow, 1995; Balaban & Waxman, 1997; Fulkerson & Waxman, 2007; Ferry, Hespos, & Waxman, 2010). These results suggest that naming invites children, as young as three months (Ferry, Hespos, & Waxman, 2013) to form conceptual categories that they would not have considered without the use of words. This learning can be postponed, and it can appear only after a sleep period. Only six- to eight-month-old infants who nap generalize the name of an object to other exemplars of the same category (Friedrich et al., 2017; Friedrich et al., 2015).

Labeling an object with a word also allows infants to represent several objects. Before the age of twelve months, infants have difficulty maintaining the simultaneous representation of several objects. For example, when two objects appear alternately from the back of an opaque screen, infants do not seem to expect two objects when the screen is removed (Xu & Carey, 1996). However, if each time the objects appear from behind the screen they are named by two different words, the infants are surprised when only one object is revealed. They are not surprised when the objects are designated by the same generic word "toy," or accompanied by two separate musical notes or sounds (Xu, 2002). The fact of naming each object specifically allows the individualization of the two objects.

Finally, labeling makes it possible to maintain more objects in working memory. In an experiment by Feigenson and Halberda (2008), four identical objects were hidden one by one in a box. The fourteen-month-old child must subsequently recover them, but two objects were surreptitiously removed by the experimenter. The time spent by the child searching for the two missing objects is then measured. In a first condition, the first two objects are named differently from the last two—"Look, a dax," then "Look, a blicket"—while in a second condition, each object is generically called "Look at this!" Children spend more time searching for missing objects when the experimenter has separated the four physically identical objects into two groups using two separate words than when he designates them with the same generic sentence. Young children can therefore use words to push the limits of their memory storage and thus memorize the four hidden objects, an ability that only appears much later in the absence of a name.

In these experiments, infants combine two interesting properties. They use speech as a source of valuable information about the world and use the label provided by speech to help them distinguish and memorize objects categories. What is the function of this label? Is it a powerful attention grabber because speech is a common and rewarding stimulus thanks to the social context in which it is embedded? Or does the label have a symbolic value, which enables it to represent a category in a very compressed form that can be more easily handled in the internal working space?

Are Words Symbols?

In a symbolic system, there is an equivalence relationship between the set of symbols and the set of objects the symbols stand for. Thus, unlike associative learning—in which if A is followed by B, it does not mean than B is followed by A—in symbolic learning there is no direction between A and B because the object A and the symbol B point toward the same representation. Ekramnia and Dehaene-Lambertz (2019) trained four-month-old infants in a naming task for two categories of images— "fribbles" versus flying birds. The pseudo-word "kafon" was presented, followed 1 second later by one of 180 images belonging to one category (e.g., birds) whereas "pauvou" was paired in the same way to the other category (e.g., fribbles).

After thirty trials, 10 percent of incongruent trials with a mispairing were introduced to verify the infants' learning. But also 20 percent of reversed trials were introduced, in which the object was presented first followed by the name (in 10 percent of cases, the pairing was correct and in the other 10 percent, incorrect). Incorrect pairs, whether in canonical or reversed trials, induced ERP responses of surprise. Because infants have built an equivalence between the category and the name, reversed and canonical pairs are the two sides of the same coin, a process very different from associative learning, which is directional.

Kabdebon and Dehaene-Lambertz (2019) went farther and showed that infants are also able to name algebraic rules. In a series of experiments, five-month-olds were trained to associate ever-changing trisyllabic nonce words characterized by the location of the repetition of a syllable (either immediate: AAB words; or on the edges: ABA words) with an image (a fish or a lion). In the test, infants were surprised by incongruent pairings both in canonical and reversed trials.

This immediate generalization to reversed trials without further training is not observed in animals (Medam, Marzouki, Montant, & Fagot, 2016), not even in chimpanzees (Kojima, 1984). Usually animals must learn separately both directions. This does not mean that symbols are not accessible to animals. For example, macaques can learn to represent quantities by abstract visual shapes, and they can even add these symbols and associate the result with the correct quantity (Livingstone et al., 2014; Srihasam, Mandeville, Morocz, Sullivan, & Livingstone, 2012). However, it is a very slow process. By contrast, the speed at which children learn these pairs and the spontaneous bidirectional mapping indicate another learning mechanism than simple slow associative learning, as was previously assumed in infants (Nazzi & Bertoncini, 2003).

These recent experiments suggest that human infants assume an isomorphism between an internal symbolic space and the external world. Therefore, if the experimenter uses two words, the infant assumes that she must discover two kinds, whereas one word implies that all the objects presented can be grouped together. Furthermore, infants' errors in the studies by Xu (2002) and Feigenson and Halberda (2008) may suggest that they are more attentive to symbolic representations than to sensory representations, probably because of the simpler manipulation and memorization relative to the overwhelming richness of sensation. It might also reveal that the format of representations in the central workspace is symbolic, and that in explicit tasks infants have access only to this format.

Speech, an Information Tool for the World

If external information can be summarized by a symbol, what can be considered a symbol by infants? Waxman and collaborators showed, first, that it is not enough to couple an image with any sound for the sound to represent the image category; and second, that younger babies are more tolerant and accept a greater variety of sounds than are older babies: If speech and lemur vocalizations (but neither tones nor backward speech) are relevant for three-month-old infants, this is no longer the case for lemur vocalizations at six months (Ferry et al., 2013). At this older age, English-speaking babies are also not helped by a distant foreign language (such as Cantonese) as opposed to a closer language (such as German) (Perszyk & Waxman, 2019). It therefore seems that they are progressing not in their symbolic competence but in their understanding of the communication medium accepted in their cultural environment.

Indeed, infants discover very early on that speech conveys information. In an eye-tracking experiment, Marno and colleagues (2015) presented four-month-old infants with a video of an experimenter fixating on them, who then directed his gaze to the right (or left) where an object would subsequently appear. Infants eyes moved more quickly toward the object when the experimenter was talking compared with a silent video or with a video accompanied by backward speech.

Martin and colleagues (2012) presented a brief situation of interaction between two experimenters and one-year-old children. During familiarization, a first person chooses one of the two objects presented, clearly indicating her preference for that object. In the next scene, the second experimenter faces the same two objects and interacts indifferently with each of them, without showing any preference. Finally, in the test phase, both people are present, but the objects are out of reach of the first person. She turns to the second and coughs for a first group of children or says "koba" for a second group. The second experimenter then gives her either her favorite object or the distractor. In the *word* but not in the *cough* condition, children were surprised when the second experimenter did not hand over the first experimenter's favorite object and therefore looked at them significantly longer. One-year-old children thus seem to expect that information about the target object was conveyed between the two experimenters by a word, unlike the cough noise. These results were subsequently replicated in six-month-old infants (Vouloumanos, Martin, & Onishi, 2014).

Even with an attention grabbing and highly natural activity such as singing, six- to eight-month-olds display fewer communicative behaviors (vocalization, visual contact, body movements, and synchrony of these behaviors with maternal interactive behavior) in face-to-face interactions than they do with a talking mother (Arias & Pena, 2016). Infants therefore perceive the communication dimension of speech and its role as providing information about the world.

If children have inferred that speech is a privileged channel of communication, given their daily experience, they can also accept another medium of communication if social exchanges have depicted its use. For example, if they see two women exchanging with ostensive social signals but one speaking and the other "beeping," six-month-old infants become able to use the "beeps" to identify a "dinosaur" category as opposed to "fish," unlike babies who have heard the same audio file but not correlated with the social exchanges (Ferguson & Waxman, 2016). This experiment illustrates the three key elements that support pedagogy in human infants: social cognition to figure out the communication medium, an ability to sort objects in categories, and a symbolic system to label any category. These three ingredients rely on different neural bases and codevelop during infancy. They allow infants to take advantage of other people's knowledge to identify relevant information in the environment and thus boost their learning.

If infants in the laboratory easily accept different types of labels (images, words, beeps, etc.) to represent a category, in everyday life the situation is much more complex. Infants must find out how their cultural group communicates (for example, oral or sign language), analyze this efficient but rich and complex communication system, and at the same time be attentive to the pertinent cues in their environment and correctly assign the proper label to the right object. The huge complexity of the task explains the apparent slowness of progress during the first years of life, and masks infants' early possibilities for symbolic representations.

Although early communication needs in infants make language the first domain in which symbols are used, symbolic representations go beyond language in adults. It is interesting to note that mathematical knowledge and verbal knowledge are clearly separated in the adult brain (Amalric & Dehaene, 2016), which calls into question the general capacity for symbolic representations versus an extension from an initial verbal domain to other domains.

Conclusion

To conclude, we have proposed that teaching, an important activity for both parents and children from the first few months of life, requires summarizing information in a compressed abstract format for an effective sharing between individuals, which thus implies symbolic representations. Although our hypothesis of symbolic representations in infants requires further experimental evidence, it parsimoniously explains several experimental results in the literature and is consistent with the brain imaging observations that have revealed a stronger continuity than previously thought in the functional cerebral architecture between infants and adults. The question of whether symbols are initially limited to the linguistic domain remains open, but we may postulate that symbols extend beyond language and might represent the required representation of information in a conscious workspace.

Acknowledgments: This work was supported by the Fondation de France, Fondation NRJ-Institut de France and the European Research Council (ERC) under the European Union's Horizon 2020 research and innovation program (grant agreement No. 695710)

References

Amalric, M., & Dehaene, S. (2016). Origins of the brain networks for advanced mathematics in expert mathematicians. *Proceedings of the National Academy of Sciences of the United States of America, 113*(18), 4909–4917.

Arias, D., & Pena, M. (2016). Mother-infant face-to-face interaction: The communicative value of infant-directed talking and singing. *Psychopathology, 49*(4), 217–227.

Bahrami, B., Olsen, K., Latham, P. E., Roepstorff, A., Rees, G., & Frith, C. D. (2010). Optimally interacting minds. *Science, 329*(5995), 1081–1085.

Balaban, M. T., & Waxman, S. R. (1997). Do words facilitate object categorization in 9-month-old infants? *Journal of Experimental Child Psychology, 64,* 3–27.

Baldwin, D. A., & Markman, E. M. (1989). Establishing word-object relations: A first step. *Child Development, 60*(2), 381–398.

Bergelson, E., & Swingley, D. (2012). At 6–9 months, human infants know the meanings of many common nouns. *Proceedings of the National Academy of Sciences of the United States of America, 109*(9), 3253–3258.

Chaplin, T. A., Yu, H. H., Soares, J. G., Gattass, R., & Rosa, M. G. (2013). A conserved pattern of differential expansion of cortical areas in simian primates. *Journal of Neuroscience, 33*(38), 15120–15125.

Coqueugniot, H., Hublin, J. J., Veillon, F., Houet, F., & Jacob, T. (2004). Early brain growth in *Homo erectus* and implications for cognitive ability. *Nature, 431*(7006), 299–302.

Csibra, G., & Gergely, G. (2009). Natural pedagogy. *Trends in Cognitive Sciences, 13*(4), 148–153.

D'Errico, F., & Banks, W. E. (2015). The archaeology of teaching: A conceptual framework. *Cambridge Archaeological Journal, 25*(4), 859–866.

Dehaene-Lambertz, G., Dehaene, S., & Hertz-Pannier, L. (2002). Functional neuroimaging of speech perception in infants. *Science, 298*(5600), 2013–2015.

Dehaene-Lambertz, G., Hertz-Pannier, L., Dubois, J., Meriaux, S., Roche, A., Sigman, M., & Dehaene, S. (2006). Functional organization of perisylvian activation during presentation of sentences in preverbal infants. *Proceedings of the National Academy of Sciences of the United States of America, 103*(38), 14240–14245.

Dehaene-Lambertz, G., Montavont, A., Jobert, A., Allirol, L., Dubois, J., Hertz-Pannier, L., & Dehaene, S. (2010). Language or music, mother or Mozart? Structural and environmental influences on infants' language networks. *Brain and Language, 114*(2), 53–65.

Dehaene, S., & Naccache, L. (2001). Towards a cognitive neuroscience of consciousness: Basic evidence and a workspace framework. *Cognition, 79*(1–2), 1–37.

DeSilva, J., & Lesnik, J. (2006). Chimpanzee neonatal brain size: Implications for brain growth in *Homo erectus. Journal of Human Evolution, 51*(2), 207–212.

Ekramnia, M., & Dehaene-Lambertz, G. (2019). Naming is more than creating an association: Four-month-olds create equivalence relation between words and categories of objects. Manuscript in preparation.

Emberson, L. L., Richards, J. E., & Aslin, R. N. (2015). Top-down modulation in the infant brain: Learning-induced expectations rapidly affect the sensory cortex at 6 months. *Proceedings of the National Academy of Sciences of the United States of America, 112*(31), 9585–9590.

Feigenson, L., & Halberda, J. (2008). Conceptual knowledge increases infants' memory capacity. *Proceedings of the National Academy of Sciences of the United States of America, 105*(29), 9926–9930.

Ferguson, B., & Waxman, S. R. (2016). What the [beep]? Six-month-olds link novel communicative signals to meaning. *Cognition, 146,* 185–189.

Ferry, A. L., Hespos, S. J., & Waxman, S. R. (2010). Categorization in 3- and 4-month-old infants: An advantage of words over tones. *Child Development, 81*(2), 472–479.

Ferry, A. L., Hespos, S. J., & Waxman, S. R. (2013). Nonhuman primate vocalizations support categorization in very young human infants. *Proceedings of the National Academy of Sciences of the United States of America, 110*(38), 15231–15235.

Friedrich, M., & Friederici, A. D. (2011). Word learning in 6-month-olds: Fast encoding-weak retention. *Journal of Cognitive Neuroscience, 23*(11), 3228–3240.

Friedrich, M., & Friederici, A. D. (2017). The origins of word learning: Brain responses of 3-month-olds indicate their rapid association of objects and words. *Developmental Science, 20*(2), e12357.

Friedrich, M., Wilhelm, I., Born, J., & Friederici, A. D. (2015). Generalization of word meanings during infant sleep. *Nature Communications, 6,* 6004.

Friedrich, M., Wilhelm, I., Molle, M., Born, J., & Friederici, A. D. (2017). The sleeping infant brain anticipates development. *Current Biology, 27*(15), 2374–2380.e2373.

Fulkerson, A. L., & Waxman, S. R. (2007). Words (but not tones) facilitate object categorization: Evidence from 6- and 12-month-olds. *Cognition, 105*(1), 218–228.

Gogate, L. J. (2010). Learning of syllable-object relations by preverbal infants: The role of temporal synchrony and syllable distinctiveness. *Journal of Experimental Child Psychology, 105*(3), 178–197.

Gogate, L. J., Prince, C. G., & Matatyaho, D. J. (2009). Two-month-old infants' sensitivity to changes in arbitrary syllable-object pairings: The role of temporal synchrony. *Journal of Experimental Psychology: Human Perception and Performance, 35*(2), 508–519.

Gunz, P., Neubauer, S., Maureille, B., & Hublin, J. J. (2010). Brain development after birth differs between Neanderthals and modern humans. *Current Biology, 20*(21), R921–922.

Hermer-Vazquez, L., Spelke, E. S., & Katsnelson, A. S. (1999). Sources of flexibility in human cognition: Dual-task studies of space and language. *Cognitive Psychology, 39*(1), 3–36.

Kabdebon, C., & Dehaene-Lambertz, G. (2019). Symbolic labeling in 5-month-old human infants. *Proceedings of the National Academy of Sciences of the United States of America, 116*(12), 5805–5810.

Kojima, T. (1984). Generalization between productive use and receptive discrimination of names in an artificial visual language by a chimpanzee. *International Journal of Primatology, 5*(2), 161–182.

Kouider, S., Long, B., Le Stanc, L., Charron, S., Fievet, A.-C., Barbosa, L. S., & Gelskov, S. V. (2015). Neural dynamics of prediction and surprise in infants. *Nature Communications, 6,* 8537.

Kouider, S., Stahlhut, C., Gelskov, S. V., Barbosa, L. S., Dutat, M., de Gardelle, V., …Dehaene-Lambertz, G. (2013). A neural marker of perceptual consciousness in infants. *Science, 340*(6130), 376–380.

Livingstone, M. S., Pettine, W. W., Srihasam, K., Moore, B., Morocz, I. A., & Lee, D. (2014). Symbol addition by monkeys provides evidence for normalized quantity coding. *Proceedings of the National Academy of Sciences of the United States of America, 111*(18), 6822–6827.

Marno, H., Farroni, T., Vidal Dos Santos, Y., Ekramnia, M., Nespor, M., & Mehler, J. (2015). Can you see what I am talking about? Human speech triggers referential expectation in four-month-old infants. *Scientific Reports, 5,* 13594.

Martin, A., Onishi, K. H., & Vouloumanos, A. (2012). Understanding the abstract role of speech in communication at 12 months. *Cognition, 123*(1), 50-60.

Medam, T., Marzouki, Y., Montant, M., & Fagot, J. (2016). Categorization does not promote symmetry in Guinea baboons (*Papio papio*). *Animal Cognition, 19*(5), 987–998.

Mesulam, M. M. (1998). From sensation to cognition. *Brain, 121,* 1013–1052.

Nazzi, T., & Bertoncini, J. (2003). Before and after the vocabulary spurt: Two modes of word acquisition? *Developmental Science, 6*(2), 136–142.

Neubauer, S., Gunz, P., & Hublin, J. J. (2010). Endocranial shape changes during growth in chimpanzees and humans: A morphometric analysis of unique and shared aspects. *Journal of Human Evolution, 59*(5), 555–566.

Neubauer, S., Hublin, J. J., & Gunz, P. (2018). The evolution of modern human brain shape. *Science Advances, 4*(1), eaao5961.

Neubert, F. X., Mars, R. B., Thomas, A. G., Sallet, J., & Rushworth, M. F. (2014). Comparison of human ventral frontal cortex areas for cognitive control and language with areas in monkey frontal cortex. *Neuron, 81*(3), 700–713.

Nieder, A. (2009). Prefrontal cortex and the evolution of symbolic reference. *Current Opinion in Neurobiology, 19*(1), 99–108.

Perszyk, D. R., & Waxman, S. R. (2019). Infants' advances in speech perception shape their earliest links between language and cognition. *Scientific Reports, 9*(1), 3293.

Price, D. J., Kennedy, H., Dehay, C., Zhou, L., Mercier, M., Jossin, Y., … Molnar, Z. (2006). The development of cortical connections. *European Journal of Neuroscience, 23*(4), 910–920.

Sakai, T., Hirata, S., Fuwa, K., Sugama, K., Kusunoki, K., Makishima, H., … Takeshita, H. (2012). Fetal brain development in chimpanzees versus humans. *Current Biology, 22*(18), R791–792.

Sakai, T., Mikami, A., Tomonaga, M., Matsui, M., Suzuki, J., Hamada, Y., … Matsuzawa, T. (2011). Differential prefrontal white matter development in chimpanzees and humans. *Current Biology, 21*(16), 1397–1402.

Srihasam, K., Mandeville, J. B., Morocz, I. A., Sullivan, K. J., & Livingstone, M. S. (2012). Behavioral and anatomical consequences of early versus late symbol training in macaques. *Neuron, 73*(3), 608–619.

Tincoff, R., & Jusczyk, P. W. (1999). Some beginnings of word comprehension in 6-month-olds. *Psychological Science, 10*(2), 172–175.

Tincoff, R., & Jusczyk, P. W. (2012). Six-month-olds comprehend words that refer to parts of the body. *Infancy, 17*(4), 432–444.

Vouloumanos, A., Martin, A., & Onishi, K. H. (2014). Do 6-month-olds understand that speech can communicate? *Developmental Science, 17*(6), 872–879.

Wang, L., Uhrig, L., Jarraya, B., & Dehaene, S. (2015). Representation of numerical and sequential patterns in macaque and human brains. *Current Biology, 25*(15), 1966–1974.

Waxman, S. R., & Markow, D. B. (1995). Words as invitations to form categories: Evidence from 12- to 13-month-old infants. *Cognitive Psychology, 29*, 257–302.

Xu, F. (2002). The role of language in acquiring object kind concepts in infancy. *Cognition, 85*(3), 223–250.

Xu, F., & Carey, S. (1996). Infants' metaphysics: The case of numerical identity. *Cognitive Psychology, 30*(2), 111–153.

8 Developing a Theory of Mind: Are Infants Sensitive to How Other People Represent the World?

Dora Kampis, Frances Buttelmann, and Ágnes Melinda Kovács

Overview

A crucial question that has intrigued researchers for decades is how children arrive at an understanding that others' behavior is not only guided by the observable reality but also by their (unobservable) mental states. Theory of mind (ToM), the ability to attribute mental states, and its development have been of interest to many domains in cognitive science. Here we review some of the central findings on ToM abilities in children and infants, and discuss the related theoretical proposals. We (a) introduce research on children's false-belief understanding, (b) describe why understanding that the same reality may be represented under different subjective descriptions (understanding the aspectuality of beliefs) may be a critical milestone in ToM development, (c) overview findings about infants' false-belief understanding, and (d) present evidence indicating that infants understand that another person may encode the world under a different aspect. We conclude by pointing to exciting new advancements in the field and potential directions for future research.

Introduction

In their first years children have to acquire a great deal of information about the physical and the social world, and among other things they also must learn that human behavior is guided by mental states that are not directly observable. Theory of mind (ToM) can be characterized by ascribing mental states—such as goals, intentions, beliefs, and desires—to oneself and to others, and by predicting and explaining behavior based on these. The term *theory of mind* was introduced by Premack and Woodruff (1978), who raised the question of whether other animals besides humans were capable of attributing mental states, not merely acting based on the physical constraints of a situation.

Some forty years later it is still an open question as to what degree other animals understand mental states (Call & Tomasello, 2008; Krupenye, Kano, Hirata, Call, & Tomasello, 2016; Martin & Santos, 2016), but the article's largest influence came from an argument raised in the commentaries. If an observer's mental state coincides with that of the observed person, we cannot know whether the observer attributes a mental state to the other or simply predicts the other's action based on her own mental states (e.g., Dennett, 1978). The suggested solution was to develop scenarios where the other person holds a false-belief that is thus different from what the observer herself believes about the true states of affairs.

As a result, the most prominent tool used to investigate the ability to attribute mental states—in particular, beliefs—became the *false-belief task*. In the standard change-of-location paradigm (Wimmer & Perner, 1983) children are told a story about a protagonist who puts an object in location A. Afterward he leaves the scene and a second character moves this object to location B. When the protagonist comes back, children are asked where he will look for his object. In order to answer correctly (that he will look for it in location A), children need to understand that the protagonist falsely believes the object to be at the location where he has left it and does not know that it was transferred to a new location in his absence (see also Baron-Cohen, Leslie, & Frith, 1985). This task is often referred to as an "explicit" false-belief task because children are prompted to verbally answer about the protagonist's behavior or belief with a direct question.

The significance of the false-belief task lies in the fact that reasoning about the action of the protagonist requires attributing to her a belief that differs from one's own reality representation (e.g., knowing that the ball is actually in location B, while the protagonist believes it to be in location A). Thus, such scenarios create a contrast between the unique mental perspective of the observer and that of another person.

Attributing False Beliefs about Locations

Children seem to pass explicit false-belief tasks around the age of four years, as suggested by a comprehensive meta-analysis based on one-hundred and seventy-eight studies (Wellman, Cross, & Watson, 2001). Younger children tend to show systematic errors: they do not consider the false-belief of the protagonist and say that he will search at the actual location of the toy. This developmental pattern, that preschoolers go from giving answers based on their own belief to answers that reflect others' divergent beliefs, is relatively consistent across different variants of the false-belief task (Wellman et al., 2001). These tasks have been used with typically and atypically developing children (Baron-Cohen et al., 1985) and in several different cultures (Avis & Harris, 1991; Vinden, 1996). Some researchers have argued for a universal, synchronized onset of the capacities that allow typically developing children to pass ToM tasks (Callaghan et al., 2005), but other studies have found cross-cultural variations regarding the age that children succeed in such tasks (Liu, Wellman, Tardif, & Sabbagh, 2008).

The improvement in performance on explicit false-belief tasks between three and four years of age seems relatively robust, but it is still debated what abilities are responsible for this change. The explicit false-belief tasks may involve various other capacities besides ToM (for a review, see Baillargeon, Scott, & He, 2010), such as executive function (Carlson & Moses, 2001), language (Astington & Jenkins, 1999; Happé, 1995), or pragmatic abilities (Rubio-Fernandez & Geurts, 2013). Various modifications of the task allow even younger children to succeed (Setoh, Scott, & Baillargeon, 2016). For example, by using a pragmatically less ambiguous question, such as asking the child where the protagonist will look for the object *first,* children tend to answer correctly before the age of four (Siegal & Beattie, 1991).

Other studies have found that three-year-old children succeed on a modified version of the change-of-location task in which perspective tracking is not interrupted (the protagonist does not leave the scene, merely turns away while the location change happens), and when children do not have to respond to questions but are asked to continue the story by acting it out (Rubio-Fernández & Geurts, 2013). Yet although such aids seem to help three-year-old children, these tasks heavily rely on verbal communication and thus might not be suitable for testing toddlers or infants who have limited linguistic competencies.

Attributing False Beliefs about an Object's Identity

Children pass the previously described standard change-of-location tasks around the age of four, which suggests that they grasp that others can have beliefs that are different from their own regarding an object's location. Subsequent proposals have questioned whether children at this age also understand a crucial characteristic of beliefs: their aspectuality. It was proposed that a full-fledged ToM system requires appreciating that beliefs are representations about external entities and states of affairs under a certain *aspect* or *description,* and not under others (Rakoczy, Schwarz, Bergfeld, & Fizke, 2015). For example, a person can believe that "Clark Kent lives next door," but this does not imply that this person also believes that "Superman lives next door," unless the person also knows that Superman is Clark Kent. It was argued that a correct answer in a change-of-location false-belief task does not require such understanding, as it only demonstrates children's ability to track *whether* a protagonist witnessed the relocation of an object, but not *how* (under which aspect or description) she has encoded the object.

Early studies targeting whether children understand such scenarios (e.g., Russell, 1987) exposed children to stories in which a protagonist had incomplete knowledge about a situation (e.g., he knew his watch was stolen but did not know what the thief looked like). Test questions involved asking children whether it is possible to make certain statements about the protagonist's thoughts—"Can we say that George was thinking: 'I must find the man with the curly red hair who stole my watch'?"—where the correct answer required matching the content of the statement to what the protagonist was aware of (e.g., that George knew there was a thief, but he did not know his hair color). Children tended to make many errors until around age seven, and it was argued that while around four years of age children seem to be able to disregard their own *contrasting* knowledge about the location of an object, it is not until a few years later that they can judge another person's *partial* knowledge about something.

However, the task developed by Russell (1987) differed from the change-of-location tasks in several ways. For example, instead of asking what a character would do, as is usually done in the location-change tasks, in the incomplete knowledge tasks children were asked to judge whether certain statements about the character's beliefs are warranted ("Can we say that…?"), making a comparison between the performance on the two types of tasks difficult. In an attempt to make the task demands more comparable, Apperly and Robinson (1998) designed a task where they (a) lowered the linguistic complexity (i.e., the number of embedded clauses), (b) made the task more similar in structure to a change-of-location task (i.e., they asked children whether a protagonist knows *X*), and (c) aimed at conveying partial knowledge in a clearer way (by using objects that can be referred to in two different ways, such as by their appearance or their function, while the protagonist only knows about one aspect of the object). In their task, they showed four- and five-year-old children two objects: a regular eraser and an eraser that looked like a die. Subsequently a puppet appeared who watched these two objects being put in two distinct locations—but was not told about the dual identity of the die. Finally, the children were asked where the puppet will look for an eraser. The children answered at chance between the location of the regular eraser and the eraser that looked like a die, even though when asked directly, they stated that the puppet does not know that the die is also an eraser. Thus, four- and five-year-old children did not correctly predict another person's action when this involved understanding the other person's belief about dual-identity objects.

These and similar findings were taken as evidence that children's full understanding of others' mental states develops much later compared to when they pass the change-of-location false-belief tasks. Rakoczy and colleagues (2015), however, pointed out that the tasks involving dual-identity objects may have been still more difficult than change-of-location tasks because they involved ambiguous referential expressions (e.g., the protagonist was looking for "an eraser" when two erasers were present) and thus required a reference resolution (which "eraser"?). To circumvent such possible issues, they used a single object with two different appearances (e.g., a bunny that was also a carrot when turned inside out). They found that four- to six-year-olds inferred correctly a protagonist's false-belief based on her partial knowledge about the object—which appearance she was aware of. Furthermore, the children's score on this task was correlated with their performance on a standard change-of-location false-belief task, suggesting that (a) the previous implementations of the aspectuality task where children at a similar age failed likely involved additional task demands, and (b) those children who could correctly respond to questions regarding beliefs about locations could also do so about beliefs that were argued to require understanding the aspect under which an object was encoded. (For additional evidence from four- to five-year-olds, see Gopnik & Astington, 1988.)

Together, these studies involving explicit tasks suggest that around the age of four years, when children are able to represent that others may have different beliefs about an object's location, they also understand that others may represent reality under some but not other aspects.

Can Infants Encode Others' False Beliefs?

The original view that ToM abilities do not emerge before age of four has been challenged over the past fifteen years by accumulating evidence that already preverbal infants show sensitivity to others' beliefs. Such evidence pointing to belief understanding in infants and toddlers comes from tasks that mainly use implicit measures.

In tasks involving implicit measures, the participant is not asked verbally to reflect on someone's mental state; rather, they are simply exposed to belief scenarios, and their spontaneous response is assessed (for reviews, see Baillargeon et al., 2010; Sodian, 2011). In the seminal study by Onishi & Baillargeon (2005) fifteen-month-old infants were surprised (looked longer) when a person searched for an object at a location that was incongruent with where she believed it to be. This study involved a violation-of-expectation paradigm, where infants' longer looking toward a particular outcome of events is considered to be an indication that their expectations about the scene were violated. For example, they expect the protagonist to look for an object at the location where she believed it to be, not where it really is.

Infants also showed surprise when an outcome was incongruent with a protagonist's beliefs in a one-location scenario (see figure 8.1; Kovács, Téglás, & Endress, 2010). A study using eye tracking measured the infants' anticipatory looks and found that infants correctly anticipated that the protagonist would reach for her object at the location where she falsely believed the object to be *before* she started to reach for it (Southgate, Senju, & Csibra, 2007).

Infants not only show an understanding of others' beliefs as measured via their looking patterns, but they also incorporate such attributed beliefs into their active behavior when they are involved in an interaction. For instance, they consider others' beliefs when they engage in helping them (Buttelmann, Carpenter, & Tomasello, 2009), when they communicate through pointing (Knudsen & Liszkowski, 2012), and when they disambiguate referential communication (Southgate, Chevallier, & Csibra, 2010). There are currently more than thirty studies pointing to an understanding of false beliefs among infants (Scott & Baillargeon, 2017), but some

Figure 8.1
Schema of the false-belief scenario used by Kovács et al. (2010). Seven-month-olds watched a character first seeing a ball rolling behind an occluder, which then rolled out of the scene in his absence. Afterward the character came back and witnessed an outcome that was either congruent or incongruent with his beliefs (whether the ball was/was not behind the occluder). An infant looking longer at the belief-incongruent outcome was taken to be an indication for computing the character's beliefs.

studies did not succeed in reproducing some of the earlier findings (Kulke & Rakoczy, 2018). How these can be interpreted and whether subtle methodological differences are responsible for such findings is a matter of debate (Baillargeon, Buttelmann, & Southgate, 2018; Poulin-Dubois et al., 2018), highlighting the importance of future research in uncovering the underlying processes that explain infants' performance.

Overall, research from the past decade has shown that infants display sensitivity to others' beliefs in a variety of scenarios, under a wide range of methodologies. Crucially, all of these studies involved measuring spontaneous reactions because the infant participants cannot be verbally asked to reflect on someone's mental states. These findings have challenged the view that ToM abilities emerge during the preschool years, and have provided evidence consistent with an earlier onset of mental state understanding. However, recently some proposals have challenged the existence of ToM abilities in infants, putting the aspectuality of beliefs under scrutiny again.

Such proposals argue that infants cannot understand that others may represent the same reality under a different aspect, and they "pass" implicit ToM tasks using cognitive mechanisms other than ToM. Crucially, similarly to earlier debates on children's ToM abilities, these theories questioning infant ToM claim that the critical test should be exposing infants to scenarios that require them to represent others' beliefs involving aspectuality.

One of these proposals (Perner, Mauer, & Hildenbrand, 2011) suggests that a full-fledged ToM relies on particular ToM-external capacities that do not develop until the age of four years. This critical, later-developing ability is proposed to be a general understanding of the fact that one single entity can be represented in different ways, (e.g., understanding identity statements such as "the yellow key is the green key" to refer to a key that is yellow on one side and green on the other). According to this proposal infants and young children do not appreciate that there may be different ways to encode the same entity and thus people can have different (subjective) representations about the same (objective) reality. This account claims that the majority of scenarios in infant ToM paradigms, such as tracking someone's beliefs about the location of an object, do not require attributing beliefs to others and can be solved with simple rules (e.g., "people search for objects where they last saw them").

Another theory was proposed by Apperly and Butterfill (2009; Butterfill & Apperly, 2013) and became known as the "two-systems" view on ToM. According to this view, infants use a cognitive system (referred to as System 1) that resembles rudimentary ToM abilities but does not entail belief attributions. In addition to System 1, older children and adults also possess System 2, which enables them to perform full-fledged belief attributions. System 2 is proposed to involve meta-representing others' mental contents, but System 1 operates on simple representations, such as person–object relations (called "registrations"). Registrations are proposed to approximate ToM-like reasoning in some contexts, and they can result in outcomes similar to representing beliefs about locations.

According to the two-systems view, infants in false-belief scenarios form a relation between the person and the object. This relation (in combination with some simple rules, such as registrations are formed and updated only if the events involving the object happen in the proximity of the person, approximating conditions of perception) allows them to behave as though they are tracking the person's belief about the location of this object, without actually ascribing a belief to her. Critically, registrations link the person to the object itself (not to the representation of that object under a specific aspect), and thus such relations cannot successfully deal with situations involving dual-identity objects used in aspectuality tasks. If one forms a registration between a person and an object or entity in one particular form, one would mistakenly grant that person all knowledge about *any* other appearance or form the entity may take. By contrast, if an observer using full-fledged ToM (System 2) sees a person encountering an entity of appearance A (e.g., Superman), and then later of appearance B (e.g., Clark Kent), without evidence that the person knows that the two are the same, they would attribute to this person that she must believe they are two separate entities.

Furthermore, according to this view, System 1 continues to operate across the life span, and even adults rely on it when they track belief-involving scenarios spontaneously (without being prompted to do so) or when they need to react fast (e.g., when we move the steering wheel if we suspect a pedestrian does not see us). A strong prediction that follows is that not only infants are limited in the range of scenarios their cognitive apparatus can deal with, but even adults would fail to spontaneously ascribe beliefs involving aspectuality.

Can Adults and Young Children Spontaneously Encode Others' Beliefs about Dual-Identity Objects?

Some recent studies have investigated spontaneous inferences regarding others' beliefs involving dual-identity objects where the protagonist encoded the object under one but not another aspect. Low and Watts (2013) tested three-year-olds, four-year-olds, and adults in a dual-identity task as well as a change-of-location task. They measured the participants' gaze in anticipating an agent's behavior as indicators of their implicit or spontaneous understanding, and their verbal responses as indicators of explicit understanding of others' beliefs. Both children and adults correctly anticipated the protagonist's reach as a function of his false-belief in the change-of-location task, but they did not do so in the dual-identity task, which was interpreted by the authors as a blind spot specific to the spontaneous computation of other's false-beliefs about object identity. By contrast, the four-year-olds and adults answered correctly to the questions regarding the protagonist's belief for both tasks, revealing that for their explicit responses they did compute false-beliefs in both situations.

Another study investigated two- and three-year-olds' ability to spontaneously understand others' beliefs about an object's location as well as an object's identity in an active helping task (Fizke, Butterfill, van de Loo, Reindl, & Rakoczy, 2017). In the change-of-location task (based on Buttelmann et al., 2009), the participants saw a person placing an object in box A, which was transferred to box B in the person's absence (false-belief condition) or presence (true-belief condition). Afterward he tried unsuccessfully to open box A (i.e., the empty box). The children's spontaneous helping behavior (opening one of the boxes) differed between the two conditions: they helped the person open box A in the true-belief condition and box B in the false-belief condition. This suggest that in the latter they encoded that he falsely believed the object to be in box A, assumed that his goal was to get the object, and thus helped him achieve his goal by opening box B (where the object really was).

In the dual-identity task, children were first familiarized with the dual aspect of the target object: showing that a toy carrot puppet could be turned into a toy rabbit. Afterward the children saw the person putting the rabbit into a box. Next, during her presence (true-belief) or absence (false-belief), the rabbit was taken out, turned into a carrot, and put back into the box. Then in both conditions, in the presence of the person, the carrot was taken out and placed on the floor between her and the child; then the person unsuccessfully tried to open the box. Again the children's spontaneous helping behavior was measured: whether they opened the box (correct in true-belief) or gave her the carrot (which was *also* the rabbit, the correct choice in false-belief). The results showed that the children mostly opened the box in both conditions (instead of giving the carrot), which was interpreted as an indication of their problem in understanding her belief about the object's identity (i.e., they did not understand that she was not aware that the carrot is in fact also the rabbit she was looking for).

These findings were considered evidence that children and adults do not spontaneously track others' beliefs involving dual-identity objects, but they also may be subject to alternative explanations. For example, Carruthers (2013) argued that the study by Low and Watts (2013) requires extensive executive control and working memory resources in order to perform mental rotations to understand the other's visual perspective; which may explain why the participants failed to spontaneously predict the actor's action but succeeded on the explicit questions regarding the person's behavior after they had some time for deliberation. Such task demands also apply to other tasks that use similar scenarios (Low, Drummond, Walmsley, & Wang, 2014). This raises the possibility that if the ToM-unrelated external task demands were lowered (e.g., no mental rotation is required), adults may show spontaneous tracking of beliefs about identity, and even younger infants may be able to deal with such tasks.

Relatedly, the task of Fizke and colleagues (2017) involving dual-identity objects may have been too demanding for some children, even from their own perspective. At fourteen months of age only about 70 percent of infants show evidence of understanding dual-identity objects from their own perspective (Cacchione, Schaub, & Rakoczy, 2013), and even three-year-olds show difficulties with dual-identity statements in verbal tasks (Perner et al., 2011). Infants' own understanding of a scene will also influence what beliefs they can ascribe to another person, as no individual is capable of ascribing a belief to someone else that she herself cannot potentially entertain. It will be therefore important in future tasks to ensure that children can track the events from their own point of view—for example, by including measures assessing their competency of understanding the critical aspects of the tasks.

Can Infants Spontaneously Encode Beliefs Involving Identity Relations?

Would infants be able to understand that others may represent an entity under a certain aspect and not under another if the tasks are made suitably easy for them? Recent evidence suggests that by eighteen months of age infants understand scenarios involving aspectuality: they understand that a person may mistake an object for another based on how she perceives those objects (Scott & Baillargeon, 2009; for related results with fourteen-month-olds, see Song & Baillargeon, 2008), or that others may only know about the (misleading) appearance and not the real function of an object (Buttelmann, Suhrke, & Buttelmann, 2015).

For example, Scott and Baillargeon (2009) used a violation-of-expectation paradigm with eighteen-month-olds where infants were familiarized with two toy penguins that looked alike, one composed of one piece only and the other of two pieces. Infants watched scenarios in which a protagonist had the goal of putting her key inside the two-piece penguin. In the false-belief condition, the two-piece penguin appeared the same to the protagonist as the one-piece penguin, whereas in the true-belief condition the protagonist witnessed that the two-piece penguin was assembled in such a way as to appear like the one-piece penguin. The results showed that infants expected the protagonist to mistake the two-piece penguin for the one-piece penguin in the false-belief condition but not in the true-belief condition. Thus, infants attributed a false-belief to another person based on the way she perceived the toys in question. Others found that those fourteen-months-olds, who show evidence for understanding events involving dual-identity objects from their own perspective, can also attribute beliefs about identity mistakes to another person, thus using the person's unique perspective on these objects to infer her belief about the object(s) in question (Kampis & Kovács, 2019).

In an interactive helping task involving "deceptive objects" that have a misleading appearance (e.g., a box that looks like a book) eighteen-month-olds helped the experimenter according to her belief about the object's identity (Buttelmann et al., 2015). The experimenter only witnessed the appearance of an object (which looked like a book) whereas the infants also knew about its "real" identity (that it was a box). When she asked for an object, infants gave her the one resembling what she believed it to be (a book, even though infants knew it was in fact a box). Recently it was found that even fourteen-month-olds are able to anticipate another's actions based on their false-beliefs involving such deceptive objects (Buttelmann & Kovács, 2019).

Infants also show an understanding of others' beliefs about identity in third-party observations, in which they have to consider a person's deceptive intention (Scott, Richman, & Baillargeon, 2015). Seventeen-month-old infants watched videos in which a person attempted to secretly steal a rattling toy from another person during her absence, and replaced the object with a visually identical but silent object. The results showed that infants understood the intention (to implant a false-belief about an object's identity) and the resulting actions of the deceptive person (putting a suitable substitute for the original object).

Together, these findings point to the possibility that infants from about fourteen months of age can follow others' beliefs in scenarios that involve thinking about how another person represents an object, suggesting that early on infants seem to be sensitive to the aspectual nature of beliefs.

Concluding Remarks

In sum, data from false-belief tasks involving location change or identity understanding using spontaneous measurements (e.g., looking times, gaze patterns, or helping behavior) suggest that infants do track other people's beliefs. They readily anticipate others' actions based on these beliefs and even modulate their own behavior accordingly. Evidence from studies targeting the neural correlates of these abilities corroborate such findings. For instance, it was found that the cognitive systems involved in representing the world from infants' own perspective are also recruited for encoding others' perspective and beliefs in eight-month-olds, as evidenced by gamma oscillations (Kampis, Parise, Csibra, & Kovács, 2015). Further, six-month-olds generate differential predictions in their motor cortex depending on an agent's belief (Southgate & Vernetti, 2014). Recent evidence with near-infrared spectroscopy has suggested that seven-month-olds recruit the same brain areas that are specifically involved in false-belief reasoning in adults (the right temporal-parietal junction) while processing scenes involving false-belief scenarios (Hyde, Simon, Ting, & Nikolaeva, 2018).

However, although the brain networks that support such abilities are subject to a considerable maturation and specialization throughout childhood, which are also reflected in performance improvements (Gweon, Dodell-Feder, Bedny, & Saxe, 2012; Grosse Wiesmann, Schreiber, Singer, Steinbeis, & Friederici, 2017), these ToM networks are present and functional already in young children who do not yet pass the explicit ToM tasks (Richardson, Lisandrelli, Riobueno-Naylor, & Saxe, 2018).

Infants' abilities may be restricted by various factors (such as task complexity or memory limitations). However, the abilities tackled via diverse tasks and a wide range of measurements seem to very much resemble the complex ToM abilities that can be observed in adults. Of course, over the years children come to understand more diverse scenarios and apply their ToM abilities more readily, in concert with the development of some more general abilities. For instance, a recent study targeted the separation of online, or prospective belief tracking and offline, or retrospective belief tracking. In prospective belief tracking the observers must encode and compute others' beliefs as an event unfolds; in retrospective belief tracking they must "go back in time" to retrieve relevant events and perform belief inferences after the events have taken place. Eighteen-month-olds participating in this study were successful when the situation allowed for online belief tracking, but they had difficulties with retrospective belief revisions that likely rely more heavily on episodic memory abilities. The three-year-olds performed well in both situations (Király, Oláh, Csibra, & Kovács, 2018).

Although based on theoretical grounds it is often argued that identity-based false-belief tasks are the most convincing test for full-fledged ToM, one might wonder whether the difference between a location-change false-belief task and an identity-based false-belief task is a quantitative or qualitative one. It can be argued that all false-belief tasks require, in fact, encoding another person's subjective perception of a scene. However, targeting the attribution of different belief contents involving location versus identity (or other possible contents, see Kovács, 2016) may shed light on the format of the representations and the specific processes involved.

In any case, the evidence accumulated up to now seems to suggest that young infants may possess rich ToM abilities, and understand that different people may represent the same reality under different subjective descriptions. Such abilities may play an important role in efficient social learning and may explain the unique collaborative structure of human societies.

Acknowledgments: This work was partly supported by the European Research Council under the European Union's Seventh Framework Programme (FP7/2007–2013)/ERC starting grant (284236-REPCOLLAB).

References

Apperly, I. A., & Butterfill, S. A. (2009). Do humans have two systems to track beliefs and belief-like states? *Psychological Review, 116*, 953–970. doi: 10.1037/a0016923

Apperly, I. A., & Robinson, E. J. (1998). Children's mental representation of referential relations. *Cognition, 63*, 287–309.

Astington, J. W., & Jenkins, J. M. (1999). A longitudinal study of the relation between language and theory-of-mind development. *Developmental Psychology, 35*(5), 1311–1320. doi: 10.1037/0012-1649.35.5.1311

Avis, J., & Harris, P. L. (1991). Belief-desire reasoning among Baka children. *Child Development, 62*, 460–467. doi: 10.2307/1131123

Baillargeon, R., Buttelmann, D., & Southgate, V. (2018). Invited commentary: Interpreting failed replications of early false-belief findings: Methodological and theoretical considerations. *Cognitive Development, 46,* 112–124. doi: 10.1016/j.cogdev.2018.06.001

Baillargeon, R., Scott, R. M., & He, Z. (2010). False-belief understanding in infants. *Trends in Cognitive Science, 14*(3), 110–118. doi: 10.1016/j.tics.2009.12.006

Baron-Cohen, S., Leslie, A. M., & Frith, U. (1985). Does the autistic child have a "theory of mind"? *Cognition, 21,* 37–46. doi: 10.1016/0010-0277(85)90022-8

Buttelmann, D., Carpenter, M., & Tomasello, M. (2009). Eighteen-month-old infants show false belief understanding in an active helping paradigm. *Cognition 112,* 337–342. doi: 10.1016/j.cognition.2009.05.006

Buttelmann, F., & Kovács, Á. M. (2019). 14-Month-olds anticipate others' actions through belief attribution in an unexpected-identity task. *Infancy, 24*(5), 738–751. doi: 10.1111/infa.12303

Buttelmann, F., Suhrke, J., & Buttelmann, D. (2015). What you get is what you believe: Eighteen-month-olds demonstrate belief understanding in an unexpected-identity task. *Journal of Experimental Child Psychology, 131,* 94–103. doi: 10.1016/j.jecp.2014.11.009

Butterfill, S. A., & Apperly, I. A. (2013). How to construct a minimal theory of mind. *Mind and Language, 28*(5), 606–637. doi:10.1111/mila.12036

Cacchione, T., Schaub, S., & Rakoczy, H. (2013). Fourteen-month-old infants infer the continuous identity of objects on the basis of non-visible causal properties. *Developmental Psychology, 49,* 1325–1329. doi: 10.1037/a0029746

Call, J., & Tomasello, M. (2008). Does the chimpanzee have a theory of mind? 30 years later. *Trends in Cognitive Sciences, 12*(5), 187–192. doi: 10.1016/j.tics.2008.02.010

Callaghan, T., Rochat, P., Lillard, A., Claux, M. L., Odden, H., Itakura, S., … Singh, S. (2005). Synchrony in the onset of mental-state reasoning: Evidence from five cultures. *Psychological Science, 16*(5), 378–384. doi: 10.1111/j.0956-7976.2005.01544.x

Carlson, S. M., & Moses, L. J. (2001). Individual differences in inhibitory control and children's theory of mind. *Child Development, 72,* 1032–1053.

Carruthers, P. (2013). Mindreading in infancy. *Mind and Language, 28,* 141–172. doi: 10.1111/mila.12014

Dennett, D. C. (1978). Beliefs about beliefs. *Behavioral and Brain Sciences 1*(4), 568–570. doi: 10.1017/S0140525X00076664

Fizke, E., Butterfill, S., van de Loo, L., Reindl, E., & Rakoczy, H. (2017). Are there signature limits in early theory of mind? *Journal of Experimental Child Psychology, 162,* 209–224. doi: 10.1016/j.jecp.2017.05.005

Gopnik, A., & Astington, J. W. (1988). Children's understanding of representational change and its relation to the understanding of false belief and the appearance-reality distinction. *Child Development, 59,* 26–37. doi: 10.2307/1130386

Grosse Wiesmann, C. G., Schreiber, J., Singer, T., Steinbeis, N., & Friederici, A. D. (2017). White matter maturation is associated with the emergence of theory of mind in early childhood. *Nature Communications, 8,* 14692.

Gweon, H., Dodell-Feder, D., Bedny, M., & Saxe, R. (2012). Theory of mind performance in children correlates with functional specialization of a brain region for thinking about thoughts. *Child Development, 83,* 1853–68.

Happé, F. G. E. (1995). The role of age and verbal ability in the theory of mind task performance of subjects with autism. *Child Development, 66*(3), 843–855. doi: 10.2307/1131954

Hyde, D. C, Simon, C. E., Ting, F., & Nikolaeva, J. (2018). Functional organization of the temporal-parietal junction for theory of mind in preverbal infants: A near-infrared spectroscopy study. *Journal of Neuroscience, 38*(18), 4264–4274. doi: 10.1523/JNEUROSCI.0264-17.2018

Kampis, D., & Kovács, A. M. (2019). Human infants understand the aspectuality of mental representations: Evidence from spontaneous belief tracking. Manuscript under review for publication.

Kampis, D., Parise, E., Csibra, G., & Kovács, Á. M. (2015). Neural signatures for sustaining object representations attributed to others in preverbal human infants. *Proceedings of the Royal Society B: Biological Sciences, 282*(1819), 20151683. doi: 10.1098/rspb.2015.1683

Király, I., Oláh, K., Csibra, G., & Kovács, Á M. (2018). Retrospective attribution of false beliefs in 3-year-old children. *Proceedings of the National Academy of Sciences of the United States of America, 115*(45), 11477–11482. doi: 10.1073/pnas.1803505115

Knudsen, B., & Liszkowski, U. (2012). 18-month-olds predict specific action mistakes through attribution of false belief, not ignorance, and intervene accordingly. *Infancy, 17,* 672–691. doi: 10.1111/j.1532-7078.2011.00105.x

Kovács, Á. M. (2016). Belief files in theory of mind reasoning. *Review of Philosophy and Psychology, 7*(2), 509–527. doi: 10.1007/s13164-015-0236-5

Kovács, Á.M., Téglás, E., & Endress, A.D. (2010). The social sense: Susceptibility to others' beliefs in human infants and adults. *Science, 330,* 1830–1834. doi: 10.1126/science.1190792

Krupenye, C., Kano, F., Hirata, S., Call, J., & Tomasello, M. (2016). Great apes anticipate that 638 other individuals will act according to false beliefs. *Science, 354,* 110–114.

Kulke, L., & Rakoczy, H. (2018). Implicit theory of mind—an overview of current replications and non-replications. *Data in Brief, 16,* 101–104.

Liu, D. M., Wellman, H., Tardif, T., & Sabbagh, M. A. (2008). Theory of mind development in Chinese children: A meta-analysis of false-belief understanding across cultures and languages. *Developmental Psychology, 44* (2), 523–531. doi: 10.1037 /0012-1649.44.2.523

Low, J., Drummond, W., Walmsley, A., & Wang, B. (2014). Representing how rabbits quack and competitors act: Limits on preschoolers' efficient ability to track perspective. *Child Development, 85*(4), 1519–1534. doi: 10.1111/cdev.12224

Low, J., & Watts, J. (2013). Attributing false beliefs about object identity reveals a signature blind spot in humans' efficient mind-reading system. *Psychological Science, 24*, 305–311. doi: 10.1177/0956797612451469

Martin, A., & Santos, L. R. (2016). What cognitive representations support primate theory of mind? *Trends in Cognitive Science, 20* (5), 375–382. doi: 10.1016/j.tics .2016.03.005

Onishi, K. H., & Baillargeon, R. (2005). Do 15-month-old infants understand false beliefs? *Science, 308*, 255–258. doi: 10.1126/science.1107621

Perner, J., Mauer, M. C., & Hildenbrand, M. (2011). Identity: Key to children's understanding of belief. *Science, 333*(6041), 474–477. doi: 10.1126/science.1201216

Poulin-Dubois, D., Rakoczy, H., Burnside, K., Crivello, C., Dörrenberg, S., Edwards, K.,... Perner, J. (2018). Do infants understand false beliefs? We don't know yet—a commentary on Baillargeon, Buttelmann and Southgate's commentary. *Cognitive Development, 48*, 302–315.

Premack, D., & Woodruff, G. (1978). Does the chimpanzee have a theory of mind? *Behavioral and Brain Sciences, 1*, 515–526. doi: 10.1017/S0140525X00076512

Rakoczy, H., Schwarz, I., Bergfeld, D., & Fizke, E. (2015). Explicit theory of mind is even more unified than previously assumed: Belief ascription and understanding aspectuality emerge together in development. *Child Development, 86*(2), 486–502. doi: 10.1111/cdev.12311

Richardson, H., Lisandrelli G., Riobueno-Naylor A., & Saxe R. (2018). Development of the social brain from age three to twelve years. *Nature Communications, 9*(1), 1027.

Rubio-Fernández, P., & Geurts, B. (2013). How to pass the false-belief task before your 4th birthday. *Psychological Science, 24*, 27–33. doi: 10.1177/0956797612447819

Russell, J. (1987). "Can we say...?" Children's understanding of intensionality. *Cognition, 25*, 289–308. doi: 10.1016/S0010-0277(87)80007-0

Scott, R. M., & Baillargeon, R. (2009). Which penguin is this? Attributing false beliefs about object identity at 18 months. *Child Development, 80*, 1172–1196. doi: 10.1111/j.1467-8624.2009.01324.x

Scott, R. M., & Baillargeon, R. (2017). Early false-belief understanding. *Trends in Cognitive Sciences, 21,* 237–249. doi: 10.1016/j.tics.2017.01.012

Scott, R. M., Richman, J. C., & Baillargeon, R. (2015). Infants understand deceptive intentions to implant false beliefs about identity: New evidence for early mentalistic reasoning. *Cognitive Psychology, 82,* 32–56. doi: 10.1016/j.cogpsych.2015.08.003

Setoh, P., Scott, R. M., & Baillargeon, R. (2016). Two-and-a-half-year-olds succeed at a traditional false-belief task with reduced processing demands. *Proceedings of the National Academy of Sciences of the United States of America, 113,* 13360–13365. doi: 10.1073/pnas.1609203113

Siegal, M., & Beattie, K. (1991). Where to look first for children's knowledge of false beliefs. *Cognition, 38,* 1–12. doi: 10.1016/0010-0277(91)90020-5

Sodian, B. (2011). Theory of mind in infancy. *Child Development Perspectives, 5,* 39–43. doi: 10.1111/j.1750-8606.2010.00152.x

Song, H., & Baillargeon, R. (2008). Infants' reasoning about others' false perceptions. *Developmental Psychology, 44,* 1789–95. doi: 10.1037/a0013774

Southgate, V., Chevallier, C., & Csibra, G. (2010). Seventeen-month-olds appeal to false beliefs to interpret others' referential communication. *Developmental Science, 13,* 907–912.

Southgate, V., Senju, A., & Csibra, G. (2007). Action anticipation through attribution of false belief by 2-year-olds. *Psychological Science, 18*(7), 587–592.

Southgate, V., & Vernetti, A. (2014). Belief-based action prediction in preverbal infants. *Cognition, 130*(1), 1–10.

Vinden, P. (1996). Junin Quecha children's understanding of mind. *Child Development, 67,* 1707–1716.

Wellman, H. M., Cross, D., & Watson, J. (2001). Meta-analysis of theory-of-mind development: The truth about false belief. *Child Development, 72,* 655–684.

Wimmer, H., & Perner, J. (1983). Beliefs about beliefs: Representation and constraining function of wrong beliefs in young children's understanding of deception. *Cognition, 13,* 103–128.

9 How Do Young Children Become Moral Agents?
A Developmental Perspective

Markus Paulus

Overview

In this chapter, I will present the sketch of a developmental account on how
young children become moral agents. That is, I aim to provide an answer to
the question of how normativity comes into human life and, in particular,
how children start to conceive of actions as being good or bad, or as being
obligated or prohibited. The account builds on considerations put forward
by cultural evolution theory and philosophy of language, as well as constructivist and social-interactionist developmental approaches. I will argue
that only normative language represents an unequivocal empirical indicator
for the presence of a moral stance and that becoming a moral agent means
to become a player in the moral language game. I conclude that a full understanding of moral development requires us to acknowledge the dynamic
interplay between our biological basis as social beings, the socializing force
of our social environment, and the active role of the child itself.

Introduction

Morality concerns the rules and principles that differentiate between good
and bad behavior or, in other words, between right and wrong. Given that
good and bad are not natural entities but normative evaluations, these distinctions are constituted by rules and principles. Morality describes obligations that we have toward each other and, in some moral theories, toward
ourselves (e.g., Kant, 1785/2018). One central function of moral rules is
to regulate social behavior and, in consequence, to allow for cooperation
and coexistence of individuals in groups. In other words, they allow for
the existence and maintenance of social groups and social institutions that

are the backbone of cooperation and survival of its members (Tomasello, 2016). Moral rules are an efficient manner of social cooperation as they are in principle flexible and allow for changes in their content (e.g., think about how norms of child upbringing have changed in Western societies over only a few generations)—different from the more limited behavioral programs that subserve animal behavior. One key aspect of moral norms (to be discussed in more detail later in this chapter) is that they consist of and only exist in language. Nature itself does not contain normativity and obligations (cf. Moore, 1903).

In the following paragraphs, I will sketch an account of the emergence of moral stances—that is, the capacity to judge actions as good or bad, right or wrong. For the purposes of this chapter, I will rely on a broader conception of morality that does not draw a sharp distinction between, for example, moral and conventional norms (e.g., Turiel, 1983). I do not want to contend that this distinction is not meaningful, but rather that it is the consequence of a developmental process in which children learn to differentiate the severity of different norm violations as well as their scope (e.g., context-dependency versus unconditional validity). First, I will focus on the ontogeny of normative stances in general. Given the complexity of the topic and the brevity of the format, this piece will necessarily be restricted to a few central lines of thinking.

Is Morality Innate?

One key question for developmental science is to explain the emergence of moral stances. How do young children come to develop a concern for right and wrong? How does the experience of obligations emerge in development? One prominent approach has focused on the existence of social preferences in infants and toddlers (Bloom, 2013; Hamlin, 2013). In a programmatic series of studies, it has been examined whether and to what extent young children display preferences for prosocial others and aversion against antisocial others. For example, it has been reported that six- and ten-month-old infants would rather reach for a protagonist who previously helped another agent than a protagonist who hindered that agent (Hamlin, Wynn, & Bloom, 2007). Moreover, the ten-month-old infants looked longer when the agent approached the hinderer than when it approached the helper. Similar results have been obtained in other studies (for review, see Baillargeon, Setoh, Sloane, Jin, & Bian, 2014). In a related vein, infants' instrumental helping behaviors have been interpreted as evidence for an altruistic motivation (Warneken & Tomasello, 2006). The findings of these studies have been taken as evidence for an innate moral core (Hamlin, 2013) and that already babies are moral beings (Bloom, 2013).

While the extent to which the single findings are replicable is a matter of ongoing debate (e.g., Scarf, Imuta, Colombo, & Hayne, 2012; Scola, Holvoet, & Arciszewski, 2015), these studies have led to an increased interest on the early ontogeny of social perception. However, one needs to be aware of one crucial point. The studies examined social preferences in young children, yet a social preference is logically different from a moral evaluation, an appreciation of rules and principles, and a feeling of normative obligations (for a similar argument, see Dahl, 2014). We all have preferences for one or the other individual, but these do not constitute moral judgments. The social preferences for particular others might be the consequence of a moral judgment (e.g., I might prefer to interact with people whom I judge to be morally good), they might be behavioral inclinations that feed into moral judgments (e.g., I might be more inclined to judge someone as morally good if I prefer to be with this person), or they might be unrelated (e.g., I prefer someone because it feels good to be with this person). These points make clear that social preferences are not equivalent with moral evaluations. Likewise, I can predict other agents' future behaviors due to a variety of cues, but action prediction is not equivalent with an appreciation of moral evaluations.

If we mix up social preferences with moral evaluations, we gloss over the very characteristics that define morality. It is a key aspect of the concept of moral obligation that it may interfere with someone's preferences (Dahl & Paulus, 2019). Likewise, with respect to early helping, it has been suggested that these behaviors are rather evidence for a motivation (one could also say a social preference) to interact with others than a genuine altruistic motivation (Paulus, 2019). It is certainly possible that these early preferences are a factor in the emergence of moral stances (Tan, Mikami, & Hamlin, 2018), yet they should not be mistaken for normative stances in itself. Indeed, recent neuroscientific evidence suggests that infants' preferences could be related to a domain-general approach-avoidance system rather than a specific moral module (Decety & Cowell, 2018). This relates well to recent theoretical proposals that humans start with a rather unspecific arousal-valence system while specific emotions are constructed in the course of development (Barrett, 2017). Overall, there are empirical and conceptual reasons that infant preferences are not moral stances.

Empirical Indicators of Normative Stances

One key question thus concerns what are appropriate empirical indicators for moral stances? When do we attribute a moral stance to someone? What are clear indicators that children conceive of an action as being obligatory? One could think that empathic concern might fulfill this role. Indeed, a classical view that can be related to David Hume proposes that empathic concerns are constitutive for moral judgments (for discussion, see Prinz, 2011). Yet empathic concern and the sympathetic actions that might follow from it are feelings and behavioral inclinations. Empirical research has demonstrated that empathy can interfere with moral judgment (Decety & Cowell, 2014). Moreover, empathy is partial, preferring those who are more similar to us or closer to us (Gutsell & Inzlicht, 2012). Most important for the present purpose is the point that empathic concern is conceptually different from judgments and evaluations, and thus constitutes no indicator for morality.

One traditional measure of moral stances is explicit judgments about good and bad, and the permissibility of an action (Turiel, 1983). These judgments are often complemented with further justifications in which children can explain their reasoning. Hereby, children are usually presented with a story or a vignette about morally relevant behavior and are asked to judge the goodness and permissibility of an action. The engagement in moral justifications and, in particular, the use of normative vocabulary is a clear indication of a moral stance whereas weaker evidence is provided when children merely respond with Yes and No to moral question. Such a developmental increase in the use of normative vocabulary and moral justifications has been observed in the early preschool period (e.g., Smetana & Brags, 1990; cf. Killen & Smetana, 2015).

More recently, the judgment measures have been complemented by assessments of spontaneous protest and spontaneous affirmation (e.g., Rakoczy, Kaufmann, & Lohse, 2016; Wörle & Paulus, 2018). In these studies, children are usually presented with a protagonist's behavior, and it is assessed whether they spontaneously comment or intervene by utterances that indicate a normative orientation—that is, whether they use normative vocabulary. The advantage of these measures is that children's reactions are more spontaneous than their answers to an experimenter's questions. In these third-party scenarios a protagonist does something to a recipient (usually both displayed by puppets) and children themselves have no benefit from (omission of) the action. The findings that children intervene in such contexts are an indicator for a strong normative stance and the application of agent-neutral norms (Rakoczy & Schmidt, 2013). The disadvantage of this measure is that usually only a minority of children engage in such spontaneous behaviors and that they need to be familiarized with the setup in order to raise the likelihood of engagement. Yet, taken together, the findings suggest that by two to three years children develop normative stances and that these stances are expressed in normative language.

Notably, also acts of punishment and reward have been interpreted as indicating a normative stance. Yet these behaviors are not in themselves unequivocal indicators of normativity (unless they are verbally assessed; e.g., Killen, Mulvey, Richardson, Jampol, & Woodward, 2011). Punishment and reward shape another agent's behavior by decreasing or increasing the probability of its occurrence; the occurrence of these actions can therefore most easily be explained by their function to change another agent's behavior. In line with these considerations, a recent study reported that chimpanzees—who do not show any other form of moral evaluation (e.g., moral reasoning or normative protest)—do punish others when they steal their food but do not engage in third-party punishment (Riedl, Jensen, Call, & Tomasello, 2012). Yet even when chimpanzees would have demonstrated third-party punishment, it would not have constituted an unequivocal indicator of normativity. Behaving in line with a norm is not equivalent with behaving on basis of the acceptance of a norm.

This overview suggests that moral development is closely linked to language development. In fact, the most widely accepted measures of moral stances are based on language. The few measures that do not directly rely on verbal utterances (punishment or reward by giving/removing resources) are behaviors that are usually embedded in a language-based context in which the meaning of the particular action is explained or in which the action is embedded into other language-based assessments of moral stances (Wörle & Paulus, 2018). If this were not the case, even this measure would be difficult to interpret. That is, it would not be clear, for example, whether a negative item was given as a means of moral punishment or mere dislike. In other words, we would not be able to distinguish between mere liking or antipathy toward another person and a clear representation of a moral stance.

Language and Morality

Influential philosophical approaches have pointed to the foundational role of language for morality (Habermas, 1983, 1991). Only language provides a format for concepts such as obligation, right or wrong, ought, moral deed, and many more. Given that these concepts do not represent natural entities, they cannot be found as items in our physical world. This relates to David Hume's famous assertion that one cannot derive an ought from an is. Thus, becoming a genuine moral agent means to get into the language game of morality.

Importantly, one needs to realize that we are not dealing with an empirical question. We cannot run another study and then "find" that there are other measures of morality or detect that one can derive an ought from an is. Here, we are concerned with conceptual and definitional questions on what we mean by morality. In other words, we need to reflect on the behavioral criteria that we accept as clear indicators of normative stances. These questions have to be dealt with before running an empirical study. Otherwise, we risk a fallacy of confusing empirical and conceptual questions (Racine & Slaney, 2013)—which leads to severe problems for theoretical progress in psychology (Bennett & Hacker, 2003).

The question of what is morally relevant and how to recognize normative stances builds on an already existing understanding of these concepts in ordinary language. The community of researchers can only communicate about these concepts because we share a (fuzzy, but still sufficiently overlapping) understanding of morality. The conceptual question of what we mean by morality and what indicators we accept has to start by our shared understanding of these concepts. One could object that he or she has a different definition of morality and therefore comes to different measures. Yet one cannot change definitions and meanings of concepts ad libitum. The definitions of our central terms are socially shared and are part of a semantic network of concepts. It is not possible to change conceptual definitions without changing our entire net of concepts and, therefore, in effect our language.

The Emergence of Normative Stances

How do young children thus develop normative stances? How does the ought become an aspect of their life? In their own interactions with significant others children first experience that some of their own behaviors are not good. Parental permissions and prohibitions (e.g., to not hit their sibling) are their first encounters with the moral world. As long as these rules merely change children's behavior (e.g., by being experienced as punishment and reward) they remain on a behavioral level.

Yet with their increasing language abilities children start to represent (in the sense of remembering and learning) the moral utterances of their caregivers. They internalize them, and, as a consequence, future situations may evoke the internalized rule of conduct and guide children's behavior. This process of internalization is a process of social learning—something young children are very keen for and are able to do from early on (for review, see Paulus, 2014). Indeed, empirical studies have demonstrated a relation between toddlers' imitative tendencies and later conscience development (Forman, Aksan, & Kochanska, 2004; Kochanska, Forman, & Aksan, 2005). Moreover, it is well-known that in pretend-play and role-play scenarios young children reenact the rules of their social world. This process of role-taking serves children's understanding of the claims and entitlements humans impose on each other, and represents a basis for the appreciation of the universality of moral norms (cf. Habermas, 1992; Joas, 1989). To the extent that young children express moral language in these situations, this indicates that they became moral agents.

Next to the direct verbal instructions and imitation, there is an affluence of social cues that signal (moral) norms to young children. Children overhear other persons expressing normative views and normative vocabulary (e.g., expressions with verbs such as ought, must, should). This not only entails close relatives and nonparental caregivers, but also conversations with the neighbor, of people in the supermarket, and (nowadays) in various media. Besides verbal utterances, caregivers also rely on nonverbal cues such as tone of voice to signal moral transgressions (e.g., Dahl & Tran, 2016). Here, it has to be closely distinguished whether young children merely react to particular social signals because they feel more threatened by an aversive voice typical for moral transgressions and/or because they have learned that this tone of voice usually indicates further punishment—or because they infer the presence of a moral norm. Again, the only unequivocal criterion for the latter case (that is, that children inferred a moral norm) would be a child's verbal utterance. To the extent that this is the case, these cues support children's acquisition of moral norms.

What is the psychological basis for children's engagement in the normative world? There are several factors that are important. Here, I would like to highlight three points: children's dependency on others, their increasing capability of mastering language, and their pleasure in social interactions.

First, one key point is that young children are highly dependent on others, in particular on their caregivers. Young children would not survive if they were not to be taken care of, and they crucially depend on others for self-regulation (Ainsworth, Blehar, Waters, & Wall, 1978). They need their caregivers to provide them with love, food, and toys. Making their caregivers mad thus has unfortunate consequences for their life. Therefore, learning to maintain relationships with their parents entails understanding and following (some) of their norms.

Second, young children have a genuine proclivity to become independent agents. They find pleasure in exploring and acting out their abilities (von Hofsten, 2004). One of these abilities is language. With language they become able to communicate more efficiently. Moreover, the social world encourages young children to, for example, learn labels for objects (Moore, 2006). Young children show a strong interest in learning language by asking their caregivers what things are and how they are labeled (Gopnik & Meltzoff, 1997). Mastering language becomes thus a genuine interest of young children. Language entails normative vocabulary, so learning how to master language therefore entails learning how to become a player of the moral language game.

Third, children experience joy in interacting with others. Playing with their caregivers and, later, with other children constitutes an activity of high interest. Here, two aspects are important: one concerns the content of children's play and the other the formal aspect of cooperative activities. With respect to the content, it is well known that children often reenact social roles and appropriate the rules of their social world (Vygotsky, 1978). Normative development is thus fostered by children's appropriation of norms in their play activities. With respect to the formal aspect, engaging in an enduring play interaction entails to coordinate one's own interest with the interest of others and to find a mutual agreement in order to keep the interaction going (Habermas, 1985). Some developmental studies demonstrate an appreciation of the normative structure of agreements. For example, by three years children appreciate the normative obligations of a joint commitment to play a game (Kachel, Svetlova, & Tomasello, 2018). Relatedly, preschool children negotiate social norms with their peers and establish agreement on rules (Kalish, 2005; Köymen, Rosenbaum, & Tomasello, 2014; Piaget, 1932). Moreover, a longitudinal study found that children's enjoyment of interactions with their mothers predicted moral conduct and moral cognition (Kochanska et al., 2005).

The important point here is that children's interest in social exchange and their joy in social interactions is a driving force of normative development. In order to keep social interactions and social relationships (that is, consolidated forms of social interactions) going, they need to find joint agreements on how to organize these interactions. Taken together, children's dependency on others, their motivation to master language, and their joy in social interactions are three motivational forces that could support their engagement in the normative world.

It is important to note that children do not need to invent norms from the scratch. They do not need to invent novel normative concepts about what is good or bad. Rather, they develop within rich normative cultures that already provide the normative vocabulary and normative orientations at several levels (Bronfenbrenner, 1979). That is, normative development is different from the ontogeny of, for example, body morphology and sensory abilities in which each child (to some extent) independently develops what has been inherited through biological evolution. Normative development capitalizes on the social institutions and social forms of life—that is, cultural evolution. (For related thoughts, see Richerson & Boyd, 2005.)

To the extent that humans are a social species, one could also call the three aforementioned points to be the biological basis of normative development. In other words, if we were not social beings, if our survival did not depend on others, if we were not enjoying social interactions with others, and if we were not endowed with the ability to master language, we would not enter the world of normative obligations and negotiations. Yet in this view the biological basis does not entail that morality is inborn or that the content of normative orientations is evolutionary inherited. Rather, it sets the starting point for a cascade of developmental dynamics that eventually make children moral agents.

The Developmental Construction of Moral Norms and the Active Role of the Child

How can we best characterize young children's normative development? The previous considerations make clear that there is no biological substrate of morality (although the specifics of our social nature provide the basis for the acquisition of norms). Is it then only a matter of mere internalization or simple copying of others' behavior? This approach falls short of the characteristics of early moral development as well. Children are not passive beings; they are not containers that are filled with information. Rather, children are active agents who try to make sense of their experiences (Carey, 2011).

For one, children reconstruct the rules and regularities that underlie their experiences. This has been extensively studied in research on early language development. For example, when young children acquire the basics of grammar, they inflect irregular verbs in the same manner as regular verbs. This demonstrates that they try to extract the underlying rules and apply them to novel contexts. It exemplifies that children do not merely copy what they experience, but that development is a constructive process. Importantly, this is not only true for language development but also holds for the evaluation of actions (e.g., Oppenheim, Emde, Hasson, & Warren, 1997). It has been demonstrated that three-year-old children infer the presence of a norm even when merely observing an incidental action (Schmidt, Butler, Heinz, & Tomasello, 2016). In the same study, children even enforced this (assumed) norm from a third party. This finding relates well to work demonstrating that preschool-aged children also use nonverbal cues to infer a moral transgression (Dahl & Tran, 2016). Taken together, we see that young children infer the presence of norms by a variety of verbal, behavioral, and nonverbal cues. Moreover, young children reconstruct the underlying rules and apply them to novel contexts (Göckeritz, Schmidt, & Tomasello, 2014).

It has to be noted that in order to infer a norm and in order to apply it to a novel context, children already need to have an appreciation of norms. I suggest that this is possible because children have already entered the normative world and acquired the necessary normative vocabulary. Indeed, in the study by Schmidt, Butler, Heinz, and Tomasello (2016), children subsequently verbally protested against an agent that conducted the action differently. I argue that this is possible because children already have an understanding that social interactions are guided by rules and because the underlying norm is in principle accessible in a verbal format.

A second aspect of normative development is young children's well-known propensity to ask for justifications and reasons (Chouinard, 2007). If young children are presented with rules (particularly those they do not want to obey), they ask for reasons and try to negotiate. In disputes, they also use justifications for their claims (Dunn & Munn, 1987). Even when children do not directly ask for reasons, caregivers tend to explain to their children why it is important to follow a particular rule (Dunn, 1988). Notably, the inclusion of reasons into moral evaluations might in itself be a developmental outcome. In fact, young preschool children quite often provide irrelevant reasons when asked to justify their moral decisions, whereas older preschoolers are more likely to give adequate reasons (Mammen, Köymen, & Tomasello, 2018). Thus, in early ontogeny, normative claims are increasingly related to reasons that are supposed to support a claim. This reserves a foundational role for reasoning in moral development (May, 2018).

Moreover, classical developmental theories (Piaget, 1932) and recent accounts (Carpendale, Hammond, & Atwood, 2013) suggest that coordination with peers—that is, with others of equal status—plays an important role in young children's moral development. Indeed, the propensity to ask for justifications and to give justifications is higher when children interact with peers than with their mothers (Mammen, Köymen, & Tomasello, 2019). Children learn that they need to give reasons in order to make their point, and to accept reasons in order to keep the relationship going.

Overall, this represents the basis for moral discourse—that is, the giving and taking of reasons in moral debates (Habermas, 1985). We accept normative views to the extent that we are convinced of the reasons that support such a view. This sets the starting point for moral development as children become increasingly able to follow more complex lines of reasoning that take different factors into account. For example, whereas young preschool children mainly consider an equal split to be an appropriate norm for fair sharing (Rakoczy et al., 2016), older preschoolers might also consider the preexisting inequalities between different recipients and have a norm that one should give more to a poor than to a rich other: a norm of equity (Wörle & Paulus, 2018). This progress has been suggested to be related to children's increasing social experiences (Paulus & Leitherer, 2017), their perspective-taking abilities (Li et al., 2017), and an increase in information-integration capacities (Kienbaum & Wilkening, 2009). Importantly, these norms also align with children's actions (Elenbaas, Rizzo, Cooley, & Killen, 2016; Paulus, Nöth, & Wörle, 2018).

Finally, with the development of a self-concept and the attribution of stable characteristics to the self in early childhood (Harter, 2012), a moral self-concept emerges. That is, young children start to consider themselves as being someone who more or less likes to engage in other-oriented behaviors and who is more or less good or bad (Sengsavang & Krettenauer, 2015; Sticker, Christner, Pletti, & Paulus, 2019). For some philosophers, the relation between moral action and the self constitutes even a central aspect of human morality (Foucault, 1984; Korsgaard, 1996). Recent developmental accounts have proposed that this self-concept is an important factor in translating normative stances into concrete actions (Hardy & Carlo, 2011).

Conclusion

Taken together, in this chapter I provided a developmental account on how young children become moral agents. Normative development capitalizes on the (biological) basis that humans depend on each other, enjoy social interactions, and are endowed with the capability for language. This paves the way for an interest in ongoing social interactions. With the acquisition of language, children enter the social world of normativity. This social world consists of a vast amount of normative rules, and it demands that children from early on follow the rules. It sets the starting point for normative development and allows children to understand obligations and to make normative claims. Children actively reconstruct rules from their experiences and apply them to novel circumstances, highlighting the active role of the child in moral development. At the same time, the social environment provides reasons for their claims, and children ask for reasons, allowing children to enter the game of giving and taking reasons. An interest in maintaining social interactions requires children to negotiate with each other. This, again, sets another starting point for further moral development and the differentiation of principles and norms exemplified in children's normative stances.

Thus, a full understanding of normative development requires the recognition of three factors: our potentially inherited dependence on others and language ability; the fact that children grow up in a social environment that provides them with a variety of social norms from early on and requires children to follow them; and the active role of children who reconstruct the rules underlying their experience and their ability to shape norms by participating in the game of giving and taking reasons.

With each child who enters the moral community of individuals who reconstruct social rules and assess their generalizability, and engage in the game of giving and taking reasons, there thus comes a new perspective. It is this new perspective that contributes to the ongoing dynamics and change of human societies. And it is this perspective that raises doubt on claims that morality is either biologically inherited or merely internalized. It is this active role of the individual that led political and moral philosopher Hanna Arendt (1958) to introduce the term "natality" in order to describe the possibility of a novel beginning with each individual who enters the stages of this world.

Acknowledgments: For valuable feedback I am thankful to the members of the Munich social development laboratory. Preparation of this chapter was supported by an ERC Starting Grant (MORALSELF, no. 679000) and a James S. McDonnell Foundation 21st Century Science Initiative in Understanding Human Cognition–Scholar Award (No. 220020511).

References

Ainsworth, M. D. S., Blehar, M. C., Waters, E., & Wall, S. (1978). *Patterns of attachment: A psychological study of the strange situation.* Hillsdale, NJ: Erlbaum.

Arendt, H. (1958). *The human condition.* Chicago: University of Chicago Press.

Baillargeon, R., Setoh, P., Sloane, S., Jin, K., & Bian, L. (2014). Infant social cognition: Psychological and sociomoral reasoning. In M. S. Gazzaniga & G. R. Mangun (Eds.), *The cognitive neurosciences* (pp. 7–14). Cambridge, MA: MIT Press.

Barrett, L. F. (2017). *How emotions are made: The secret life the brain.* New York: Houghton-Mifflin-Harcourt.

Bennett, M. R., & Hacker, P. M. S. (2003). *Philosophical foundations of neuroscience.* Malden, MA: Blackwell.

Bloom, P. (2013). *Just babies: The origins of good and evil.* New York: Crown.

Bronfenbrenner, U. (1979). *The ecology of human development: Experiments by nature and design.* Cambridge, MA: Harvard University Press.

Carey, S. (2011). *The origin of concepts.* New York: Oxford University Press.

Carpendale, J. I. M., Hammond, S. I., & Atwood, S. (2013). A relational developmental systems approach to moral development. In R. M. Lerner & J. B. Benson (Eds.), *Embodiment and epigenesis: Theoretical and methodological issues in understanding the role of biology within the relational developmental system* (pp. 105–133). Burlington, MA: Academic Press.

Chouinard, M. M. (2007). Children's questions: A mechanism for cognitive development. *Monographs of the Society for Research in Child Development, 72,* 1–129.

Dahl, A. (2014). Definitions and developmental processes in research on infant morality. *Human Development, 57,* 241–249.

Dahl, A., & Paulus, M. (2019). From interest to obligation: The gradual emergence of human altruism. *Child Development Perspectives, 13,* 10–14. doi: 10.1111/cdep.12298

Dahl, A., & Tran, A. Q. (2016). Vocal tones influence young children's responses to prohibitions. *Journal of Experimental Child Psychology, 152,* 71–91.

Decety, J., & Cowell, J. M. (2014). The complex relation between morality and empathy. *Trends in Cognitive Sciences, 18,* 337–339. doi: 10.1016/j.tics.2014.04.008

Decety, J., & Cowell, J. M. (2018). Interpersonal harm aversion as a necessary foundation for morality: A developmental neuroscience perspective. *Development and Psychopathology, 30,* 153–164.

Dunn, J. (1988). *The beginnings of social understanding.* Cambridge, MA: Harvard University Press.

Dunn, J., & Munn, P. (1987). Development of justification in disputes with mother and sibling. *Developmental Psychology, 23,* 791–798.

Elenbaas, L., Rizzo, M., Cooley, S., & Killen, M. (2016). Rectifying social inequalities in a resource allocation task. *Cognition, 155,* 176–187

Forman, D. R., Aksan, N., & Kochanska, G. (2004). Toddlers' responsive imitation predicts preschool-age conscience. *Psychological Science, 15,* 699–705.

Foucault, M. (1984). *Histoire de la sexualité. Vol. 3: Le souci de soi.* Paris: Gallimard.

Göckeritz, S., Schmidt, M. F. H., & Tomasello, M. (2014). Young children's creation and transmission of social norms. *Cognitive Development, 30,* 81–95.

Gopnik, A., & Meltzoff, A. N. (1997). *Words, thoughts and theories.* Cambridge, MA: Bradford/MIT Press.

Gutsell, J. N., & Inzlicht, M. (2012). Intergroup differences in the sharing of emotive states: neural evidence of an empathy gap. *Social Cognitive and Affective Neuroscience, 7,* 596–603.

Habermas, J. (1983). *Moralbewußtsein und kommunikatives Handeln* [*Moral consciousness and communicative action*]. Frankfurt: Suhrkamp.

Habermas, J. (1985). *The theory of communicative action: Vol. 1, Reason and rationalization of society.* Boston: Beacon.

Habermas, J. (1991). *Erläuterungen zur Diskursethik.* Frankfurt: Suhrkamp.

Habermas, J. (1992). *Nachmetaphysisches Denken. Philosophische Aufsätze.* Frankfurt: Suhrkamp.

Hamlin, J. K. (2013). Moral judgment and action in preverbal infants and toddlers: Evidence for an innate moral core. *Current Directions in Psychological Science, 22,* 186–193. doi: 10.1177/0963721412470687

Hamlin, J. K., Wynn, K., & Bloom, P. (2007). Social evaluation by preverbal infants. *Nature, 450,* 557–559.

Hardy, S. A., & Carlo, G. (2011). Moral identity: What is it, how does it develop, and is it linked to moral action? *Child Development Perspectives, 5,* 212–218.

Harter, S. (2012). *The construction of the self: Developmental and sociocultural foundations.* New York: Guilford Press.

Joas, H. (1989). *Praktische Intersubjektivität.* Frankfurt: Suhrkamp.

Kachel, U., Svetlova, M., & Tomasello, M. (2018). Three-year-olds' reactions to a partner's failure to perform her role in a joint commitment. *Child Development, 89,* 1691–1703.

Kalish, C. W. (2005). Becoming status conscious: Children's appreciation of social reality. *Philosophical Explorations, 8,* 245–263.

Kant, I. (2018). *Groundwork for the metaphysics of morals* (Allen W. Wood, Ed. and Trans.). New Haven, CT: Yale University Press. (Original work published 1785)

Kienbaum, J., & Wilkening, F. (2009). Children's and adolescents' intuitive judgements about distributive justice: integrating need, effort, and luck. *European Journal of Developmental Psychology, 6,* 481–498.

Killen, M., Mulvey, K. L., Richardson, C., Jampol, N., & Woodward, A. (2011). The accidental transgressor: Morally-relevant theory of mind. *Cognition, 119,* 197–215.

Killen, M., & Smetana, J. G. (2015). Origins and development of morality. In M. E. Lamb (Ed.), *Handbook of child psychology and developmental science* (Vol. 3, 7th ed.) (pp. 701–749). New York: Wiley-Blackwell.

Kochanska, G., Forman, D. R., Aksan, N., & Dunbar, S. B. (2005). Pathways to conscience: early mother-child mutually responsive orientation and children's moral emotion, conduct, and cognition. *Journal of Child Psychology and Psychiatry, 46,* 19–34.

Korsgaard, C. M. (1996). *The sources of normativity*. Cambridge: Cambridge University Press.

Köymen, B., Rosenbaum, L., & Tomasello, M. (2014). Reasoning during joint decision-making by preschool peers. *Cognitive Development, 32,* 74–85.

Li, L., Rizzo, M. T., Burkholder, A. R., & Killen, M. (2017). Theory of mind and resource allocation in the context of hidden inequality. *Cognitive Development, 43,* 25–36.

Mammen, M., Köymen, B., & Tomasello, M. (2018). The reasons young children give to peers when explaining their judgments of moral and conventional rules. *Developmental Psychology, 54,* 254–262.

Mammen, M., Köymen, B., & Tomasello, M. (2019). Children's reasoning with peers and parents about moral dilemmas. *Developmental Psychology, 55,* 2324–2335.

May, J. (2018). *Regard for reason in the moral mind*. Oxford: Oxford University Press.

Moore, C. (2006). *The development of comonsense psychology*. Mahwah, NJ: Lawrence Erlbaum Associates.

Moore, G. E. (1903). *Principia ethica*. Cambridge: Cambridge University Press.

Oppenheim, D., Emde, R. N., Hasson, M., & Warren, S. (1997). Preschoolers face moral dilemmas: A longitudinal study of acknowledging and resolving internal conflict. *The International Journal of Psycho-Analysis, 78,* 943–957.

Paulus, M. (2014). How and why do infants imitate? An ideomotor approach to social and imitative learning in infancy (and beyond). *Psychonomic Bulletin and Review, 21,* 1139–1156.

Paulus, M. (2019, February 13). Is young children's helping affected by helpees' need? Preschoolers, but not infants selectively help needy others. *Psychological Research.* doi: 10.1007/s00426-019-01148-8

Paulus, M., & Leitherer, M. (2017). Preschoolers' social experiences and empathy-based responding relate to their fair resource allocation. *Journal of Experimental Child Psychology, 161,* 202–210.

Paulus, M., Nöth, A., & Wörle, M. (2018). Preschoolers' resource allocations align with their normative judgments. *Journal of Experimental Child Psychology, 175,* 117–126.

Piaget, J. (1932). *Le jugement moral chez l'enfant*. Paris: Alcan.

Prinz, J. (2011). Against empathy. *The Southern Journal of Philosophy, 49,* 214–233.

Racine, T. P., & Slaney, K. L. (Eds.). (2013). *A Wittgensteinian perspective on the use of conceptual analysis in psychology*. Basingstoke, United Kingdom: Palgrave Macmillan.

Rakoczy, H., Kaufmann, M., & Lohse, K. (2016). Young children understand the normative force of standards of equal resource distribution. *Journal of Experimental Child Psychology, 150,* 396–403.

Rakoczy, H., & Schmidt, M. F. H. (2013). The early ontogeny of social norms. *Child Development Perspectives, 7*(1), 17–21.

Richerson, P. J., & Boyd, R. (2005). *Not by genes alone: How culture transformed human evolution.* Chicago: University of Chicago Press.

Riedl, K., Jensen, K., Call, J., & Tomasello, M. (2012). No third-party punishment in chimpanzees. *Proceedings of the National Academy of Sciences of the United States of America, 109,* 14824–14829.

Scarf, D., Imuta, K., Colombo, M., & Hayne, H. (2012). Social evaluation or simple association? Simple associations may explain moral reasoning in infants. *PloS One, 7,* e42698.

Schmidt, M. F. H., Butler, L. P., Heinz, J., & Tomasello, M. (2016). Young children see a single action and infer a social norm: Promiscuous normativity in 3-year-olds. *Psychological Science, 27,* 1360–1370.

Scola, C., Holvoet, C., Arciszewski, T., & Picard, D. (2015). Further evidence for infants' preference for prosocial over antisocial behaviors. *Infancy, 20,* 684–692.

Sengsavang, S., & Krettenauer, T. (2015). Children's moral self-concept: The role of aggression and parent-child relationships. *Merrill-Palmer Quarterly, 61,* 213–235.

Smetana, J. G., & Braeges, J. L. (1990). The development of toddlers' moral and conventional judgments. *Merrill-Palmer Quarterly, 36*(3), 329–346.

Sticker, R. M., Christner, N. M., Pletti, C., & Paulus, M. (2019). *Dimensions of the moral self-concept and prosocial behavior in preschool children.* Manuscript in preparation.

Tan, E., Mikami, A., & Hamlin, J. K. (2018). Do infant sociomoral evaluation and action studies predict preschool social and behavioral adjustment? *Journal of Experimental Child Psychology, 176,* 39–54.

Tomasello, M. (2016). *A natural history of human morality.* Cambridge, MA: Harvard University Press.

Turiel, E. (1983). *The development of social knowledge: Morality and convention.* New York: Cambridge University Press.

Von Hofsten, C. (2004). An action perspective on motor development. *Trends in Cognitive Sciences, 8,* 266–272.

Vygotsky, L. S. (1978). *Mind in society.* Cambridge, MA: Harvard University Press.

Warneken, F., & Tomasello, M. (2006). Altruistic helping in infants and young chimpanzees. *Science, 311*(5765), 1301–1303.

Wörle, M., & Paulus, M. (2018). Normative expectations about fairness: The development of a charity norm in preschoolers. *Journal of Experimental Child Psychology, 165,* 66–84.

10   Understanding Others' Minds and Morals: Progress and Innovation of Infant Electrophysiology

Caitlin M. Hudac and Jessica A. Sommerville

Overview

Within the first two years of life, infants make tremendous gains in their ability to process social information. Despite efforts to characterize the nature and developmental timing of various aspects of social cognition, interpreting infants' emerging sensitivities is challenging, particularly because it can be difficult to determine whether infants possess precursor processes that support higher-level social cognition. We explore what we know about the developmental origins of two critical aspects of social cognition and discuss the utility of electrophysiology to address outstanding questions in these areas. First, we describe existing knowledge related to theory of mind, including how infants represent goals, desires, with a special focus on infants' understanding of false-beliefs. Second, we describe existing knowledge related to infants' sociomoral cognition, with a special focus on infants' emerging fairness concerns. We argue that electrophysiology can help identify the psychological processes that underlie early theory of mind abilities and early sociomoral concerns and thus can shed light on how these sensitivities emerge as a developmental process.

Introduction

Our ability to navigate the social world hinges on our understanding of others' minds as well as our ability to interpret others' behavior according to social and moral norms. Mental state reasoning and the detection and evaluation of others' behavior according to social and moral norms were once thought to be the unique providence of adulthood and childhood. Yet research over the past two decades has suggested that the underpinnings of these abilities can be found in infancy. Methodological advances in the form of techniques that rely on infants' visual attention and behavior and infants' behavior in interactive paradigms have been at the forefront in revealing these early sensitivities.

One significant challenge in the study of infants' social knowledge (and cognition, more broadly) is how to best interpret the sensitivities that infants possess early in life, before they can express themselves verbally or engage in sophisticated motor behaviors, which limits the range of paradigms that can be employed to tap infants' knowledge. Moreover, the hypotheses that link particular infant behaviors to underlying functions cannot always be fully elaborated (Aslin, 2007). Consequently, there is still considerable debate regarding the richness of infants' social and potentially moral cognition, and whether and how such knowledge differs from that of more mature reasoners.

The goals of this chapter are twofold. First, we review the state of our knowledge about infants' ability to reason about other people's minds, in particular their ability to understand others' false-beliefs, as well as their knowledge of sociomoral norms, in particular their developing concerns regarding fairness norms. Second, we discuss the utility of electrophysiology to address outstanding questions regarding these topics; we present the design and findings from two recent experiments we conducted on precursors to false-belief understanding and developmental changes in infants' fairness concerns, as well as future directions that stem from this work.

Infants' Understanding of Others' Minds

The ability to attribute, interpret, and reason about our own and others' mental states is known as theory of mind (ToM). ToM is an umbrella term that refers to a host of psychological processes (Wellman, 1990) that range from an understanding of more basic mental states and subjective experiences, such as goals, intentions, desires and perceptions, to understanding more complex, representational mental states, such as beliefs, as well as understanding the consequences of these mental states for people's behavior across a range of contexts.

A variety of evidence suggests that infants possess basic elements of ToM. Within the first two years of life, infants understand others' goals (Robson, Lee, Kuhlmeier, & Rutherford, 2014; Woodward, 1998), preferences (Choi,

Mou, & Luo, 2018), emotional states (Barna & Legerstee, 2005; Striano & Vaish, 2006), and perceptual experiences (Luo & Beck, 2010). An open question concerns when children can represent others' beliefs about the world. It is particularly difficult to represent a person's false-belief, as children need to understand that beliefs do not merely *reflect* reality but rather *represent* reality, and therefore can be accurate or inaccurate. Classic research has documented that it is not until the age of four that children possess this ability, as marked by their success on false-belief tasks (Gopnik & Wellman, 1995). The transition between age three and age five in children's ability to use others' false-beliefs to predict, describe, and explain others' behavior is thought to signal a qualitative, conceptual shift in children's understanding of the mind (Wellman, Cross, & Watson, 2001).

Yet infants and children younger than four years of age demonstrate successful performance on implicit tasks that may provide early evidence of false-belief understanding (for a recent review, see Scott & Baillargeon, 2017). A growing body of literature capitalizing on infants' visual attention shows that infants appear to expect an agent who possesses a false-belief to act in accord with that mistaken belief (e.g., search for an object in the location where they last saw it) rather than in accord with reality (e.g., search for an object in its actual location). In one such violation-of-expectation study (Onishi & Baillargeon, 2005), infants watched as an actress hide a toy in one of two boxes. In false-belief conditions, a screen prevented the actress from knowing that the object had switched locations. In the true-belief conditions, the actor witnessed the object switch locations. In both conditions, infants expected the actress to search for the object where she believed it to be, given her perceptual experience. More specifically, in the true-belief condition, infants expected the actress to search in the location where the object actually was, whereas in the false-belief condition, they expected her to search in the object's initial location. Additional experiments using violation-of-expectation paradigms also demonstrate similar looking time patterns in as young as seven months of age (Kovács, Téglás, & Endress, 2010).

Other behaviors that rely on infants' spontaneous responses add evidence that infants may be generating predictions regarding how a person will act based upon their beliefs. Evidence from anticipatory looking studies indicates that by twenty-four months of age infants spontaneously predict that agents will search in a location consistent with a person's false-beliefs (Senju, Southgate, Snape, Leonard, & Csibra, 2011; Southgate, Senju, &

Csibra, 2007). Furthermore, experiments that rely on infants' active help-ing behavior suggest that toddlers eighteen months of age and older may appreciate others' false-beliefs (Buttelmann, Carpenter, & Tomasello, 2009; Buttelmann, Suhrke, & Buttelmann, 2015). For example, when toddlers see an experimenter try to open an empty box, infants selectively open a second box that contains the target item when the experimenter falsely believes the target item to be in the empty box, assuming her actions on the empty box reflect a desire to retrieve the hidden object. In contrast, when infants see the experimenter try to open an empty box she knows to be empty, infants instead try to help her open the empty box, suggesting that because the actor knows the box to be empty she wishes to play with the empty box (Buttelmann et al., 2009). Some have argued that converging evidence across spontaneous response paradigms suggests that the quali-tative, conceptual shift proposed to underlie children's false belief under-standing (Wellman et al., 2001) may be an artifact of the heavy inhibitory demands that explicit false-belief tasks place on children (Baillargeon, Scott, & He, 2010).

Yet despite this converging evidence suggesting that infants may possess an ability to reason about others' (false) beliefs implicitly, recent studies have failed to replicate these findings (Kulke, Reiß, Krist, & Rakoczy, 2018; Powell, Hobbs, Bardis, Carey, & Saxe, 2018). In addition, even when findings are rep-licable, some scholars have suggested that infants' performance on these tasks can be accounted for by lower-level interpretations. In one famous study, Kovács, Téglás, and Endress (2010) claimed adults and infants as young as seven months of age encode others' false-beliefs in an automatic manner, based on the fact that infants' ability to detect a target object appears to be influenced by the beliefs of another agent involved in the scene. However, a subsequent replication study with adults determined that these effects may be an artifact of the timing and sequence of the task (Phillips et al., 2015). Thus, whether infants possess the ability to reasons about people's beliefs, including their false-beliefs, remains an outstanding question.

Infants' Expectations and Evaluations of Others' Sociomoral Behavior

Understanding sociomoral norms and acting upon them are a critical part of mature moral cognition. Although there are intense debates about the content of the concerns that fall into the moral realm, there tends to be agreement that moral reasoning involves concerns about the wellness of other people and concerns about justice and fairness. Moreover, a variety of evidence has suggested that preschool age children possess these concerns and, furthermore, that they distinguish violations to moral norms as distinct from social conventions (Turiel, 1998).

Within the last decade or so, the field has turned its attention to the origins of sociomoral sensitivities in infancy. In a pivotal study, infants watched as a puppet twice attempted to climb a hill unsuccessfully. On the third try, the puppet was either supported by a "helper" who pushed the puppet from behind or thwarted by a "hinderer" who pushed the puppet farther down the hill (Hamlin, Wynn, & Bloom, 2007). Consistent with prior work (Kuhlmeier, Wynn, & Bloom, 2003), infants looked longer when the puppet approached the hinderer than the helper. Additionally, when offered to interact with the two characters, infants selectively chose to reach for the helper over the hinderer. These findings suggested that infants were evaluating the agents based on their prosocial or antisocial behavior. Subsequent studies observed consistent looking time and preference behaviors that indicate infants prefer to interact with prosocial over antisocial individuals (Holvoet, Scola, Arciszewski, & Picard, 2016; Van de Vondervoort & Hamlin, 2018).

In a similar vein, recent work suggests that infants may have a nascent understanding of distributive fairness. Specifically, infants possess expectations for how resources are typically distributed to individuals (Geraci & Surian, 2011; Schmidt & Sommerville, 2011). For example, twelve-month-old infants, but not six-month-old infants, who saw an actress distribute crackers to two recipients looked significantly longer when one recipient received more crackers than the other (i.e., three crackers to one cracker) than when both recipients received equal amounts of crackers (i.e., each recipient received two crackers) (Ziv & Sommerville, 2016). Similarly, infants appear to prefer to interact with fair over unfair individuals (Burns & Sommerville, 2014; Lucca, Pospisil, & Sommerville, 2018).

Together these two types of finding suggest that the origins of moral foundations appear to be traceable back to infancy. However, the source of these foundations is debated. One perspective is that infants' preferences reflect an "innate moral core" (Van de Vondervoort & Hamlin, 2016). Others posit that sociomoral understanding is constructed with influence from external sources (Dahl & Killen, 2018), such that the emergence of early expectations for sociomoral behavior and sociomoral preferences are not simply an innate, biological process. Indeed, there is empirical evidence to support both perspectives. On the one hand, from as young as three months of age infants appear to prefer hinderers to helpers (Hamlin et al., 2007), consistent with a nativist perspective. On the other hand, infants appear to develop a sensitivity to equitable resource distribution between six and twelve months of age, tied to the onset of their naturalistic sharing behavior (Ziv & Sommerville, 2016), consistent with a more constructivist perspective. Thus, the origins and development of infants' sociomoral sensitivities requires further study.

Advantages of Electrophysiology

Electroencephalography (EEG) is a safe, noninvasive electrophysiological technique that measures underlying electrical fluctuations at the scalp (see figure 10.1), reflecting the summation of postsynaptic discharges from bundles of neurons within the cortex. Modern electrophysiology systems record these fluctuations at the millisecond level across a high-density array of electrodes (i.e., upward of 100 or more recording sites), providing unique insight into the temporal and topographical specificity of cognitive processes. Event-related potential (ERP) components capture temporal (i.e., proximity to the onset of the stimulus) and topographic (i.e., location on the scalp) elements of neurocognitive processes. One advantage of ERP is that it is possible to delineate between discrete cognitive processes, which may illuminate the manner by which infants become proficient in aspects of higher-level social cognition. Components with early peaks (i.e., closer to stimulus onset) are thought to reflect endogenous activity such that they capture external factors (e.g., perception, perceptual encoding), whereas mid- to late-latency peaks may be driven by internal factors (e.g., attention/cognitive reorienting, evaluation).

Another critical advantage is that EEG and ERP can be used continuously across age groups to measure cognitive processes. Although neural efficiency improves as the brain matures (i.e., white and grey matter development), infant ERP components corresponding to child and adult ERP components have been identified, albeit often with greater amplitude and slower latency (Kushnerenko et al., 2002). Despite these subtle differences, discrete cognitive processes are still separable and evident in infant ERPs and, importantly,

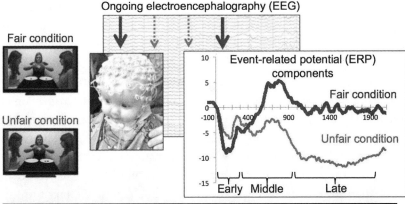

	Early	Middle	Late
Latency peak (ms)	< 300	300-600	600-900
Components	P1, P2, N1	N2, Nc, P300	Late slow waves (LSW)
Cognitive processes	Perceptual encoding	Attention, Cognitive reorienting	Evaluation, prediction of others' future behavior

Figure 10.1

Infant event-related potentials. After the electroencephalography (EEG) net is placed on the infant's head (<5 seconds for this style of net), infants sit on a parent's lap and watch vignettes while recording EEG. Event-related potentials (ERP) are time-locked to when a particular condition is shown, as indicated by the red (thin dashed, Unfair condition) and blue (thick solid, Fair condition) arrows. For example, in Hudac, Pospisil, and Sommerville (under review) infants watched videos in which a central actress distributed a resource fairly (one cracker to each) or unfairly (two crackers to one recipient). The ERP is averaged across all infants ($N = 22$, here) and all instances of the condition being shown (two times per vignette, >20 trials per infant). The ongoing changes in voltage (μV, y-axis) are characterized across time (milliseconds, x-axis) as components that reflect discrete cognitive processes, often described in reference as the point in time when the component amplitude is most positive or most negative.

can elucidate age-related changes in category discrimination. These advantages are important for exposing the emerging development and maturation of discrete cognitive processes that support infant social cognition.

Using Electrophysiology to Understand the Nature of Infants' Theory of Mind

Recent electrophysiological studies in adults have implicated several critical ERP components during ToM processes. First, primary convergence across several adult belief-state ERP studies report late slow wave effects across left frontal regions when adults are asked to make belief-based judgments (Geangu, Gibson, Kaduk, & Reid, 2013; Liu, Sabbagh, Gehring, & Wellman, 2004). These effects start approximately 300 milliseconds after stimulus onset with the peak difference occurring at approximately 800 milliseconds. Second, although not explicitly tested in these studies, visual inspection of the waveforms has indicated early perceptual components may be sensitive to belief states (N1, P1). Finally, Geangu and colleagues (2013) observed a midlatency P2 component that was larger for true rather than false belief states, which the investigators interpreted as a need for increased attention when tracking behavior aligned with a true representation of reality. Considering the paucity of work using implicit tasks for adults, it is important to note that adults recruit similar brain regions for both explicit and implicit ToM tasks (Kovács, Kühn, Gergely, Csibra, & Brass, 2014), supporting initial comparisons between infants and older individuals regardless of paradigm type. Additional support is provided by source analysis indicating that late slow wave activity is associated with regions within the ToM brain network (e.g., right inferior frontal gyrus, medial orbitofrontal cortex, and left temporal pole) during ToM processing (McCleery, Surtees, Graham, Richards, & Apperly, 2011).

The late slow wave effect is also present in children age six to eight years old (Meinhardt, Sodian, Thoermer, Döhnel, & Sommer, 2011) and appears to characterize differences between children aged four to six years who either succeed or fail at traditional false-belief tasks (Liu, Sabbagh, Gehring, & Wellman, 2009). In the latter study, a cartoon character placed two animals into separate boxes and then moved to the foreground so that the character could not see the boxes. One box opened, and the animal either moved to the other box or remained at the same location. Participants were asked questions regarding the animal's true location and the character's belief regarding the animal's location. For adults and children who responded correctly, the ERP waveform generated a larger late slow wave in response to the false-belief condition relative to the true-belief condition. There were no neural differences between conditions in children who did not pass the false-belief task, in support of the conceptual shift theory. However, while these findings reveal the neural correlates associated with successful performance on explicit false-belief tasks, to the extent that successful performance on false-belief tasks relies on ancillary abilities, they do not necessarily tell us anything about the underlying neural basis of understanding false-beliefs per se.

In a recent project, we sought to use an ERP paradigm to shed new light on infants' implicit false-belief understanding. The goals of this experiment were twofold. First, because a true understanding of beliefs requires the ability to use this understanding to form novel inferences, we sought to address whether infants are capable of understanding the affective *consequences* of one's belief being met or violated. Adults understand that if Scott expects a friend to stop by for a visit and he hears the doorbell ring, he believes that his friend is behind the door. If he opens the door to a traveling salesperson (i.e., *belief unmet*), adults would predict Scott to be disappointed, whereas if it were his friend (i.e., *belief met*), Scott would be happy. Although children's capacity for understanding affective consequences of beliefs is evident by three years of age (Wellman & Banerjee, 1991), little is known about how these skills emerge during infancy. Infants exhibit some understanding of affective consequences after goal fulfillment: infants eight and ten months of age expect an agent to be happy when it meets its goal, but do not necessary expect the agent to be unhappy if it fails to meet its goal (Skerry & Spelke, 2014).

Our second goal was to overcome one of the limitations of existing implicit false-belief paradigms. In most implicit paradigms, there is no way to verify whether infants possess the prerequisite understanding of different aspects of event scenarios upon which an understanding of false-beliefs must be built—for example, did infants encode the object switch? Do they differentiate scenarios in which the protagonist attends to the object switch from those in which the protagonist does not attend to the object switch? Thus, when infants succeed on these tasks it is assumed that they have encoded and differentiate key aspects of the vignette, and if infants fail it is difficult to know whether they failed the task because they failed to encode and differentiate this information or they lacked the ability to reason about false-beliefs per se.

Our strategy was to use an ERP paradigm to dynamically measure discrete cognitive processes involved in a false-belief scenario by revealing information in a step-wise fashion throughout the duration of a vignette. We designed a task that incrementally revealed information about an actress's mental state expectations and whether or not these expectations were met or violated in order to analyze the discrete neural indices of the outcome of the task (i.e., whether the actor's belief was met or violated), and also the

prerequisite processes that must be present for infants to perform successfully on the task (Hudac, Schutte, Webb, & Sommerville, *under review*).

We pursued these questions in the context of an object-labeling paradigm in which infants ($N=28$) watched as an actress looked for the labeled
object in one of two locations. We chose to test infants at almost nine
months of age due to evidence that as early as six months of age infants can
predict an actress will seek out a location based upon her belief (Southgate
& Vernetti, 2014). The parents were trained to teach their infant about the
object-labels within a play-based learning interaction that was conducted
twice a day for two to three days. Then, in the laboratory, the infants were
fitted with an EEG net, and they watched a series of video vignettes that
consisted of an actress (a) being told to expect an object (that was hidden
in a box), (b) finding either the matching or mismatching object (i.e., an
object that was implied by the label, or an object that was not implied by
the label), and (c) reacting with either positive or negative affect. Specifically, the actress inquired, "What is it?" and was told by a voice off-screen
that "It's a *bigu*." In the first critical event, the box opened up to reveal the
hidden object, which was either the *bigu* or a different object. In the second
critical event, the actress repeated the name with either positive or negative facial affect and prosody (i.e., "*Bigu!*" or "*Bigu?*" respectively). In this
way, we were able to evaluate whether the infants' knowledge of another
person's belief (i.e., the box contains *bigu*) leads to particular expectations,
such that the person expresses happiness when these expectations are met
(i.e., *bigu* is in the box) and confusion when these expectations are not met
(i.e., a different object is in the box).

Importantly, in our paradigm, we recorded ERPs at each of the two critical events to determine whether infants had encoded the prerequisite information to better understand infants' success or failure on our task. First,
we tested whether infants correctly paired the label to the object during
the first critical event in which it is revealed whether the box contained
the object to match the actress's belief. Our results confirmed that infants
link labels with their referent object as evidenced by a larger middle latency
frontal negative central (Nc) component in response to a matching object
and verbal label, than a mismatching object and verbal label, similar to
prior work (Reid et al., 2009). Second, we tested whether infants could distinguish positive and negative affective responses as indicated by differential brain responses between conditions. Indeed, a frontal P300 component

discriminated between emotions with greater amplitude for negative compared to positive affect. These findings suggest that infants can associate objects with their labels and distinguish positive and negative emotions, both of which are necessary prerequisites for using the actress's expectations to determine her affective reaction.

We next looked at ERP responses as a function of whether or not the actress's emotional response was congruent with her expectation (i.e., a *bigu* is in the box). No ERP components exhibited a significant interaction between object outcome (match, mismatch) and emotional response (positive, negative), suggesting that at this age infants do not sufficiently generate predictions such that affective consequences will align with others' beliefs. However, inspection of pairwise comparisons indicated that within the mismatch condition (when the object did not match the label), the P300 amplitude was greater when the actress had a negative affective response than a positive affective response. These findings may indicate an asymmetry in infants' ability to predict affective responses to met and unmet expectations: infants do not make predictions regarding affective responses to met expectations, but they may expect differential affective responses when expectations are not met.

In future work, we plan to test older infants in a paradigm that juxtaposes the infants' beliefs with that of the protagonist. In the current paradigm, infants' beliefs are consistent with that of the protagonist. For example, infants can be pre-exposed to the object within the box before the protagonist sees it, such that they (i.e., the infants) know the actual identity of object before the labeling event that sets up either accurate or inaccurate expectations for the protagonist. This manipulation will provide a stringent test of infants' ability to appreciate the affective consequences of an agent's true or false beliefs.

Using Electrophysiology to Understand the Nature and Developmental Origins of Infants' Sociomoral Cognition

A variety of functional magnetic resonance imaging (fMRI) research studies have indicated that the neural system underlying moral sensitivities is tethered to other related systems, including affective and social integration networks (Buon, Seara-Cardoso, & Viding, 2016; Decety, Michalska, & Kinzler, 2012). Considering the challenges of isolating these systems and processes, EEG and ERP methods can provide insight into discrete socio-moral cognitive processes that will be critical for assessing how these skills emerge during early development.

Much of the limited work using these methods in children and adults has targeted explicit sociomoral decision-making, although a handful of studies have focused on implicit sociomoral cognitive processes during passive observation. One study examined adult ERP responses to morally good or bad intentional action sequences (e.g., helping off the floor versus pulling hair) (Yoder & Decety, 2014). Results indicated early and middle latency components (N1, N2, late positive potential) were greater for morally good over morally bad actions, confirming that adults observed and recognized that these actions represent distinct moral features. Work in children has also implicated similar components as well as late slow wave effects. For instance, children aged three to five years discriminate when observing prosocial (e.g., sharing, helping) and antisocial (e.g., harming) intentional actions (Cowell & Decety, 2015b), such that early, middle, and late portions of the waveform (early posterior negativity, N2, late positive potential) confirm that children observe and shift attention based upon the valence of moral actions. The later waveform (late-positive potential, LPP) effects but not earlier waveform effects were directly tied to children's decision to share stickers with an unknown child stranger, highlighting differentiate the role of discrete cognitive processes in children's sociomoral sensitivities as well as their differential contributions to overt behavior.

Research with infants points to age-related developmental changes in sociomoral concerns. In comparison with older children, only the middle latency attentional Nc component significantly differentiated between prosocial and antisocial actions in infants aged twelve to twenty-four months (Cowell & Decety, 2015a). Additionally, in the classic hill climb task, six-month-old infants differentiated between helper and hinder scenarios as reflected by the P400 component, likely reflecting perceptual encoding (Gredebäck et al., 2015). Yet no differences were observed for the attentional Nc component (or other later ERP components), suggesting that younger infants may not fully evaluate this information as it relates to sociomoral expectations. These implied age-related changes in response to prososcial and antisocial actions suggest that young infants differentiate these actions, but that extracting the meaning of these actions is not evident until later in development.

We recently undertook a project to investigate whether age-related changes in infancy occur regarding infants' concerns about distributive fairness. Behavioral evidence suggests that infants may not expect an equitable division of resources until ten to twelve months of age (Surian, Ueno, Itakura, & Meristo, 2018; Ziv & Sommerville, 2016). However, it is possible that younger infants may not demonstrate expectations for equal resource distributions for reasons outside of concerns about fairness per se.

First, most existing studies have focus on relative unfairness; that is, distributions compare outcomes in which one recipient receives three resources and the other recipient receives one resource (as opposed to outcomes in which one recipient receives two resources, and the other recipient receives zero). It is possible that infants are sensitive to categorical instances of unfairness before they are sensitive to continuous instances of unfairness. Alternately, or additionally, it is possible that infants fail to notice important perceptual aspects of the event, such that in the case of the unfair distribution they fail to recognize that one recipient has received more resources than another recipient. Thus, we tested infants in an ERP paradigm that allowed us to directly compare infants' ERP response to categorical versus continuous instances of unfairness, and served to ensure that infants recognized the relevant perceptual distinctions that are necessary to detect unfairness.

We used an ERP paradigm to test infants between six months ($N = 22$) and twelve months ($N = 20$) of age (Hudac, Pospisil, and Sommerville, under review). In a violation-of-expectancy paradigm adapted from our prior work (Ziv & Sommerville, 2016), infants watched an actress distribute crackers to two recipients. Next, infants were presented with a series of static photos that revealed whether the actress had distributed the crackers fairly (i.e., equally) in either 1:1 or 2:2 distributions or unfairly (i.e., unequally) by giving all to one recipient in a 2:0 distribution or the majority to one recipient in a 3:1 distribution. Ongoing EEG was recorded and time-locked to the onset of the outcome, such that we could evaluate specific cognitive processes in response to the distribution.

Our results indicated that at twelve months infants demonstrated three different waveform components that discriminate between fair and unfair outcomes. First, P100 and P400 components varied as a function of fair versus unfair outcomes; such waveforms are typically thought to reflect encoding of visual information (Taylor, 2002), and therefore they likely indicate infants' ability to perceptually differentiate between the fair and unfair outcomes. The Nc component, which is thought to reflect differences in attentional orientation (Richards, 2003), also differed between fair and unfair outcomes for twelve-month-olds, suggesting that the unfair event yielded greater attentional orientation. Finally, there were late slow wave differences in late slow wave components, thought to reflect different evaluations of the events (de Haan, Johnson, & Halit, 2003). Thus, these findings suggest that twelve-month-old infants noticed the perceptual differences between the fair and unfair outcomes and demonstrated greater attentional orienting to the unfair event; also evaluative processes were implemented in relation to the two outcomes. These findings are consistent with behavioral work showing that twelve-month-old infants notice differences in resource distributions, possess expectations for the outcomes of these distributions, and evaluate outcomes and actors based on whether resources are distributed fairly or unfairly (Ziv & Sommerville, 2016). Also of note was the finding that the P400 and late slow wave varied according to whether the difference between fair and unfair outcomes was categorical or continuous, suggesting that infants of this age are sensitive to *degrees* of fairness.

By contrast, six-month-old infants only differentially responded to the P100, consistent with the findings of Gredebäck and colleagues (2015), indicating that while younger infants may perceive different distributions (e.g., 2:0 is different from 1:1), there is no evidence that younger infants encode this information as an evaluation of the distributor (e.g., *she was unfair*). Thus, despite the fact that infants of this age can perceptually distinguish between fair and unfair outcomes, they do not seem to perceive the difference between these outcomes as particularly noteworthy either in terms of continued enhanced attention or in terms of evaluation. Again, these findings are consistent with previously documented changes between six and twelve months of age in infants' responses to resource distribution outcomes (Ziv & Sommerville, 2016).

As a next step we plan to investigate associations between infants' experience and ERP responses to violations to distributive fairness, along with other types of sociomoral violations. Although there is evidence to suggest that infants' distributive fairness concerns are linked to the onset of their naturalistic sharing behavior and that individual differences in fairness concerns are related to the presence of siblings (Ziv & Sommerville, 2016), it is unclear how such experiences impact the underlying psychological processes that engender reactions to fair and unfair outcomes. Future studies can address this issue.

Conclusion

In this chapter, we have argued that electrophysiological approaches may shed light on the developmental origins of infants' social and moral cognition. Specifically, we believe that such approaches can help to identify the range of psychological processes that such sensitivities invoke and can help us understand the developmental trajectory of infants' theory of mind understanding and sociomoral concerns at a fine-grained level, in a manner that other paradigms may miss. Decomposing theory of mind understanding and sociomoral cognition into constituent cognitive processes is a particularly timely and important endeavor now, given the wide range of experiments that have demonstrated relative sophistication in infants' early sensitivities on the one hand, and preschooler's failures on explicit tasks that tap theory of mind and sociomoral cognition on the other hand. Adopting an electrophysiological approach holds potential for understanding the developmental process by which infants and children come to understand and navigate the social world.

References

Aslin, R. N. (2007). What's in a look? *Developmental Science, 10*(1), 48–53. doi: 10.1111/j.1467-7687.2007.00563.x

Baillargeon, R., Scott, R. M., & He, Z. (2010). False-belief understanding in infants. *Trends in Cognitive Sciences, 14*(3), 118. doi: 10.1016/j.tics.2009.12.006

Barna, J., & Legerstee, M. (2005). Nine- and twelve-month-old infants relate emotions to people's actions. *Cognition and Emotion, 19*(1), 53–67. doi: 10.1080/02699930341000021

Buon, M., Seara-Cardoso, A., & Viding, E. (2016). Why (and how) should we study the interplay between emotional arousal, theory of mind, and inhibitory control to understand moral cognition? *Psychonomic Bulletin and Review, 23*(6), 1660–1680. doi: 10.3758/s13423-016-1042-5

Burns, M. P., & Sommerville, J. A. (2014). "I pick you": The impact of fairness and race on infants' selection of social partners. *Frontiers in Psychology, 5*, 93. doi: 10.3389/fpsyg.2014.00093

Buttelmann, D., Carpenter, M., & Tomasello, M. (2009). Eighteen-month-old infants show false belief understanding in an active helping paradigm. *Cognition, 112*(2), 337–342. doi: 10.1016/j.cognition.2009.05.006

Buttelmann, F., Suhrke, J., & Buttelmann, D. (2015). What you get is what you believe: Eighteen-month-olds demonstrate belief understanding in an unexpected-identity task. *Journal of Experimental Child Psychology, 131,* 94–103. doi: 10.1016/j.jecp.2014.11.009

Choi, Y.-J., Mou, Y., & Luo, Y. (2018). How do 3-month-old infants attribute preferences to a human agent? *Journal of Experimental Child Psychology, 172,* 96–106. doi: 10.1016/j.jecp.2018.03.004

Cowell, J. M., & Decety, J. (2015a). Precursors to morality in development as a complex interplay between neural, socioenvironmental, and behavioral facets. *Proceedings of the National Academy of Sciences of the United States of America, 112*(41), 12657–12662. doi: 10.1073/pnas.1508832112

Cowell, J. M., & Decety, J. (2015b). The neuroscience of implicit moral evaluation and its relation to generosity in early childhood. *Current Biology, 25*(1), 93–97. doi: 10.1016/j.cub.2014.11.002

Dahl, A., & Killen, M. (2018). A developmental perspective on the origins of morality in infancy and early childhood. *Frontiers in Psychology, 9,* 1736. doi: 10.3389/fpsyg.2018.01736

Decety, J., Michalska, K. J., & Kinzler, K. D. (2012). The contribution of emotion and cognition to moral sensitivity: A neurodevelopmental study. *Cerebral Cortex, 22*(1), 209–220. doi: 10.1093/cercor/bhr111

De Haan, M., Johnson, M. H., & Halit, H. (2003). Development of face-sensitive event-related potentials during infancy: A review. *International Journal of Psychophysiology, 51*(1), 45–58. doi: 10.1016/S0167-8760(03)00152-1

Geangu, E., Gibson, A., Kaduk, K., & Reid, V. M. (2013). The neural correlates of passively viewed sequences of true and false beliefs. *Social Cognitive and Affective Neuroscience, 8*(4), 432–437. doi: 10.1093/scan/nss015

Geraci, A., & Surian, L. (2011). The developmental roots of fairness: Infants' reactions to equal and unequal distributions of resources. *Developmental Science, 14*(5), 1012–1020. doi: 10.1111/j.1467-7687.2011.01048.x

Gopnik, A., & Wellman, H. M. (1995). Why the child's theory of mind really is a theory. *Mind and Language, 7*(1–2), 145–171.

Gredebäck, G., Kaduk, K., Bakker, M., Gottwald, J., Ekberg, T., Elsner, C.,…Kenward, B. (2015). The neuropsychology of infants' pro-social preferences. *Developmental Cognitive Neuroscience, 12,* 106–113. doi: 10.1016/j.dcn.2015.01.006

Hamlin, J. K., Wynn, K., & Bloom, P. (2007). Social evaluation by preverbal infants. *Nature, 450*(7169), 557–U13. doi: 10.1038/nature06288

Holvoet, C., Scola, C., Arciszewski, T., & Picard, D. (2016). Infants' preference for prosocial behaviors: A literature review. *Infant Behavior and Development, 45*(B), 125–139. doi: 10.1016/j.infbeh.2016.10.008

Hudac, C. M., Pospisil, J. W., & Sommerville, J. A. (2019). The neural correlates of fairness expectations in infancy. Manuscript under review.

Hudac, C. M., Schutte, A. R., Webb, S. J., & Sommerville, J. A. (2019). Infant encoding of violation of expectations: Evidence from event-related potentials. Manuscript in preparation.

Kovács, Á. M., Kühn, S., Gergely, G., Csibra, G., & Brass, M. (2014). Are all beliefs equal? Implicit belief attributions recruiting core brain regions of theory of mind. *PLoS One, 9*(9), e106558. doi: 10.1371/journal.pone.0106558

Kovács, Á. M., Téglás, E., & Endress, A. D. (2010). The social sense: Susceptibility to others' beliefs in human infants and adults. *Science, 330*(6012), 1830–1834. doi: 10.1126/science.1190792

Kuhlmeier, V., Wynn, K., & Bloom, P. (2003). Attribution of dispositional states by 12-month-olds. *Psychological Science, 14*(5), 402–408. doi: 10.1111/1467-9280.01454

Kulke, L., Reiß, M., Krist, H., & Rakoczy, H. (2018). How robust are anticipatory looking measures of theory of mind? Replication attempts across the life span. *Cognitive Development, 46*, 97–111. doi: 10.1016/j.cogdev.2017.09.001

Kushnerenko, E., Ceponiene, R., Balan, P., Fellman, V., Huotilaine, M., & Näätäne, R. (2002). Maturation of the auditory event-related potentials during the first year of life. *Neuroreport, 13*(1), 47–51.

Liu, D., Sabbagh, M. A., Gehring, W. J., & Wellman, H. M. (2004). Decoupling beliefs from reality in the brain: An ERP study of theory of mind. *Neuroreport, 15*(6), 991–995. doi: 10.1097/01.wnr.0000123388.87650.06

Liu, D., Sabbagh, M. A., Gehring, W. J., & Wellman, H. M. (2009). Neural correlates of children's theory of mind development. *Child Development, 80*(2), 318–326. doi: 10.1111/j.1467-8624.2009.01262.x

Lucca, K., Pospisil, J., & Sommerville, J. A. (2018). Fairness informs social decision making in infancy. *PLoS ONE, 13*(2), e0192848. doi: 10.1371/journal.pone.0192848

Luo, Y., & Beck, W. (2010). Do you see what I see? Infants' reasoning about others' incomplete perceptions. *Developmental Science, 13*(1), 134–142. doi: 10.1111/j.1467-7687.2009.00863.x

McCleery, J. P., Surtees, A. D. R., Graham, K. A., Richards, J. E., & Apperly, I. A. (2011). The neural and cognitive time course of theory of mind. *Journal of Neuroscience, 31*(36), 12849–12854. doi: 10.1523/JNEUROSCI.1392-11.2011

Meinhardt, J., Sodian, B., Thoermer, C., Döhnel, K., & Sommer, M. (2011). True- and false-belief reasoning in children and adults: An event-related potential study of theory of mind. *Developmental Cognitive Neuroscience, 1*(1), 67–76. doi: 10.1016/j. dcn.2010.08.001

Onishi, K. H., & Baillargeon, R. (2005). Do 15-month-old infants understand false beliefs? *Science, 308*(5719), 255–258. doi: 10.1126/science.1107621

Phillips, J., Ong, D. C., Surtees, A. D. R., Xin, Y., Williams, S., Saxe, R., & Frank, M. C. (2015). A second look at automatic theory of mind. *Psychological Science, 26*(9), 1353–1367. doi: 10.1177/0956797614558717

Powell, L. J., Hobbs, K., Bardis, A., Carey, S., & Saxe, R. (2018). Replications of implicit theory of mind tasks with varying representational demands. *Cognitive Development, 46,* 40–50. doi: 10.1016/j.cogdev.2017.10.004

Reid, V. M., Hoehl, S., Grigutsch, M., Groendahl, A., Parise, E., & Striano, T. (2009). The neural correlates of infant and adult goal prediction: Evidence for semantic processing systems. *Developmental Psychology, 45*(3), 620–629. doi: 10.1037/a0015209

Richards, J. E. (2003). Attention affects the recognition of briefly presented visual stimuli in infants: An ERP study. *Developmental Science, 6*(3), 312–328. doi: 10.1111/1467-7687.00287

Robson, S. J., Lee, V., Kuhlmeier, V. A., & Rutherford, M. D. (2014). Infants use contextual contingency to guide their interpretation of others' goal-directed behavior. *Cognitive Development, 31,* 69–78. doi: 10.1016/j.cogdev.2014.04.001

Schmidt, M. F. H., & Sommerville, J. A. (2011). Fairness expectations and altruistic sharing in 15-month-old human infants. *PLoS One, 6*(10), e23223.

Scott, R. M., & Baillargeon, R. (2017). Early false-belief understanding. *Trends in Cognitive Sciences, 21*(4), 237–249. doi: 10.1016/j.tics.2017.01.012

Senju, A., Southgate, V., Snape, C., Leonard, M., & Csibra, G. (2011). Do 18-month-olds really attribute mental states to others?: A critical test. *Psychological Science, 22*(7), 878–880. doi: 10.1177/0956797611411584

Skerry, A. E., & Spelke, E. S. (2014). Preverbal infants identify emotional reactions that are incongruent with goal outcomes. *Cognition, 130*(2), 204–216. doi: 10.1016/j.cognition.2013.11.002

Southgate, V., Senju, A., & Csibra, G. (2007). Action anticipation through attribution of false belief by 2-year-olds. *Psychological Science, 18*(7), 587–592. doi: 10.1111/j.1467-9280.2007.01944.x

Southgate, V., & Vernetti, A. (2014). Belief-based action prediction in preverbal infants. *Cognition, 130*(1), 1–10. doi: 10.1016/j.cognition.2013.08.008

Striano, T., & Vaish, A. (2006). Seven- To 9-month-old infants use facial expressions to interpret others' actions. *British Journal of Developmental Psychology, 24*(4), 753–760. doi: 10.1348/026151005X70319

Surian, L., Ueno, M., Itakura, S., & Meristo, M. (2018). Do infants attribute moral traits? fourteen-month-olds' expectations of fairness are affected by agents' antisocial actions. *Frontiers in Psychology, 9,* 79–5. doi: 10.3389/fpsyg.2018.01649

Taylor, M. J. (2002). Non-spatial attentional effects on P1. *Clinical Neurophysiology, 113*(12), 1903–1908. doi: 10.1016/S1388-2457(02)00309-7

Turiel, E. (1998). The development of morality. In W. Damon & N. Eisenberg (Ed.), *Handbook of child psychology: Social, emotional, and personality development* (pp. 863–932). Hoboken, NJ: John Wiley & Sons.

Van de Vondervoort, J. W., & Hamlin, J. K. (2016). Evidence for intuitive morality: Preverbal infants make sociomoral evaluations. *Child Development Perspectives, 10*(3), 143–148. doi: 10.1111/cdep.12175

Van de Vondervoort, J. W., & Hamlin, J. K. (2018). The early emergence of sociomoral evaluation: Infants prefer prosocial others. *Current Opinion in Psychology, 20,* 77–81. doi: 10.1016/j.copsyc.2017.08.014

Wellman, H. (1990). Theory of mind. In R. A. Scott and S. M. Kosslyn (Eds.), *Emerging trends in the social and behavioral sciences.* New York: Wiley Online Library. doi: 10 .1002/9781118900772.etrds0360

Wellman, H. M., & Banerjee, M. (1991). Mind and emotion: Children's understanding of the emotional consequences of beliefs and desires. *British Journal of Developmental Psychology, 9*(2), 191–214. doi: 10.1111/j.2044-835X.1991.tb00871.x

Wellman, H. M., Cross, D., & Watson, J. (2001). Meta-analysis of theory-of-mind development: The truth about false belief. *Child Development, 72*(3), 655–684.

Woodward, A. L. (1998). Infants selectively encode the goal object of an actor's reach. *Cognition, 69*(1), 1–34.

Yoder, K. J., & Decety, J. (2014). Spatiotemporal neural dynamics of moral judgment: A high-density ERP study. *Neuropsychologia, 60*(C), 39–45. doi: 10.1016/j .neuropsychologia.2014.05.022

Ziv, T., & Sommerville, J. A. (2016). Developmental differences in infants' fairness expectations from 6 to 15 months of age. *Child Development, 289,* 128. doi: 10.1111/ cdev.12674

11 Cognitive and Neural Correlates of Children's Spontaneous Verbal Deception

Xiao Pan Ding and Kang Lee

Overview

Verbal deception—lying—is a common behavior in childhood, the scientific study of which dates back to the nineteenth century. Much of the scientific work for over 100 years has focused on children's lying behavior itself. Only in the last two decades has researchers begun to explore the cognitive and neural correlates of the development of verbal deception. In this chapter, we will review the current evidence regarding the cognitive and neural factors related to children's spontaneous lying. First, we introduce three common paradigms used to study children's spontaneous lying. Second, we review the growing evidence regarding the role of theory of mind understanding and executive functioning in the development of spontaneous lying. Third, we present emerging evidence from recent neuroimaging studies about children's lying in relation to the more extensive evidence about the neural correlates of lying in adults. Finally, we will discuss and outline the gaps in current knowledge and future directions of research to elucidate the cognitive and neural mechanisms underlying verbal deception.

Introduction

Lying is a common human behavior. Its developmental origin has fascinated scientists and laypeople alike. The earliest scientific report of spontaneous lying in early childhood was by Charles Darwin (1877). Darwin observed his son telling lies at two years of age to cover up the fact that he had stolen and eaten sugar from the kitchen. This observation was confirmed by a series of studies in our laboratories and others' about a century later. Indeed, we found that children begin to tell spontaneously various types of lies as early as 2.5 years of age (Evans & Lee, 2013).

Researchers began to take a keen interest in children's deception in the late 1990s following early work on children's lying (Darwin, 1877; Hartshorne & May, 1928; Piaget, 1932). This was due to the realization that children's lying is not only related to moral development but also strongly associated with cognitive development—especially the development of theory of mind (ToM) and executive function (EF). Researchers and theorists have recognized that children's lying can be a unique window into understanding cognitive and moral development. Additionally, with the development of different neuroimaging tools, it became possible to investigate the neural mechanisms underlying children's lying.

In this chapter, we first describe the cognitive factors underlying children's spontaneous lying and then discuss the neuroimaging work related to their deceit. Before discussing these issues, we will first introduce three widely used behavioral paradigms for studying children's spontaneous lying: the temptation resistance paradigm (TRP), the hide-and-seek paradigm, and the undesirable gift paradigm.

Temptation Resistance Paradigm

TRP is used to measure whether children would tell a lie after a minor transgression. It was initially used to test lying behavior in preschoolers (Talwar, Lee, Bala, & Lindsay, 2002), but researchers modified the paradigm to be suitable even for children in late childhood to middle adolescence as well (Evans & Lee, 2011).

During the paradigm (figure 11.1), children play a guessing game with an experimenter. For the first two trials, the experimenter plays a sound (e.g., barking) associated with a toy (e.g., a dog), and children are asked to guess what they think the toy is based on the sound. After two successful trials of the game, the experimenter makes an excuse to leave the room but tells children not peek at the toy. The experimenter places a bear on the table but plays an unrelated sound so that children are not able to guess the toy without peeking at it. All children are then left alone in the room with the toy for a minute. The experimenter returns to the room, covers the toy, and asks children whether they had peeked at the toy. If a child peeked at the toy but does not admit to it, he or she is coded as a "lie-teller."

Figure 11.1
The temptation resistance paradigm with hidden video cameras.

The experimenter then asks the child, "What do you think the toy is?" The purpose of this question is to assess lie-tellers' ability to maintain their lie. Children are coded as "revealers" if they respond with the correct identity of the toy (i.e., a bear). Conversely, children are coded as "concealers" if they either feign ignorance to the toy's true identity (i.e., "I don't know") or respond with a different toy (e.g., "music box").

Hide-and-Seek Paradigm

The hide-and-seek paradigm (figure 11.2) is a method to measure children's strategic deceptive behavior (Ding, Wellman, Wang, Fu, & Lee, 2015). Unlike TRP, it not only measures whether children can lie, but also the frequency of their lies.

Before the game begins, child chooses their most desired stickers from a box. These stickers provide an incentive for the child to lie. During game, the experimenter tells the child, "I will close my eyes. You need to hide the sticker under either of these two cups." After the child indicates that he or she has hidden the sticker, the experimenter opens her eyes and asks, "Where did you hide the sticker?" If the child points to the actual cup, the child is telling the truth. If the child points to the empty cup, the child is

Figure 11.2
The hide-and-seek paradigm.

telling a lie. The experimenter will always search for the sticker in the cup that the child indicates. The total deceptive score is the cumulative number of times that the child points to the empty cup.

Undesirable Gift Paradigm

The most common procedure to test prosocial lying is the undesirable gift paradigm. Before the formal game, children are shown a dozen different items and asked which item they like the most (desirable gift) and which they like the least (undesirable gift). Upon identifying the desirable and undesirable gift items, the experimenter makes an excuse to leave the room; another experimenter (confederate) then comes in and plays a game with the child. After the game, the confederate tells the child that she would like to give the child a gift because of the child's good performance on the game. Contrary to the child's expectations, the confederate gives the undesirable item to the child. After the child inspects the gift, the confederate asks the child whether he or she likes the gift. After the child has responded, the confederate leaves the room, and the first experimenter enters the room again. If the child indicates that he or she likes the gift, the experimenter asks, "I just heard from my friend that you told her that you liked this gift, but you told me earlier that you didn't like the gift. Do you really like it?"

If the children respond that they do not like the gift, they are considered to have told a white lie.

Part I: Cognitive Correlates Underlying Children's Lying

Theory of Mind

Theory of mind refers to the ability to reason about mental states (Wellman, 2014). ToM is one of the core components of cognitive development because extensive research has shown that ToM can predict children's social and cognitive development, such as children' prosocial behavior, interpersonal interaction, and popularity with peers (Mizokawa & Koyasu, 2015; Slaughter, Imuta, Peterson, & Henry, 2015; Watson, Nixon, Wilson, & Capage, 1999).

Lying is considered to be ToM in action. In order to lie and to lie successfully, children need to be equipped with an understanding of other's mental states (Lee, 2013). Specifically, ToM plays an important role in three levels of children's lying.

Knowledge access and lying.   As an early precursor of ToM, knowledge access understanding enables children to understand the similarities and differences between their own mental states and those of other people. They begin to understand that others may not have the same knowledge as them. In turn, this allows children to make appropriate decisions about whether to lie and what to lie about (Lee, 2013). This is because children know that others do not know the truth, and, as a result, they can tell a lie.

Two studies provided the first evidence to support the relation between knowledge access and lying; these studies found that two- to three-year-old children's simple denials of a transgression are related to knowledge access understanding (first-order ignorance: others do not know the truth) (Leduc, Williams, Gomez-Garibello, & Talwar, 2017; Ma, Evans, Liu, Luo, & Xu, 2015). The study by Sai, Ding, Gao, Fu (2018) examined the cognitive factors that contribute to children's second-order lying (i.e., children's ability to tell *truth* to deceive others). They also found that children's ability to tell second-order lies was positively related with their knowledge access understanding.

First-order false-belief understanding and lying.   As the most popular topic in ToM research, first-order false-belief understanding refers to the realization that it is possible for others to hold a false-belief about events in the world (Miller, 2009). To lie successfully, first-order false-belief understanding provides an important cognitive tool to allow children to deliberately instill a false-belief in the mind of the lie recipient (Talwar & Lee, 2008). Extensive studies have provided evidence to support the close relation between lying and ToM. For example, Talwar and Lee (2008) found that children with better first-order false-belief understanding tend to lie more in the temptation resistance paradigm. This is also supported by other studies using different paradigms. For example, Fu, Sai, Yuan, and Lee (2017) found that the greater the children's first-order false-belief abilities, the more frequently they lied in the hide-and-seek paradigm.

Second-order false-belief understanding and lying. Second-order false-belief understanding refers to the ability to realize that it is possible for others to hold a false-belief about someone else's belief (Miller, 2009). Second-order false-belief understanding also needs to provide the further recursive reasoning about mental states to help children to maintain lies (Talwar & Lee, 2008). Children with better second-order false-belief understanding tend to be better at maintaining their lies (Talwar, Gordon, & Lee, 2007). Additionally, second-order false-belief understanding can predict a child's ability to maintain prosocial lies (Williams, Moore, Crossman, & Talwar, 2016).

Executive Function

As a core cognitive ability for children, the development of EF is also believed to be related to children's lying. EF refers to a set of higher order psychological processes that are involved in goal-oriented behavior (Zelazo & Müller, 2002). EF includes cognitive skills such as inhibitory control, planning, cognitive flexibility, and working memory (Diamond, 2006; Testa, Bennett, & Ponsford, 2012). Previous studies have consistently found that children with better EF are more likely to lie spontaneously (Alloway, McCallum, Alloway, & Hoicka, 2015; Evans & Lee, 2011, 2013; Leduc et al., 2017; Williams et al., 2016). This positive association between EF and lying is expected because EF's constituent cognitive skills are necessary for engaging in deception. First, children need to inhibit revealing the truthful information, which relies on *inhibitory control* (Carlson, Moses, & Hix, 1998; Evans & Lee, 2013). Second, children need *cognitive flexibility* to help them switch between telling the truth and coming up with a false response (Ding, Omrin, Evans, Fu, Chen, & Lee, 2014). Lastly, children need *working memory* to keep track of both truthful and false information (Alloway et al., 2015).

Recent Training Studies

From the previous studies, we find that children's lying is related closely to their cognitive abilities: children with better ToM and EF lie earlier and lie better. Though previous studies focus on correlations, recent training studies provide new evidence to support the causal relation that ToM and EF have with lying.

Theory of mind training to improve children's lying. Ding and her colleagues (2015) trained young children's ToM ability for ten days and found that children who initially cannot tell lies can readily learn to tell lies. In this study, they selected children who were unable to lie in the pretest. Half of the children were assigned to a ToM training condition in which they learned to reason about various mental states in different situations. The other half was assigned to a control training condition in which they learned to reason about properties of physical objects. After participating in the ToM training to learn about mental state concepts, the children who had initially been unable to lie began to lie. Although the training only lasted for ten days, this training effect lasted for more than a month. In contrast, the children who participated in control training to learn about physical concepts were significantly less inclined to lie than the ToM-trained children. These findings support the causal role of ToM in the development of deception in early childhood.

Lying training to improve cognitive abilities. Previous studies have shown that practicing competitive lying games can improve individual's cognitive abilities in adults (Ybarra, Winkielman, Yeh, Burnstein, & Kavanagh, 2011). Ding and colleagues (2018b) extended this finding to young children and examined whether it was also possible that training children to lie would in turn lead to the improvement of their cognitive abilities. In this study, they invited three-year-old children to play a competitive hide-and-seek game. After the pretest, children who could not deceive in the game were randomly assigned to one of two conditions. Those in the experimental condition were instructed on how to deceive the opponent to win the game each day during the training phase, while those in the control condition did not receive any instructions about how to win the game. Children in both conditions then played the exact same game on four consecutive days. Training effects on ToM and EF skills were assessed during a pretest and a posttest. The results showed that practicing the deceptive games led to significant improvements in children's ToM and EF skills. By engaging in deception, children practice and improve these cognitive skills concurrently.

Whether children can discover how to lie spontaneously. Ding, Heyman, Fu, Zhu, and Lee (2018a) adopted a microgenetic method to observe how children who initially cannot lie learn how to lie in ten days. In each session, children played the hide-and-seek game with experimenters. Although children originally showed little ability to deceive, most of them discovered how to lie within the ten days. Further, children with better ToM and EF learned how to tell lie faster than their peers. These results were the first to provide evidence for the importance of cognitive skills and social experience in the discovery of deception over time in early childhood.

The existing evidence taken together suggest strong associations between ToM, EF, and children's lying. The cognitive abilities facilitate emergence of children's lying, whereas the practicing of lying can promote the development of their cognitive abilities. It is worth noting that the three studies above adopted the hide-and-seek paradigm to assess children's ability of lying; we do not know whether this training effect would transfer to other kinds of lying, such as lying to conceal transgressions or prosocial lying. The answer to this question awaits future research.

Part II: The Neural Development of Children's Spontaneous Lying

Although extensive behavioral research has examined deception in children and adults for nearly a century (Hartshorne & May, 1928; Lee, 2013), only recently have researchers begun to examine the neural basis of deceptive behaviors. Most of the existing studies have only involved adults and focused on identifying the focal brain areas related to dishonesty (for a review, see Abe, 2011). In the past decades, researchers have used various neuroimaging methodologies to investigate the neural correlates underlying deception behavior, including functional magnetic resonance imaging (fMRI), positron emission tomography, functional near-infrared spectroscopy(fNIRS), and transcranial magnetic stimulation. With the recent development of neuroimaging techniques, researchers have begun to analyze the functional neutral connectivities underlying deception. Given the fact that there have been few studies about the neural development of lying, we introduce the neural correlates underlying adult's deception first and then describe the studies about the neural development of children's lying.

The Brain Regions Involved in Lying

The first motivation to study the neural correlates underlying lying comes from the area of deception detection. With the development of the neuro-imaging studies, scientists begin to adopt various of techniques to detect deception based on the orienting theory.

The first wave of neuroimaging studies about lying focused on the cognitive process of deception, and they mainly adopted the instructed lying paradigm. For example, one pioneer study used the guilty knowledge test paradigm (Langleben et al., 2002). In this paradigm, neural markers of deception can be found by comparing different brain activities between deceptive and control items. Later, researchers realized the instructed lying paradigm is artificial and does not reflect the lying in real life where liars tell lies out of their own volition.

From Greene's pioneer work, researchers started to adopt spontaneous lying paradigms (Greene & Paxton, 2009). In the spontaneous lying paradigms, researchers create a scenario in which participants are not only motivated to lie spontaneously but also actively choose different strategies to tell lies (for example, participants can use truth-telling to tell lies).

Most of existing studies involve adult participants. They have found that deception usually results in much broader brain activations than truth-telling behaviors, and these activations mainly occur in the prefrontal cortex (PFC) (Christ, Van Essen, Watson, Brubaker, & McDermott, 2009). Because the PFC plays a central role in cognitive control (Cocchi, Zalesky, Fornito, & Matting-ley, 2013; Osaka et al., 2004), these findings support the idea that, compared to honest behavior, dishonest behavior is a more intensive cognitive control task (Sip, Roepstorff, McGregor, & Frith, 2008; Vrij, Fisher, Mann, & Leal, 2006). Additionally, activating and formulating a lie is more intentional and deliberate than activating the truth (Walczyk, Harris, Duck, & Mulay, 2014). Although instructed lying and spontaneous lying share some common brain areas, there are different brain activation patterns in the dorsal PFC between spontaneous and instructed lying. To lie spontaneously, individuals need the network of working memory and cognitive flexibility to generate alternative responses additionally (Ding, Gao, Fu, & Lee, 2013). Taken together, these adult studies reveal that spontaneous deception recruits the cognitive control system including network of inhibitory control, working memory, and cognitive switching.

Despite the important role that the cognitive control system plays in deception, researchers also recognize the role of reward system (Abe &

Greene, 2014; Ding et al., 2013; Hu, Pornpattananangkul, & Nusslock, 2015). Heightened activity is found in the frontal area (the right superior frontal gyrus and left middle frontal gyrus) for winning compared to losing. Additionally, participants who begin lying sooner have greater neural response differences between winning and losing the trials. This result suggests the involvement of the cortical reward system in deception (Ding et al., 2013).

There is more direct evidence showing how the reward system is involved in honest and dishonest behavior. Abe and Greene (2014) found that participants who were sensitive to rewards were more inclined to lie and the dishonest group had a stronger activation in the dorsolateral PFC when they were not lying, which suggests that they may have actively resisted temptation. Most recently, Liang, Ding, Lee, and Fu (2018) found that the reward system not only showed different patterns during the different phases of deceptive behavior but also predicted whether the participants would switch their deceptive strategy in the next trial. These studies suggest that reward system may moderate the decision and the execution of honest and dishonest behaviors.

The Functional Connectivities of the Neural Networks for Lying

The majority of the previous studies found that the PFC is critically involved in an individual's deceptive behavior, especially the dorsolateral PFC, the ventrolateral PFC, the dorsomedial PFC, and the inferior frontal gyrus. However, given that these regions do not work in isolation, it is likely that the deception not only elicits activations of each brain area, but also changes the connections among those regions. Several fMRI and fNIRS studies explored the functional connectivities of deception. For example, Zhang and colleagues (2016) were among the first to examine the differences in brain networks between spontaneous deception and instructed deception. They found that the brain networks of spontaneous deception showed different small world proprieties compared with instructed deception. Likewise, Pornpattananangkul, Zhen, and Yu (2018) found that the functional connectivities between the reward system and cognitive control system played a key role in registering the social-related goal of dishonest decisions.

The Neural Development of Children's Lying

Although deception is common in children, there have only been a few of neuroimaging studies on children's deception to date. Yokota et al. (2013)

and Thijssen et al. (2017) reported the first two neuroimaging studies on deception in children aged seven to eleven years old. Instead of the activation in the frontal lobes which have been found in adult studies, they found that the inferior parietal lobe, the anterior cingulate cortex, and the frontal pole play an important role in children's deception. Because the frontal lobes continue to mature even after adolescence, it is easy to understand that deception in children may rely on spatially distinct, or morphologically different, cortical regions compared with adults.

Liang, Hu, Ding, and Fu (2019) explored the role of ToM in three- to six-year-old children's deceptive behavior using fNIRS. The results confirmed that activation of the right temporal parietal junction (RTPJ) was stronger in the lying condition than the truth-telling condition. They further found that when children told lies the activation of the RTPJ was less in children with better ToM ability than in children with poorer ToM ability. That is, children with better ToM ability recruited fewer neural resources when they were lying.

In another study, Ding, Wu, Liu, Fu, and Lee (2017) examined how different brain regions interact with each other during spontaneous deception in seven- to twelve-year-olds. They found that both the global and local efficiencies of children's functional neural networks decreased during lying, suggesting that lying disrupts the efficiency of children's cortical network. This means that deception taxes the brain, and young children are more vulnerable to the network disruption.

Challenges

The findings from the previous studies have laid the foundation for understanding the cognitive and neural mechanisms underlying children's verbal deception. They have established close relations between children's verbal deception, ToM and EF development, and involvement of the cognitive control and reward system brain networks in the processing of verbal deception. However, there exist considerable limitations in the current literature. We will address these limitations and discuss future directions of research.

The first challenge is the lack of neural evidence for the development of verbal deception. Although there are numerous studies examining the neural patterns underlying adults' deception, as well as behavioral studies focusing on children's verbal deception, only a few of them have investigated the neural development of children's lying. This is because it is extremely difficult to conduct neuroimaging studies on young children.

Two studies used fMRI to investigate the neural mechanisms underlying children's deceptive behavior; however, they only tested children who were above seven years of age (Yokota et al., 2013; Thijssen et al., 2017). As mentioned previously, children start to lie around 2.5 years of age, and their neutral activation patterns during may be different from those of older children because ToM and EF change dramatically during the preschool years. It is vital to study the neural development of verbal deception in three- to six-year-old children.

Neuroimaging studies with young children can provide more direct information on how the cognitive control brain network and reward system brain network develop to support children's deceptive behavior. The gold standard in neuroimaging studies is fMRI, but several of its characteristics limit its usage with young children. For example, the fMRI machine is noisy, and children may be uncomfortable lying down in a confined space.

Recently developed neuroimaging tools, such as fNIRS, have become a feasible solution for researchers to study brain development in young children. In fact, fNIRS can help us to investigate the brains of children as young as newborns (Aslin, Shukla, & Emberson, 2015). We have conducted two studies with three- to six-year-olds (Liang et al., 2019) and six- to twelve-year-olds (Ding et al., 2017), which showed that fNIRS is a feasible and reliable technique to study the neural correlates of young children's deception. Because it is an inexpensive and noninvasive neuroimaging tool, it also allows for conducting longitudinal studies of neural correlates of children's deception that to date has never been attempted.

The second challenge is how to study children's deception in a social context. Verbal deception is an important social skill and always involves other parties. Children may lie to their parents or other people for various purposes. However, most of the previous studies—especially the neuroimaging studies—have asked children to play games with computers, which made the experimental setting unnatural. Additionally, most deception studies only have focused on lying for personal gain (self-benefiting lying). Yet verbal deception takes many forms and can have many different purposes. Children can lie not only for self-interest but for self-protection, and for helping others. Thus, it is essential to study children's deception and neural mechanisms in much broader social contexts than has been done to date.

Future Directions

We suggest several future directions of research to advance our current understanding of the development of lying and its cognitive and neural mechanisms.

The neural correlates underlying children's emergence of lying. Using microgenetic methods, we observed that young children discover how to tell self-benefiting lies spontaneously in the hide-and-seek game (Ding, Heyman, Fu, Zhu, & Lee, 2018a). Although we found that children with better EF and ToM discover how to lie more quickly, we know little about how children's discovery of lying manifests itself at the neural level. Future studies can combine microgenetic and neuroimaging methods to observe more closely the process through which children discover how to lie. It would shed light on the relations between the cortical regions involve in ToM and EF and those involved in lying.

The neural correlates underlying adolescents' lying behavior. Previous studies mainly examined children and young adults' deceptive behavior, with few studies focusing on adolescents. We noticed that there is an inverted U-shaped development of children's lying: from three years onward, they begin to lie increasingly more often; however, children's lies become increasingly less obvious as they reach school age (Evans & Lee, 2011). As revealed by adult studies, the reward system (motivation) may also play an important role in children's deception behavior, especially during adolescence (Blakemore, 2008). It is possible that individuals recruit the cognitive control and reward systems differently when they are of different ages. For young children, they may need cognitive control to inhibit the truth in order to tell a lie whereas older children may need cognitive control for them to inhibit the reward system to tell the truth. It will be interesting to see how the reward system and cognitive control system contribute to adolescents' deception compared with children and adults.

Short-term and long-term behavioral and neural effects of children's deception. Several behavioral training studies have shown that children's deceptive behavior has short- and long-term effects on their cognitive development. Ding and her colleagues (2017) found that deception caused temporary changes in brain connectivities . It is not entirely clear whether lying leads to long-term changes in neural connectivities as well. Furthermore, recent behavioral studies have demonstrated the possibility that training children to deceive confers short-term cognitive benefits. It is unknown whether such training would have collateral neutral effects or whether such effects, if any exist, are beneficial or detrimental to brain development. It is possible that deception training leads to cognitive and neural short-term gains but not long-term efficiency. These intriguing possibilities await sustained research in the near future.

Verbal deception in social contexts.   Previous studies found that parenting factors affect children's lying (Ma et al., 2015; Wang et al., 2017). At the same time, a few studies showed that parent–child brain-to-brain synchrony is an indicator of children's social emotional development. Brain-to-brain synchrony may enable children to take another's perspective and understand others' actions (Nummenmaa et al., 2012). Future studies can establish a link between parent–child brain synchrony during parent–child interaction, ToM, and children's lying behavior. We can also extend the existing deception paradigms and examine how children incorporate social norms and prosocial behavior into deception.

Conclusion

The current chapter reviewed three common paradigms used to study children's spontaneous verbal deception, as well as the growing evidence regarding the role of ToM understanding and executive functioning in the development of verbal deception. We also presented the emerging evidence from recent neuroimaging studies about children's lying in relation to the more extensive evidence about the neural correlates of lying in adults. In future, the systematic investigation of lying behavior in broad social contexts—with a life span development perspective—is necessary to elucidate the cognitive and neural mechanisms underlying verbal deception in children and adults.

References

Abe, N. (2011). How the brain shapes deception: An integrated review of the literature. *The Neuroscientist, 17*(5), 560–574.

Abe, N., & Greene, J. D. (2014). Response to anticipated reward in the nucleus accumbens predicts behavior in an independent test of honesty. *Journal of Neuroscience, 34*(32), 10564–10572.

Alloway, T. P., McCallum, F., Alloway, R. G., & Hoicka, E. (2015). Liar, liar, working memory on fire: Investigating the role of working memory in childhood verbal deception. *Journal of Experimental Child Psychology, 137*, 30–38.

Aslin, R. N., Shukla, M., & Emberson, L. L. (2015). Hemodynamic correlates of cognition in human infants. *Annual Review of Psychology, 66,* 349–379.

Blakemore, S. J. (2008). The social brain in adolescence. *Nature Reviews Neuroscience, 9*(4), 267.

Carlson, S. M., Moses, L. J., & Hix, H. R. (1998). The role of inhibitory processes in young children's difficulties with deception and false belief. *Child Development, 69*(3), 672–691.

Christ, S. E., Van Essen, D. C., Watson, J. M., Brubaker, L. E., & McDermott, K. B. (2009). The contributions of prefrontal cortex and executive control to deception: Evidence from activation likelihood estimate meta-analyses. *Cerebral Cortex, 19*(7), 1557–1566.

Cocchi, L., Zalesky, A., Fornito, A., & Mattingley, J. B. (2013). Dynamic cooperation and competition between brain systems during cognitive control. *Trends in Cognitive Sciences, 17*(10), 493–501.

Darwin, C. (1877). A biographical sketch of an infant. *Mind, 2,* 285–294.

Diamond, A. (2006). The early development of executive function. In E. Bialystok & F. I. M. Craik (Eds.), *Lifespan cognition: Mechanisms of change* (pp. 70–95). New York: Oxford University Press.

Ding, X. P., Gao, X., Fu, G., & Lee, K. (2013). Neural correlates of spontaneous deception: A functional near-infrared spectroscopy (fNIRS)study. *Neuropsychologia, 51*(4), 704–712.

Ding, X. P., Heyman, G. D., Fu, G., Zhu, B., & Lee, K. (2018a). Young children discover how to deceive in 10 days: A microgenetic study. *Developmental Science, 21*(3), e12566.

Ding, X. P., Heyman, G. D., Sai, L., Yuan, F., Winkielman, P., Fu, G., & Lee, K. (2018b). Learning to deceive has cognitive benefits. *Journal of Experimental Child Psychology, 176,* 26–38.

Ding, X. P., Omrin, D. S., Evans, A. D., Fu, G., Chen, G., & Lee, K. (2014). Elementary school children's cheating behavior and its cognitive correlates. *Journal of Experimental Child Psychology, 121,* 85–95.

Ding, X. P., Wellman, H. M., Wang, Y., Fu, G., & Lee, K. (2015). Theory of mind training causes honest young children to lie. *Psychological Science, 26*(11), 1812–1821.

Ding, X. P., Wu, S. J., Liu, J., Fu, G., & Lee, K. (2017). Functional neural networks of honesty and dishonesty in children: Evidence from graph theory analysis. *Scientific Reports, 7*(1), 12085.

Evans, A. D., & Lee, K. (2011). Verbal deception from late childhood to middle adolescence and its relation to executive functioning skills. *Developmental Psychology, 47*(4), 1108–1116.

Evans, A. D., & Lee, K. (2013). Emergence of lying in very young children. *Developmental Psychology, 49*(10), 1958–1964.

Fu, G., Sai, L., Yuan, F., & Lee, K. (2017). Young children's self-benefiting lies and their relation to executive functioning and theory of mind. *Infant and Child Development, 27*(1), e2051. doi: 10.1002/icd.2051

Greene, J. D., & Paxton, J. P. (2009). Patterns of neural activity associated with honest and dishonest moral decisions. *Proceedings of the National Academy of Sciences of the United States of America, 106*(30), 12506–12511.

Hartshorne, H., & May, M. S. (1928). *Studies in the nature of character: Studies in deceit.* New York: Macmillan.

Hu, X., Pornpattananangkul, N., & Nusslock, R. (2015). Executive control- and reward-related neural processes associated with the opportunity to engage in voluntary dishonest moral decision making. *Cognitive, Affective, and Behavioral Neuroscience, 15*(2), 475–491.

Langleben, D. D., Schroeder, L., Maldjian, J. A., Gur, R. C., McDonald, S., Ragland, J. D.,…Childress, A. R. (2002). Brain activity during simulated deception: An event-related functional magnetic resonance study. *NeuroImage, 15*(3), 727–732.

Leduc, K., Williams, S., Gomez-Garibello, C., & Talwar, V. (2017). The contributions of mental state understanding and executive functioning to preschool-aged children's lie-telling. *British Journal of Developmental Psychology, 35*(2), 288–302.

Lee, K. (2013). Little liars: Development of verbal deception in children. *Child Development Perspectives, 7*(2), 91–96.

Liang, Y., Ding, X. P., Lee, K., & Fu, G. (2018) The role of reward system in dishonest behavior: A functional near-infrared spectroscopy study. fNIRS 2018, Tokyo.

Liang, Y., Hu, Z., Ding, X. P., & Fu, G. (2019). The relation between children's simple deception and theory of mind: A functional near-infrared spectroscopy study. Unpublished manuscript.

Ma, F., Evans, A. D., Liu, Y., Luo, X., & Xu, F. (2015). To lie or not to lie? The influence of parenting and theory-of-mind understanding on three-year-old children's honesty. *Journal of Moral Education, 44*(2), 198–212.

Miller, S. A. (2009). Children's understanding of second-order mental states. *Psychological Bulletin, 135*(5), 749–773.

Mizokawa, A., & Koyasu, M. (2015). Digging deeper into the link between socio-cognitive ability and social relationships. *British Journal of Developmental Psychology, 33*(1), 21–23.

Nummenmaa, L., Glerean, E., Viinikainen, M., Jääskeläinen, I. P., Hari, R., & Sams, M. (2012). Emotions promote social interaction by synchronizing brain activity across individuals. *Proceedings of the National Academy of Sciences of the United States of America, 109*(24), 9599–9604.

Osaka, N., Osaka, M., Kondo, H., Morishita, M., Fukuyama, H., & Shibasaki, H. (2004). The neural basis of executive function in working memory: An fMRI study based on individual differences. *NeuroImage, 21*(2), 623–631.

Piaget, J. (1932). *The Moral Judgement of the Child*. London: Routledge & Kegan Paul.

Pornpattananangkul, N., Zhen, S., & Yu, R. (2018). Common and distinct neural correlates of self-serving and prosocial dishonesty. *Human Brain Mapping, 39*(7), 3086–3103.

Sai, L., Ding, X. P., Gao, X., & Fu, G. (2018). Children's second-order lying: Young children can tell the truth to deceive. *Journal of Experimental Child Psychology, 176*, 128–139.

Sip, K. E., Roepstorff, A., McGregor, W., & Frith, C. D. (2008). Detecting deception: The scope and limits. *Trends in Cognitive Sciences, 12*(2), 48–53.

Slaughter, V., Imuta, K., Peterson, C. C., & Henry, J. D. (2015). Meta-analysis of theory of mind and peer popularity in the preschool and early school years. *Child Development, 86*(4), 1159–1174.

Talwar, V., Gordon, H. M., & Lee, K. (2007). Lying in the elementary school years: Verbal deception and its relation to second-order belief understanding. *Developmental Psychology, 43*(3), 804–810.

Talwar, V., & Lee, K. (2008). Social and cognitive correlates of children's lying behavior. *Child Development, 79*(4), 866–881.

Talwar, V., Lee, K., Bala, N., & Lindsay, R. C. L. (2002). Children's conceptual knowledge of lying and its relation to their actual behaviors: Implications for court competence examinations. *Law and Human Behavior, 26*(4), 395–415.

Testa, R., Bennett, P., & Ponsford, J. (2012). Factor analysis of nineteen executive function tests in a healthy adult population. *Archives of Clinical Neuropsychology, 27*, 213–224.

Thijssen, S., Wildeboer, A., van IJzendoorn, M. H., Muetzel, R. L., Langeslag, S. J. E., Jaddoe, V. W. V.,…White, T. (2017). The honest truth about deception: Demographic, cognitive, and neural correlates of child repeated deceptive behavior. *Journal of Experimental Child Psychology, 162*, 225–241.

Vrij, A., Fisher, R., Mann, S., & Leal, S. (2006). Detecting deception by manipulating cognitive load. *Trends in Cognitive Sciences, 10*(4), 141–142.

Walczyk, J. J., Harris, L. L., Duck, T. K., & Mulay, D. (2014). A social-cognitive framework for understanding serious lies: Activation-decision-construction-action theory. *New Ideas in Psychology, 34*, 22–36.

Wang, L., Zhu, L., & Wang, Z. (2017). Parental mind-mindedness but not false belief understanding predicts Hong Kong children's lie-telling behavior in a temptation resistance task. *Journal of Experimental Child Psychology, 162,* 89–100.

Watson, A. C., Nixon, C. L., Wilson, A., & Capage, L. (1999). Social interaction skills and theory of mind in young children. *Developmental Psychology, 35*(2), 386–391.

Wellman, H. M. (2014). *Making minds: How theory of mind develops.* New York: Oxford University Press.

Williams, S., Moore, K., Crossman, A. M., & Talwar, V. (2016). The role of executive functions and theory of mind in children's prosocial lie-telling. *Journal of Experimental Child Psychology, 141,* 256–266.

Ybarra, O., Winkielman, P., Yeh, I., Burnstein, E., & Kavanagh, L. (2011). Friends (and sometimes enemies) with cognitive benefits: What types of social interactions boost executive functioning?. *Social Psychological and Personality Science, 2*(3), 253–261.

Yokota, S., Taki, Y., Hashizume, H., Sassa, Y., Thyreau, B., Tanaka, M., & Kawashima, R. (2013). Neural correlates of deception in social contexts in normally developing children. *Frontiers in Human Neuroscience, 7,* 206.

Zelazo, P. D., & Müller, U. (2002). The balance beam in the balance: Reflections on rules, relational complexity, and developmental processes. *Journal of Experimental Child Psychology, 81*(4), 458–465.

Zhang, J., Lin, X., Fu, G., Sai, L., Chen, H., Yang, J., … & Yuan, Z. (2016). Mapping the small-world properties of brain networks in deception with functional near-infrared spectroscopy. *Scientific Reports, 6,* 25297.

III Prosocial Behavior

12 Multiple Mechanisms of Prosocial Development

Jean Decety and Nikolaus Steinbeis

Overview

Natural selection has equipped the human brain with a set of innate predispositions that motivate us to be social, cooperative, and altruistic. These
predispositions also provide the criteria by which we evaluate the behavior
of others. Prosociality encompasses many distinct motivations, proximate
mechanisms, and behaviors, which together represent important adaptive
elements for social cohesion and cooperation. Developmental social neuroscience contributes to charting out the mechanisms involved in social
decision-making and prosociality, and how they gradually mature in interaction with the social and cultural environments.

Introduction

Prosocial behavior is commonly used to refer to any action that is intended
to benefit another. This definition is deceptively simple, and misleads lay
people and some academics to think of prosociality as a single capacity
through the distorted lens of the opposition between selfishness versus self-
lessness, inherited from classic philosophical views such as those developed
by Plato versus Aristotle, or Hobbes versus Rousseau. Things have changed:
a growing body of interdisciplinary research in anthropology, evolutionary biology, psychology, behavioral economics, and social neuroscience has
resulted in a better attempt to define and investigate prosociality and related
behaviors across domains. Work among these various academic disciplines
indicates that prosocial behavior is an umbrella concept and includes many
different types of behaviors such as helping, cooperating, sharing, comforting, rescuing, and informing. These various forms of prosocial behaviors

have distinct underlying motivations, such as caring, fairness, reputation management, group loyalty, reciprocity, and social rewards. Furthermore, there has been a shift from a dispositional stance toward prosocial behavior and an understanding of the contextual factors that interact with individual dispositions to give rise to it.

Prosocial behavior takes multiple forms, such as helping, comforting others in distress, or sharing resources such as food or money. These forms of prosociality have been documented in infancy (Dunfield, Kuhlmeier, O'Connell, & Kelley, 2011), suggesting early and deep-seated ontogenetic roots (Warneken, 2016). While much attention has been dedicated to understanding the origins of prosociality, much less is known about how these behaviors develop throughout childhood and how they are underpinned by the maturation of dedicated neural circuitry in interaction with the social environment.

This chapter marshals new empirical findings on the development of prosociality from psychology, behavioral economics, and cognitive neuroscience. Developmental research is critical to understanding the foundations of the social mind and identifying the mechanisms that guide social decision-making and prosociality. Knowing the psychological mechanisms and the ultimate causes of why prosociality has evolved is important for promoting and fostering prosocial behavior and moral development. We believe that an interdisciplinary approach to understanding the emergence and development of prosociality is necessary to studying its underlying mechanisms.

The chapter begins with an evolutionary perspective on the ultimate and proximate levels of explanation of human prosocial behavior. It also introduces the life history theory, which seeks to explain how natural selection has shaped the timing of sociomoral development. Early sociomoral evaluations and preferences are then discussed, followed by sections on comforting, helping and cooperation, and sharing. It concludes with a number of suggestions to move the field of developmental social neuroscience of prosocial behavior forward, calling for more cultural diversity in the populations that are investigated with a consideration of contextual and socioeconomic differences that shape the developing social brain, as well as employing neurocomputational models of the affective and cognitive processes underlying social behavior to move beyond the qualitative models of prosociality.

Evolutionary Roots of Human Prosociality

Homo sapiens is an ultrasocial species. Humans are interdependent on each other to an extent unseen in other species. The majority of resources for activities including hunting, foraging, building shelter, and child rearing are obtained by cooperation with genetically nonrelated conspecifics. This interdependence creates a motivation for individuals to help others and to be concerned with each other's welfare, directly benefitting fitness (Tomasello, 2014). Prosociality thus encompasses many distinct behaviors: products of our evolutionary history with specific ecological contingencies that together represent important adaptive elements for social cohesion and cooperation. It comprises some basic mechanisms that we share with other species, such as caring for offspring, as well a set of species-unique proximate mechanisms, both motivational and cognitive, that enable human individuals to survive and thrive in uniquely cooperative social arrangements.

Evolutionary theory provides a theoretical framework to explain why humans are predisposed to sociality, acting to benefit others even at some cost to themselves. It assumes that prosocial capacities are adaptations that contributed to the fitness and the reproductive success of our forebears.[1] Adaptations promote the replication of genes. Inclusive fitness recognizes that there is a reproductive advantage for the transmission of one's genes to future generations by not only engaging in seemingly selfish behaviors but also by acting to ensure the survival of one's kin (kin selection). Moreover, natural selection has promoted prosociality with nonkin as well, via reciprocal altruism, both direct and indirect, and most specially mutualism, which facilitates cooperative exchanges and interaction with unrelated individuals (Tomasello, 2016). Thus, cooperation can be maintained as an evolutionary stable strategy when it is not detrimental to the reproductive fitness of the organisms involved. Some have argued that morality—the prescriptive norms concerning others' welfare, rights, fairness, and justice— is a consequence of cooperative interactions and emerged to guide the

---

1. The organization of the human brain is product of natural selection, differential reproductions accumulated over millions of years under the pressure of various socioecological conditions. As a result, multiple intertwined layers and diverse systems have been gradually selected for their survival value, without guarantee of optimality (Jacob, 1977).

distribution of gains resulting from these interactions (Baumard, André, & Sperber, 2013).

When it comes to properly explaining the emergence of prosociality, we must obtain both ultimate explanations and proximate explanations that achieve that functionality. These two types of explanations are distinct, yet complementary. Ultimate explanations refer to the fitness consequences of a behavior or a trait, while proximate explanations refer to its mechanisms and motivations. Furthermore, adaptations need not operate at a conscious level. We are not aware of the ultimate explanations of behaviors. People are not consciously trying to increase their evolutionary fitness; thus, ultimate and proximate causes may be decoupled. For instance, a baby does not cry because she knows it generates heat (the thermoregulatory function of infant's crying) or because she knows that if her mother comes to nurse her she is more likely to survive and then reproduce. But infants who do not cry when in need of assistance are less likely to survive. Crying elicits care and defense from caregivers. This is the ultimate explanation because it appeals to the fitness benefits of the behavior in terms of reproductive success.

Proximate explanations include the external triggers of crying, such as physical separation from the caregiver, cold, or a lack of food, and the neural circuits in the brainstem and the endogenous opioids involved in the cessation of crying. To extend that example, in many species caring for offspring is a biological necessity. The proximate motivational mechanisms for parental care rely on specific subcortical nuclei and circuits in the brainstem and hypothalamus as well as valuation processes in the orbitofrontal cortex. Importantly, caring is rewarding, and it is associated with activation of the mesolimbic dopaminergic systems (Rilling & Young, 2014).

Another important point that we have learned from evolutionary theory is that behaviors with similar outcomes can be generated by different proximate mechanisms. For instance, ants, like humans, are often at war. In some species, ants rescue their injured comrades (Frank, Wehrhahn, & Linsenmair, 2018). This is a case where similar selection pressures can result in a convergent adaptation. The behavior in both species may share the same ultimate explanation: helping the armies or colonies stay strong enough to fight another day. Nevertheless, they may not necessarily share the same proximate mechanisms. In ants, the rescue behavior is triggered by pheromones. In humans, the decision to rescue might be motivated by empathy and concern for social norms. Alternatively, such a decision to help might be prompted by more selfish, egoistic concerns, such as improving one's reputation (Silver & Shaw, 2018).[2] Individuals can be motivated to help others, because helping brings rewards to themselves, including reputational benefits or future material gains. It also may prevent punishment, including material sanctions, or may reduce the aversive arousal that comes from observing others in need.

The evolutionary approach provides a rich variety of deductions and predictions about what kinds of prosocial behavior to expect, how they emerge in infants, how they develop in children, and how social context shapes them. For instance, seventeen-month-old infants possess an abstract expectation of in-group support. They view helping as expected among individuals from the same group but as optional otherwise (Jin & Baillargeon, 2017). This capacity to categorize the social world into "us and them" allows individuals to parse the social environment into groups and calibrate the importance of particular group membership in a given circumstance (Hirschfeld, 2013). Infants are sensitive to cues of social status and dominance. By six months of age, they are capable of detecting dominance relations when provided with an ecologically relevant cue such as social group size (Pun, Birch, & Baron, 2016). This ability to detect dominance relationships is essential for survival because it helps individuals weigh the potential costs and benefits of engaging in a physical competition. These findings suggest that infants quickly build on early social biases with a presumably evolutionarily ancient origin to represent social dominance relations.

Life history theory is an important contribution to understanding sociomoral development. It seeks to explain aspects of organisms' anatomy and behavior with reference to the way that their life histories—including their reproductive development and behaviors, life span, and postreproductive behavior—have been molded by natural selection (Kaplan & Gangestad, 2005). Some authors have proposed to apply life history theory to the domain of human social behavior and have proposed that the timing of the development of moral behavior has been influenced by when moral

---

2. Around age five, children begin to recognize that their actions can signal important information about their desirability to potential social partners, and they will vary their behavior based on audience and social context. Indeed, five-year-old children are consistently more generous when they know they are being observed (Shaw, Montinari, Piovesan, Olson, Gino, & Norton, 2014).

behavior is likely to produce a net gain in resources (Sheskin, Chevallier, Lambert, & Baumard, 2014). The life history theory suggests that, in contrast with moral behavior, social evaluation of others requires little cost and can be beneficial to infants. As a result, natural selection has favored an earlier developmental emergence of third-party social evaluation than of moral behavior. This theory provides a framework for understanding why moral behavior develops years after children understand what morality requires, and accounts for the observed asymmetry between cognition and behavior. However, like for language learning, comprehension and production draw on distinct cognitive processes with their specific timing. The gap between cognition and behavior is gradually bridged by the maturation of dedicated mechanisms and circuits in constant interaction with the social and cultural environment.

Overall, natural selection has equipped the human brain with a set of innate predispositions which motivate prosociality. These adaptations also provide the criteria by which infants evaluate the behavior of others at some basic levels. Importantly, this view of prosociality as a set of evolved adaptations challenges the traditional approaches that have dominated the development of morality largely based on socialization and internalization of social norms. However, it is unlikely that these innate predispositions develop in the absence of relevant experience.

One major characteristic of the development of the human brain is that it is prolonged for fifteen years after birth without significant change in the total number of neurons. The human baby is born with a brain weighing five times less than that of the adult, whereas the chimpanzee is born with a brain 40 percent of its adult size. More than half of the synapses of the human cerebral cortex are formed after birth (Changeux, 2017). A Darwinian selection occurs during which the state of brain activity, whether spontaneous or evoked by interaction with the environment, controls the stabilization of certain connections and elimination of others during multiple critical stages of postnatal development. This development occurs at an extremely rapid rate. It has been estimated that, on average, 10 million synapses are formed per second in the baby's cerebral cortex (Lagercrantz, 2010). Essentially, this Darwinian selection is under the control of the spontaneous and evoked activity of neural networks. In other words, learning is eliminating. As suggested by Vygotsky (1978), the physical, social, and cultural environment of the newborn becomes internalized in his brain from birth. Evolution has selected the genome of a species that spends a good portion of its life building its brain in a social environment compatible with its survival.

Early Sociomoral Evaluations and Preferences

One way to approximate how evolution has shaped the basic cognitive architecture that underlies prosociality is to examine the extent to which infants are born with predispositions that provide them with a skeletal framework for evaluating the social environment around them. Uncovering infants' expectations about how people should act toward each other before infants have any experience of their own is strong evidence for social cognitive adaptations. In support of this view, a number of rudimentary elements necessary for prosociality emerge very early during human ontogeny.

The infant mind is cognitively predisposed to interpret the world in terms of agents and objects whose behaviors are constrained by different sets of principles (Leslie, 1994). Young infants during the first year of life seem to possess expectations about social interactions among two or more agents and have the cognitive capacity to assign value of valence and magnitude to social actions. Since the seminal work of Premack and Premack (1997), a growing body of evidence has demonstrated that, before their first birthday, infants assign valence when perceiving third-party interactions. Infants determine whether actions have beneficial or detrimental effects on others, distinguishing actions such as helping and comforting from those like hitting and hindering (Baillargeon et al., 2015). At nine months of age, infants prefer individuals who act prosocially over those who act antisocially and expect others to share their preferences (Hamlin, 2015).

Sensitivity to fairness appears to emerge at the end of the first year (Sommerville et al., 2013). Infants aged eighteen months also expect people to display in-group favoritism when allocating resources (Bian et al., 2018). Third-party experiments demonstrate that thirteen-month-olds expect an adult to attend to a crying baby (Johnson et al., 2010) even though they themselves show no affective concern for a crying infant nor are they particularly interested in it (Nichols et al., 2015). Around fifteen months, infants expect reciprocal actions to match initial actions in valence, in accordance with the reciprocity principle, one of the fundamental principles guiding human social interactions (Baillargeon et al., 2015).

Although these key constituents of morality emerge very early in ontogeny and are often described as innate and modular, they are probably best conceived in terms of *canalization,* or the degree to which the development of a trait is robust across typical environmental variations (Cummins & Cummins, 1999). The development of these predispositions depends on tightly coupled transactions between neurobiological predispositions and environmental inputs. This view fits well with the Bayesian models that incorporate both nativism and learning in the same processes (Carrurthers, Laurence, & Stich, 2005).

Developmental neuroscience research has begun to identify specific neural computations underpinning early sociomoral evaluations and their relation to moral preferences. One study examined the neural underpinnings of third-party moral scenarios in infants and toddlers, aged twelve to twenty-four months (Cowell & Decety, 2015a). Continuous electroencephalography (EEG), time-locked event-related potentials (ERP), and gaze fixation were recorded while babies watched cartoon characters engaging in prosocial and antisocial actions. Overall, infants and toddlers expressed preferential looking toward the prosocial characters versus the antisocial characters. Relative overall alpha asymmetry for left versus right was stronger for the perception of antisocial than prosocial behaviors, which reflects greater withdrawal/avoidance. Moreover, relatively automatic differences (300–500 milliseconds) after observing characters helping or hindering each other were detected. Children with greater negativity in this time window for the perception of prosocial characters compared with antisocial characters also tended to prefer prosocial characters over the antisocial characters in a preferential reaching task.

To examine what neural circuits support third-party moral evaluations and how they change with age, a cross-sectional neurodevelopmental study using functional magnetic resonance imaging (fMRI) and eye-tracking included a large number of participants aged four to thirty-seven years who were shown visual scenarios depicting interpersonal intentional and accidental harmful actions (Decety, Michalska, & Kinzler, 2012). Perceived intentional harm (as opposed to accidental harm) was associated with increased activation in brain regions sensitive to *Theory of Mind,* or the understanding of others' mental states, such as the medial prefrontal cortex (mPFC), right posterior superior temporal sulcus/temporoparietal junction (pSTS/TPJ), regions processing the valence of these actions (amygdala and insula), and social valuation (ventromedial prefrontal cortex, vmPFC). Age was negatively correlated with empathic sadness for the victim of harm in the scenarios, with the youngest participants reporting the greatest levels of sadness, which was predictive of the neural response in the insula, thalamus, and subgenual prefrontal cortex. The response in the amygdala followed a curvilinear function, such that the hemodynamic signal was highest at the youngest ages, decreased rapidly through childhood and early adolescence, and reached an asymptote in late adolescence through adulthood. Conversely, the neurohemodynamic signal in older participants increased in the mPFC and vmPFC, regions that are associated with valuation and moral decision-making (see figure 12.1 for a map of brain regions).

Patterns of functional connectivity—such as temporal coupling in neural response between spatially distinct regions during the perception of intentional harm relative to accidental harm—provides complementary evidence for a gradual developmental integration between the prefrontal cortex and amygdala. The oldest participants showed significant coactivation in these regions during the perception of intentional harm, whereas the youngest only exhibited a significant covariation between the vmPFC and periaqueductal gray in the brainstem. Furthermore, adult participants showed the strongest connectivity between vmPFC and pSTS/TPJ while viewing morally laden actions, suggestive of a developmental change in functional integration within the mentalizing and valuation systems. Conversely, activity in regions of the medial and vmPFC that are reciprocally connected with the amygdala and that are involved in social decision-making and valuation increased with age as they became functionally coupled. This pattern of developmental change was also reflected in explicit moral evaluations, which require the capacity to integrate a representation of the mental states and intentions of others together with the consequences of their actions.

Overall, infants seem to possess a set of predispositions for third-party sociomoral evaluations as well as preferences that guide their expectations of others in relation to fairness, empathic concern, reciprocity, and group affiliation. There is neuroscientific and behavioral evidence that infants readily distinguish between prosocial and antisocial acts, assign valence and values to perceived actions, and tend to approach the prosocial and avoid the antisocial. Importantly, these preferences do not necessarily manifest in infants' own behavior, as predicted by the life history theory. For instance, although infants expect adults to comfort a baby in distress, they

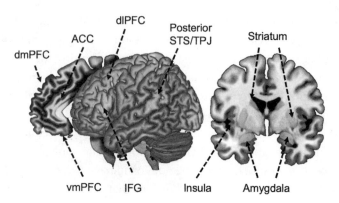

Figure 12.1
Brain regions involved in social decision-making and prosociality. Functional neuroimaging studies reveal that the anterior insula and the anterior cingulate are conjointly
activated during the experience of negative emotion and during the perception of
negative emotion in others. The insula provides a foundation for the representation
of subjective bodily feelings, which substantiates emotional awareness. The anterior
cingulate cortex (ACC) can be divided anatomically based on dorsal (cognitive) and
ventral (emotional) regions. The dorsal part is reciprocally connected with the prefrontal cortex and parietal cortex as well as the motor system, making it a central
station for processing top-down and bottom-up stimuli and assigning appropriate
control to other areas in the brain. By contrast, the ventral ACC is connected with
the amygdala (a structure involved in assigning affective significance to stimuli), striatum, hypothalamus, and anterior insula, and is involved in assessing the salience
of emotion and motivational information. Many functions are attributed to ACC,
such as error detection, anticipation of tasks, motivation, and modulation of emotional responses. The medial prefrontal cortex (mPFC) and posterior superior temporal sulcus (pSTS)/temporoparietal junction (TPJ) are associated with Theory of Mind
processes and understanding emotion. The ventromedial prefrontal cortex (vmPFC)
is involved in decision-making and representing the affective value of reinforcers.
In particular, the vmPFC is thought to regulate planning behavior associated with
sensitivity to reward and punishment, and is closely connected to the anterior insula
and amygdala. The vmPFC is implicated in encoding expectations of future reward
during value-based decision making and integrates inputs from regions involved in
social cognition, in particular the anterior insula and pSTS, during altruistic decisions. It is also a critical hub for empathic concern and caring behavior. Finally, the
inferior frontal gyrus (IFG) with the anterior insula is involved in elaborate attentional and working memory processing in low-predictable environments (Tops &
Boksem, 2011).

themselves do not show any affective concern for a crying infant (Nichols et al., 2015) or for an adult simulating pain after slamming her finger in a door (Dunfield et al., 2011).

Children's behavior in the beginning of the second year shows they do not yet view it as wrong to harm others (Dahl & Freda, 2017). Other-oriented empathic responses start to manifest in the second year of life and more reliably at 3.5 to 4.5 years of age. The evolutionary approach considers prosociality a set of biological strategies for solving problems of survival, cooperation, and conflict. Such capacities allow infants to predict others' social behaviors as well as facilitate their affiliation with similar others.

Infants seem to understand harm, fairness, and reciprocity in third-party situations. Yet children require years of development before they apply this understanding to their own interactions with others. The mismatch between competent infant social evaluations and relatively limited early childhood ability to act prosocially is explained by a life history analysis of when behaving morally becomes beneficial (Sheskin et al., 2014). The human species is characterized by an extended period of juvenile dependence during which cooperation with nonkin is mostly superfluous. Later, as children age, moral behaviors supporting cooperation become increasingly beneficial.

Consolation and Comforting

Consolation and comforting behaviors have been documented in nonhuman species including great apes and, more recently, prairie voles. Consolation is an affiliative behavioral response aimed at providing physical and emotional comfort to a distressed individual. In chimpanzees, postconflict affiliative interaction directed from a third party to the recipient of aggression has a stress-alleviating function and is more likely between individuals with more valuable relationships (Fraser, Stahl, & Aureli, 2008). In prairie voles—a highly social and monogamous species that mate for life and form deep familial bonds—a study demonstrated that when individuals who had been separated from one another were reunited, they would soothe the one that had been electrically shocked by licking it (Brukett et al., 2016). Furthermore, blocking the voles' receptors for oxytocin, a neurohormone that facilitates bonding and social recognition, in the anterior cingulate cortex (ACC) stopped the consoling behavior. Such studies show that the ability to be sensitive to the emotional states of others, coupled with a motivation to care for their well-being, relies on biological mechanisms highly conserved across species (Decety, Ben-Ami Bartal, Uzefovsky, & Knafo-Noam, 2016). In particular, the ACC plays a critical role in regulating empathic behavior in voles as in humans.

Babies appear sensitive to signals of distress early in development. Neonates possess a neural mechanism for vocal affective discrimination, as demonstrated by a mismatch EEG response over the right hemisphere in response to emotionally laden syllables, particularly fearful and happy ones, within the first few days of life (Cheng, Lee, Chen, Wang, & Decety, 2012). This affective discrimination is selectively driven by voice processing rather than low-level acoustical features, and relies on the cerebral specialization for human voice and emotion processing that emerges over the right hemisphere during the first days of life. In infants aged three- to seven-month-olds, sad vocalizations are associated with a selective increase of hemodynamic activity in brain regions such as the orbitofrontal cortex and insula that are involved in processing affective stimuli (Blasi et al., 2011).

While this negativity bias is clearly observed in the first days of life for the vocal modality (Cheng, Chen, & Decety, 2012), it emerges later for facial expressions. Only in the first year of life do infants begin to pay more attention to negative emotions. For instance, seven-month-olds looked longer at fearful than happy faces (de Haan, Belsky, Reid, Volein, & Johnson, 2004), and they exhibited greater amplitudes in a negative component over the frontal and central electrodes around 700 milliseconds labeled negative central (Nc) in response to fearful rather than to happy faces. The Nc is usually interpreted to reflect infants' allocation of attention. Another study measured EEG/ERP responses while children aged three to nine years watched visual stimuli depicting physical injuries to people. Irrespective of the age, the results demonstrated both an early automatic component (N200), which reflects attention to emotionally salient stimuli (affective arousal or affect sharing), and a late-positive potential (LPP), indexing cognitive reappraisal or more complex processing of emotional stimuli. Only the LPP response showed an age-related differentiation between painful and neutral scenes (Cheng et al., 2014).

Naturalistic observations have provided some evidence that by the end of the first year some children comfort others in distress through simple, nonverbal, mostly gestures and postures such as touching, patting, hugging, and affectionately leaning in. These prosocial acts become increasingly frequent, differentiated, and focused on victims' circumstances and problems over the second and third years of life (Zahn-Waxler, Radke-Yarrow, Wagner, & Chapman, 1992). More recent laboratory-controlled work indicates, however, that twelve-month-olds are in fact not especially interested nor concerned by their peers in distressed (Nichols et al., 2015). Eighteen- and twenty-four-month-olds, by contrast, were generally similar in their positive social interest and attentiveness to the infant in distress, but twenty-four-month-olds exhibited greater affective concern. Moreover, affective concern was rare at eighteen months, with only 25 percent of children at that age exhibiting any concern at all. In contrast, by twenty-four months of age, two-thirds of children exhibited affective concern for the crying infant, and 10 percent displayed high levels of concern. Children's ability to respond prosocially to others' distress seems to emerge more clearly during the second year of life, and this is associated with an increased understanding of other people as psychological agents with internal states that differ from the child's own (Sveltova, Nichols, & Brownell, 2010).

Emotional empathy (aka affect sharing) has been assumed to be a primary driving force for eliciting comforting others in distress. Even at early ages, emotional empathy appears as a disposition, relatively stable across time and consistent across contexts, and predicts subsequent comforting behaviors (Light & Zhan-Waxler, 2011). However, as early as eighteen months of age, children show concerned looks and subsequent comforting behaviors for an adult stranger who is in a hurtful situation but shows no overt emotion (Vaish Carpenter, & Tomasello, 2009). This demonstrates that the capacity to empathize does not require overt emotional cues, nor does an emotional distress reaction.

Overall, there is limited solid empirical evidence for signs of concern and comforting behavior before the first year of life. The neural pathways (in the brainstem and amygdala) that allow affect sharing are in place early in ontogeny, but activation alone of such circuits is not sufficient to elicit a comforting response (Decety & Cowell, 2018). Neural maturation and perhaps prefrontal regulation in dynamic interaction with learning and socialization seem requisite to give rise to such behaviors.

Sharing

Sharing with nonkin is prevalent in all human societies and facilitates reciprocity and cooperation. Young children's prosocial proclivities are apparent in their sharing behaviors, and these are combined with a sense of fairness at least in a sense of equality (e.g., Huppert et al., 2019). With development, sharing becomes increasingly selective (Martin & Olson, 2015).

Sharing valuable resources can be observed in toddlers as young as fifteen months (Schmidt & Sommerville, 2011). At this age sensitivity toward equal distributions and fairness norms also emerges (Geraci & Surian, 2011; McAuliffe, Blake, Steinbeis, & Warneken, 2017). Around three years, children state that sharing equally is the norm; from then on, they increasingly follow such sharing norms with their actual behavior (Smith, Blake, & Harris, 2013). Complying with social norms constitutes a long-term goal, which conflicts with the more immediate satisfaction of reward maximization (Buckholtz, 2015). Resolving such conflict in favor of sharing according to the norm requires behavioral control.

There is by now increasing evidence that behavioral control is positively correlated to sharing in both preschoolers (Aguilar-Pardo, Martinez-Arias, & Colmenares, 2013; Paulus, Licata, Kristen, Thoermer, Woodward, & Sodian, 2015) and school children (Blake, Piovesan, Montinari, Warneken, & Gino, 2015; Smith et al., 2013) using a variety of measures of behavioral inhibition (i.e. day-night task and questionnaires). More recently, these correlative findings have been extended to show that explicit experimental manipulations of behavioral control impact sharing directly. For instance, in one study it was shown that children aged six to nine years shared less after having engaged in a behavioral motor control task compared with sharing after a speeded reaction time task (Steinbeis, 2018). In another study children aged six to nine years shared more after having listened to stories priming behavioral control compared with stories that did not (Steinbeis & Over, 2017).

There is a sizeable body of literature suggesting that behavioral control aids sharing in accordance with the prevailing social norms during middle childhood. It has also been shown that children as young as five years alter their sharing behavior as a function of being watched by others (Engelmann, Herrmann, & Tomasello, 2012) and more so when this other is an in-group as opposed to an out-group member (Engelmann, Over, Herrmann, & Tomasello, 2013). Findings of such selective, context-specific social behavior suggest the operation of a behavioral regulation mechanism that allows for the optimal titration of social behavior (Steinbeis & Crone, 2016).

Other mechanisms related to the development of sharing have been linked with social abilities such as empathy and Theory of Mind. Both three-year-olds and five-year-olds show increased sharing behavior in a resource allocation task. This increase in generosity is predicted by their ratings of the emotional state of a needy protagonist but not with their own emotional state, suggesting that empathic concern, not personal distress, is the primary motivation (Williams, O'Driscoll, & Moore, 2014). The link to Theory of Mind seems less clear cut, with evidence of a positive (Cowell et al., 2017; Takagishi, Kameshima, Schug, Koizumi, & Yamagishi, 2010) as well as a negative relationship with sharing (Cowell, Samek, List, & Decety, 2015) in three- to five-year-old children. Interestingly, a meta-analysis that included over 6,000 children between two and twelve years of age found that, collapsed across all studies, Theory of Mind was a small but significant predictor of prosocial behavior including helping, comforting, and cooperating, but not for sharing (Imuta, Henry, Slaughter, Selcuk, & Ruffman, 2016).

In adults, it has been shown that social norm compliance relies on lateral prefrontal cortical brain regions (Spitzer, Fischbacher, Herrnberger, Gron, & Fehr, 2007). Activity in the dorsolateral prefrontal cortex (dlPFC) was positively correlated with sharing under threat of punishment (Spitzer et al., 2007), while disrupting activity in right dlPFC reduced such sharing (Ruff, Ugazio, & Fehr, 2013). It has been argued that the top-down modulation from dlPFC of subjective value signals in the vmPFC is key for both implementing and complying with social norms (Baumgartner, Knoch, Hotz, Eisenegger, & Fehr, 2011). Lateral prefrontal cortical areas are among the brain regions that undergo the most protracted age-related loss of gray matter volume throughout childhood and adolescence (Lenroot & Giedd, 2006). Further, linear age-related increases in structural connectivity are most delayed in white matter bordering the prefrontal cortex (Lebel, Gee, Camicioli, Wieler, Martin, & Beaulieu, 2012), which in turn impacts the extent of functional connectivity (Hagmann et al., 2010). Lateral portions of the prefrontal cortex are involved in actively maintaining task goals, biasing attention, and implementing behaviors (Miller & Cohen, 2001). The maturation of the lateral prefrontal cortex also underpins the development of these functions in children (Bunge, Dudukovic, Thomason, Vaidya, & Gabrieli, 2002; Rubia, Smith, Taylor, & Brammer, 2007), which makes it a suitable candidate region for supporting the emergence of the behavioral control necessary for social norm compliance.

In a study that investigated the neurocognitive mechanisms facilitating sharing, Cowell and Decety (2015b) measured ERPs and eye movements in three- to five-year-old children in response to movies depicting social interactions between two characters engaging in prosocial behavior such as helping, sharing, and comforting, or antisocial behavior such as hitting, tripping, and shoving.[3] It was found that observing helpful and harmful actions differentially affected early and late ERP components, namely the so-called early posterior negativity as well as the N2 and LPP. Early and late potentials are interpreted as reflecting automatic and controlled processing respectively. Interestingly, it was only late ERPs—namely, the LPP over left frontal sites—that predicted sharing in the Dictator Game (i.e., the number of stickers given to another anonymous child).

Another study found that sharing compliant with social norms increased between the ages of six and thirteen years, which correlated directly with an independent measure of behavioral motor control (Steinbeis, Bernhardt, & Singer, 2012). Simultaneously recorded neural activity showed that this age-related increase in social norm compliance correlated positively with activity in the left dlPFC. The age-dependent increase in activity in this region mediated the developmental increase in behavioral motor control, which in turn predicted the increase in social norm compliance through-out childhood. Connectivity analyses were not conducted in this study, but it is likely that decisions to share are supported by increased functional coupling between dlPFC and brain areas that compute the value of deci-sions, such as the vmPFC. Such a mechanism has been shown to support decisions in favor of long-term goals in similar scenarios both in adult-hood (Baumgartner et al., 2011; Hare, Hakimi, & Rangel, 2014) and during middle childhood (Steinbeis, Haushofer, Fehr, & Singer, 2016).

This interpretation is buttressed by the findings of an ERPs study that showed an increase of regulatory processes in bringing about sharing dur-ing childhood (Meidenbauer, Cowell, Killen, & Decety, 2018). Thus, in older children the P3, a component reflecting behavioral control mechanisms, predicted the equal sharing of resources. In younger children, however, this was predicted by the early posterior negativity, an early component

3. This Chicago Moral Sensitivity task was created by Decety and was used, with dif-ferent stimuli, with children, adolescents, and adults in several functional magnetic resonance imaging and high-density EEG studies, including source localization (e.g. Decety, Michalska & Kinzler, 2012; Yoder & Decety, 2014a, 2014b; Yoder, Harenski, Kiehl, & Decety, 2015).

reflecting affective evaluation. Further, a brain morphology study, conducted with 464 children between the ages of six and nine years, found that cortical thickness in the left superior frontal cortex and parts of the dorsal medial prefrontal cortex and midcingulate cortex was related to prosocial behavior assessed by parents ratings (Thijssen et al., 2015).

Overall, it appears that the development of top-down behavioral control mechanisms, supported by the maturation of function and connectivity of prefrontal cortical circuitry, account for the observed changes in sharing during childhood. Such a mechanism helps to shift decisions away from the immediate desires of reward maximization, toward compliance with social norms of equal sharing.

Helping and Cooperation

Many studies on the ontogeny of prosociality have predominantly focused on instrumental helping. Toddlers show early signs of internally motivated helping behavior. For instance, fourteen- to eighteen-month-olds fetch desired objects that appear out of reach for an experimenter, and help to complete household chores (Warneken & Tomasello, 2007). Young children also help by informing, where children as young as twelve months point toward objects that an experimenter is searching for (Dunfield et al., 2011).

Most findings come from studies looking at need-based or empathic helping. Two key motivations have been argued to underlie such helping: a selfish desire to reduce one's own distress, also known as *personal distress;* and an altruistic desire to reduce the other's distress, also known as *empathic concern* (Batson, 2012). Personal distress leads to helping only when there is no other way of terminating one's own distress, such as fleeing the situation. Empathic concern, however, leads to helping across a range of situations. In children, only indicators of empathic concern were shown to predict helping (Eisenberg & Eggum, 2009). It has been argued that empathic concern arises out of the interplay of an emotional response to the need of another and a sufficiently strong regulation of this emotional response (Decety & Lamm, 2009). Thus, the literature on the development of helping during childhood suggests that those children high in both emotional response and emotion regulation are the ones most likely to help. Support for this comes from studies using parental questionnaires, parent and teacher ratings, and psychophysiological indicators, suggesting that empathic concern is a good predictor of helping behavior in childhood.

In terms of the neurocognitive mechanisms underlying helping in adults, it has been shown that observing the painful or unpleasant experience of another person activates circuitry that is also recruited when undergoing this same experience oneself (for a meta-analysis, see Lamm, Decety & Singer, 2011). This circuitry comprises the anterior insula and ACC. Activity in the anterior insula was shown to correlate positively with empathic concern ratings, and was predictive of helping behavior, albeit only to in-group members (Hein, Silani, Preuschoff, Batson, & Singer, 2010). It has been argued that the activity in anterior insula influences value computations in the vmPFC related to prosocial behavior (Hare, Camerer, Knoepfle, & Rangel, 2010). More recently, it was shown that functional connectivity between the ACC and the anterior insula was positively related to prosocial behavior following an empathy induction (Hein, Morishima, Leiberg, Sul, & Fehr, 2016).

The ACC has been implicated in top-down regulation and control of negative emotions and processing of emotional conflict (Etkin, Egner, & Kalisch, 2011), and in this process might function as an affective regulation mechanism to produce empathic concern, in turn leading to prosocial behavior. One neuroimaging study found that individuals high in dispositional empathic concern were more likely to engage in costly altruism, and this relationship was supported by neural activity in the vmPFC (FeldmanHall et al., 2015).

Several studies on the neurocognitive development of empathic concern and helping have been performed in children. Two longitudinal studies have linked neural predictors with the emergence of helping during early childhood. One study looked at the relationship between infants' EEG resting state brain activity at fourteen months, instrumental helping at eighteen months, and comforting at twenty-four months. It was found that hemispheric asymmetries in temporal and frontal brain regions at fourteen months predicted helping and comforting at eighteen and twenty-four months, respectively (Paulus, Kühn-Popp, Licata, Sodian, & Meinhardt, 2013).

Another recent longitudinal study examined whether processing faces at age seven months was predictive of helping behavior at fourteen months of age (Grossmann, Missana, & Krol, 2018). It was found that helping at fourteen months of age was predicted both by an initial attentional bias and subsequent disengagement to fearful faces at seven months and by associated activity in left dlPFC, as measured by functional near-infrared spectroscopy. These effects were specific to processing fearful faces over happy or angry faces. Both of these studies suggest that prefrontal regions are potentially functionally linked to the emergence of empathic motivated helping. Given the role of prefrontal regions in top-down cognitive control and regulation, these findings suggest that, at the youngest ages, such control mechanisms might already be required for helping behaviors to emerge.

Functional neuroimaging with children aged seven to twelve years has shown activation of the anterior insula and ACC when the children observe the pain and distress of another (Decety, Michalska, & Akitsuki, 2008). Similar activation patterns can already be identified by four years of age (Michalska, Kinzler, & Decety, 2013). A study testing seven- to forty-year-old participants showed that activity in the amygdala in response to seeing others' pain decreased with age, while activity in the lateral prefrontal cortex increased. These findings suggest a potential decrease in brain functions related to distress and a potential increase those related to emotion regulation (Decety & Michalska, 2010). A more recent study also shows that age-related changes in empathically driven helping were related to emotional clarity, which was linked to regulatory success in an emotion regulation task. Each of these were in turn strongly linked to increased connectivity between the right anterior insula and lateral and medial prefrontal cortical areas (Hoffmann, Grosse Wiesmann, Singer, & Steinbeis, 2019).

Overall, the neurocognitive mechanisms of helping during childhood comprise affective responding to the emotional state of another, as coded in the anterior insula, in combination with regulatory mechanisms of the experienced affect, instantiated in prefrontal cortical brain regions.

Future Directions

There is still a dearth of systematic developmental research outside North American and Western European cultural contexts. A lack of attention to cultural variation risks yielding incomplete and potentially inaccurate conclusions (Apicella & Barrett, 2016). After all, human nature does not imply strict phenotypic uniformity, but rather a set of evolved developmental resources that are responsible for building psychological phenotypes, which may vary depending on the local cultural environment (Tooby &

Cosmides, 1992). To understand how prosociality emerges from the interaction between innate predispositions and input from local cultural environments, there is a need for empirical research on the development of moral cognition and its relation to prosociality across different cultures. This will help to determine the extent to which basic social decision-making processes and motivations are universal and to examine variation and stability in core domains of human psychology and prosociality.

For instance, the extent to which social norms are integrated into fairness considerations, and how they influence social preferences regarding equality and equity, was examined in a large sample of children ages four to eleven in Argentina, Canada, Chile, China, Colombia, Cuba, Jordan, Mexico, Norway, South Africa, Taiwan, Turkey, and the United States (Huppert et al., 2019). Social decision-making in distributive justice games revealed universal developmental shifts from equality-based to equity-based distribution decisions across cultures. However, differences in levels of individualism and collectivism between the countries predicted the age and extent to which children favored equity for recipients differing in terms of wealth, merit, and physical suffering. When recipients differed in regards to wealth and merit, children from the most individualistic cultures endorsed equitable distributions to a greater degree than children from more collectivist cultures. Children from the more individualistic cultures also favored equitable distribution at an earlier age than those from more collectivist cultures. These results provide insights into theories positing that fairness is a universal moral concern, and that humans naturally favor fair distributions, not equal ones (Baumard et al., 2013; Starmans, Sheskin, Bloom, Christakis, & Brown, 2017).

Many theories of prosociality and its development are qualitative. They have been invaluable for structuring research, but they are limited in their capacity to predict behavior and determine how different factors interact in a given social context. One way to move things forward is to employ neurocomputational models of affective and cognitive processes underlying social behavior. Some of these models help formally integrate both personal and situational characteristics into a model of the dynamic control of social behavior. Studies using computational models not only can capture interindividual differences in resource distribution choices and the underlying brain structures but also can predict how a given individual will respond to changes in the cost of altruistic behaviors (Hutcherson, Bushong, & Rangel,

2016). Moreover, the fact that different motives can simultaneously impact perceived utility helps clarify the relationship between different internal motives underlying distribution behaviors.

References

Aguilar-Pardo, D., Martinez-Arias, R., & Colmenares, F. (2013). The role of inhibition in young children's altruistic behaviour. *Cognitive Processes, 14*(3), 301–307.

Apicella, C. L., & Barrett, H. C. (2016). Cross-cultural evolutionary psychology. *Current Opinion in Psychology, 7,* 92–97.

Baillargeon, R., Scott, R. M., He, Z., Sloane, S., Setoh, P., Jin, K.-S., ... Bian, L. (2015). Psychological and sociomoral reasoning in infancy. In M. Mikulincer & P. Shaver (Eds.), *Handbook of personality and social psychology* (Vol. 1, pp. 79–150). Washington DC: American Psychological Association.

Batson, C. D. (2012). The empathy-altruism hypothesis: Issues and implications. In J. Decety (Ed.), *Empath—From bench to bedside* (pp. 41–53). Cambridge, MA: MIT Press.

Baumard, N., André, J. B., & Sperber, D. (2013). A mutualistic approach to morality: The evolution of fairness by partner choice. *Behavioral and Brain Sciences, 36*(1), 59–78.

Baumgartner, T., Knoch, D., Hotz, P., Eisenegger, C., & Fehr, E. (2011). Dorsolateral and ventromedial prefrontal cortex orchestrate normative choice. *Nature Neuroscience, 14*(11), 1468-U1149.

Blake, P. R., Piovesan, M., Montinari, N., Warneken, F., & Gino, F. (2015). Prosocial norms in the classroom: The role of self-regulation in following norms of giving. *Journal of Economic Behavior and Organization, 115,* 18–29.

Blasi, A., Mercure, E., Lloyd-Fox, S., Thomson, A., Brammer, M., Sauter, D., ... Murphy, D. G. M. (2011). Early specialization for voice and emotion processing in the infant brain. *Current Biology, 21,* 1–5.

Bian, L., Sloane, S., & Baillargeon, R. (2018). Infants expect ingroup support to override fairness when resources are limited. *Proceedings of the National Academy of Sciences of the United States of America, 115*(11), 2705–2710.

Buckholtz, J. W. (2015). Social norms, self-control, and the value of antisocial behavior. *Current Opinion in Behavioral Sciences, 3,* 122–129.

Bunge, S. A., Dudukovic, N. M., Thomason, M. E., Vaidya, C. J., & Gabrieli, J. D. (2002). Immature frontal lobe contributions to cognitive control in children: Evidence from fMRI. *Neuron, 33*(2), 301–311.

Burkett, J. P., Andari, E., Johnson, Z. V., Curry, D. C., de Waal, F. B., & Young, L. J. (2016). Oxytocin-dependent consolation behavior in rodents. *Science, 351*(6271), 375–378.

Carrurthers, P., Laurence, S., & Stich, S. (2005). *The innate mind*. New York: Oxford University Press.

Changeux, J. P. (2017). Climbing brain levels of organization from genes to consciousness. *Trends in Cognitive Sciences, 21*(3), 168–181.

Cheng, Y., Chen, C., & Decety, J. (2014). An EEG/ERP investigation of the development of empathy during early childhood. *Developmental Cognitive Neuroscience, 10,* 160–169.

Cheng, Y., Lee, S. Y., Chen, H. Y., Wang, P., & Decety, J. (2012). Voice and emotion processing in the human neonatal brain. *Journal of Cognitive Neuroscience, 24*(6), 1411–1419.

Cowell, J., & Decety, J. (2015a). Precursors to morality in development as a complex interplay between neural, socio-environmental, and behavioral facets. *Proceedings of the National Academy of Sciences of the United States of America, 112*(41), 12657–12662.

Cowell, J. M., & Decety, J. (2015b). The neuroscience of implicit moral evaluation and its relation to generosity in early childhood. *Current Biology, 25*(1), 93–97.

Cowell, J. M., Lee, L., Malcolm-Smith, S., Selcuk, B., Zhou, X., & Decety, J. (2017). The development of generosity and moral cognition across five cultures. *Developmental Science, 20*(4), e12403.

Cowell, J. M., Samek, A., List, J., & Decety, J. (2015). The curious relation between theory of mind and sharing in preschool age children. *PLoS One, 10*(2), e0117947.

Cummins, D. D., & Cummins, R. (1999). Biological preparedness and evolutionary explanation. *Cognition, 73,* B37-B53.

Dahl, A., & Freda, G. F. (2017). How young children come to view harming others as wrong: A developmental analysis of research from western communities. In J. A. Sommerville & J. Decety (Eds.), *Social cognition: Development across the life span* (pp. 151–184). New York: Routledge.

De Haan, M., Belsky, J., Reid, V., Volein, A., & Johnson, M. H. (2004). Maternal personality and infants' neural and visual responsivity to facial expressions of emotion. *Journal of Child Psychology and Psychiatry, 45,* 1209–1218

Decety, J., Ben-Ami Bartal, I., Uzefovsky, F., & Knafo-Noam, A. (2016). Empathy as a driver of prosocial behavior: Highly conserved neurobehavioral mechanisms across species. *Proceedings of the Royal Society B: Biological Sciences, 371,* 20150077.

Decety, J., & Cowell, J. M. (2018). Interpersonal harm aversion as a necessary foundation for morality: A developmental neuroscience perspective. *Development and Psychopathology, 30,* 153–164.

Decety, J., & Lamm, C. (2009). Empathy versus personal distress: Recent evidence from social neuroscience. In J. Decety & W. Ickes (Eds.), *The social neuroscience of empathy* (pp. 199–213). Cambridge, MA: MIT Press.

Decety, J., & Michalska, K. J. (2010). Neurodevelopmental changes in the circuits underlying empathy and sympathy from childhood to adulthood. *Developmental Science, 13*(6), 886–899.

Decety, J., Michalska, K. J., & Akitsuki, Y. (2008). Who caused the pain? An fMRI investigation of empathy and intentionality in children. *Neuropsychologia, 46*(11), 2607–2614.

Decety, J., Michalska, K. J., & Kinzler, K. D. (2012). The contribution of emotion and cognition to moral sensitivity: A neurodevelopmental study. *Cerebral Cortex, 22,* 209–220.

Dunfield, K., Kuhlmeier, V. A., O'Connell, L., & Kelley, E. (2011). Examining the diversity of prosocial behavior: Helping, sharing, and comforting in infancy. *Infancy, 16,* 227–247.

Eisenberg, N., & Eggum, N. D. (2009). Empathic responding: Sympathy and personal distress. In J. Decety & W. Ickes (Eds.), *The social neuroscience of empathy* (pp. 71–83). Cambridge, MA: MIT Press.

Engelmann, J. M., Herrmann, E., & Tomasello, M. (2012). Five-year olds, but not chimpanzees, attempt to manage their reputations. *PLoS One, 7*(10) e48433.

Engelmann, J. M., Over, H., Herrmann, E., & Tomasello, M. (2013). Young children care more about their reputation with ingroup members and potential reciprocators. *Developmental Science, 16*(6), 952–958.

Etkin, A., Egner, T., & Kalisch, R. (2011). Emotional processing in anterior cingulate and medial prefrontal cortex. *Trends Cognitive Sciences 15*(2), 85–93.

FeldmanHall, O., Dalgleish, T., Evans, D., & Mobbs, D. (2015). Empathic concern drives costly altruism. *Neuroimage, 105,* 347–356.

Frank, E. T., Wehrhahn, M., & Linsenmair, K. E. (2018). Wound treatment and selective help in a termite-hunting ant. *Proceedings of the Royal Society B: Biological Sciences, 285*(1872), 20172457.

Fraser, O. N., Stahl, D., & Aureli, F. (2008). Stress reduction through consolation in chimpanzees. *Proceedings of the National Academy of Sciences of the United States of America, 105*(25), 8557–8562.

Geraci, A., & Surian, L. (2011). The developmental roots of fairness: Infants' reactions to equal and unequal distributions of resources. *Developmental Science, 14*(5), 1012–1020.

Grossmann, T., Missana, M., & Krol, K. L. (2018). The neurodevelopmental precursors of altruistic behavior in infancy. *PLoS Biology 16*(9), e2005281.

Hagmann, P., Sporns, O., Madan, N., Cammoun, L., Pienaar, R., Wedeen, V. J., ...Grant, P. E. (2010). White matter maturation reshapes structural connectivity in the late developing human brain. *Proceedings of the National Academy of Sciences of the United States of America, 107*(44), 19067–19072.

Hamlin, J. K. (2015). The infantile origins of our moral brains. In J. Decety & T. Wheatley (Eds.), *The moral brain: Multidisciplinary perspectives* (pp. 105–122). Cambridge, MA: MIT Press.

Hare, T. A., Camerer, C. F., Knoepfle, D. T., & Rangel, A. (2010). Value computations in ventral medial prefrontal cortex during charitable decision-making incorporate input from regions involved in social cognition. *Journal of Neuroscience, 30*(2), 583–590.

Hare, T. A., Hakimi, S., & Rangel, A. (2014). Activity in dlPFC and its effective connectivity to vmPFC are associated with temporal discounting. *Frontiers in Neuroscience, 8,* 50.

Hein, G., Morishima, Y., Leiberg, S., Sul, S., & Fehr, E. (2016). The brain's functional network architecture reveals human motives. *Science, 351*(6277), 1074–1078.

Hein, G., Silani, G., Preuschoff, K., Batson, C. D., & Singer, T. (2010). Neural responses to ingroup and outgroup members' suffering predict individual differences in costly helping. *Neuron, 68*(1), 149–160.

Hirschfeld, L. A. (2013). The myth of mentalizing and the primacy of folk sociology. In M. Banaji & S. A. Gelman (Eds.), *Navigating the social world: What infants and other species can teach us* (pp. 101–106). New York: Oxford University Press.

Hoffmann, F., Grosse Wiesmann, C., Singer, T., & Steinbeis, N. (2019). Maturation of prefrontal-insular connectivity predicts developmental shift towards altruistic helping in childhood. Manuscript in preparation.

Huppert, E., Cowell, J. M., Cheng, Y., Contreras, C., Gomez-Sicard, N., Gonzalez-Gaeda, L. M.,…Decety, J. (2019). The development of children's preferences for equality and equity across 13 individualistic and collectivist cultures. *Developmental Science, 22*(2), e12729.

Hutcherson, C. A., Bushong, B., & Rangel, A. (2015). A neurocomputational model of altruistic choice and its implications. *Neuron, 87*(2), 451–462.

Imuta, K., Henry, J. D., Slaughter, V., Selcuk, B., & Ruffman, T. (2016). Theory of mind and prosocial behavior in childhood: A meta-analytic review. *Developmental Psychology, 52,* 1192–1205.

Jacob, F. (1977). Evolution and tinkering. *Science, 196,* 1161–1166.

Jin, K. S., & Baillargeon, R. (2017). Infants possess an abstract expectation of ingroup support. *Proceedings of the National Academy of Sciences of the United States of America, 114*(31), 8199–8204.

Johnson, S. C., Dweck, C. S., Chen, F. S., Stern, H. L., Ok, S.-J., & Barth, M. E. (2010). At the intersection of social and cognitive development: Internal working models of attachment in infancy. *Cognitive Science, 34,* 807–825.

Kaplan, H. S., & Gangestad, S. W. (2005). Life history theory and evolutionary psychology. In D. Buss (Ed.), The handbook of evolutionary psychology (pp. 68–95). New York: Wiley.

Lagercrantz, H. (2010). *The newborn brain.* Cambridge: Cambridge University Press.

Lamm, C., Decety, J., & Singer, T. (2011). Meta-analytic evidence for common and distinct neural networks associated with directly experienced pain and empathy for pain. *NeuroImage, 54*(3), 2492–2502.

Lebel, C., Gee, M., Camicioli, R., Wieler, M., Martin, W., & Beaulieu, C. (2012). Diffusion tensor imaging of white matter tract evolution over the lifespan. *NeuroImage, 60*(1), 340–352.

Lenroot, R. K., & Giedd, J. N. (2006). Brain development in children and adolescents: Insights from anatomical magnetic resonance imaging. *Neuroscience and Biobehavioral Reviews, 30*(6), 718–729.

Leslie, A. M. (1994). ToMM, ToBy, and Agency: Core architecture and domain specificity. In L. Hirschfeld & S. Gelman (Eds.), *Mapping the mind: Domain specificity in cognition and culture* (pp. 119–148). New York: Cambridge University Press.

Light, S., & Zahn-Waxler, C. (2011). Nature and forms of empathy in the first years of life. In J. Decety (Ed.), Empathy: From bench to bedside (pp. 109–130). Cambridge, MIT Press.

Martin, A., & Olson, K. R. (2015). Beyond good and evil: What motivations underlie children's prosocial behavior? *Perspectives on Psychological Science, 10*(2), 159–175.

McAuliffe, K., Blake, P. R., Steinbeis, N., & Warneken, F. (2017). The developmental foundations of human fairness. *Nature Human Behaviour, 1*(2), 0042.

Michalska, K. J., Kinzler, K. D., & Decety, J. (2013). Age-related sex differences in explicit measures of empathy do not predict brain responses across childhood and adolescence. *Developmental Cognitive Neuroscience, 3*, 22–32.

Miller, E. K., & Cohen, J. D. (2001). An integrative theory of prefrontal cortex function. *Annual Review of Neuroscience, 24*, 167–202.

Meidenbauer, K. L., Cowell, J. M., Killen, M., & Decety, J. (2018). A developmental neuroscience study of moral decision-making regarding resource allocation. *Child Development, 89*(4), 1177–1192.

Nichols, S. R., Svetlova, M., & Brownell, C. A. (2015). Toddlers' responses to infants' negative emotions. *Infancy, 20*(1), 70–97.

Paulus, M., Licata, M., Kristen, S., Thoermer, C., Woodward, A., & Sodian, B. (2015). Social understanding and self-regulation predict preschoolers' sharing with friends and disliked peers: A longitudinal study. *International Journal of Behavioral Development, 39*, 53–64.

Paulus, M., Kühn-Popp, N., Licata, M., Sodian, B., & Meinhardt, J. (2013). Neural correlates of prosocial behavior in infancy: Different neurophysiological mechanisms support the emergence of helping and comforting. *Neuroimage, 66,* 522–530.

Premack, D., & Premack, A. J. (1997). Infants attribute value to the goal-directed actions of self-propelled objects. *Journal of Cognitive Neuroscience, 9*(6), 848–856.

Pun, A., Birch, S. A., & Baron, A. S. (2016). Infants use relative numerical group size to infer social dominance. *Proceedings of the National Academy of Sciences of the United States of America, 113*(9), 2376–2381.

Rilling, J. K., & Young, L. J. (2014). The biology of mammalian parenting and its effect on offspring social development. *Science, 345*(6198), 771–776.

Rubia, K., Smith, A. B., Taylor, E., & Brammer, M. (2007). Linear age-correlated functional development of right inferior fronto-striato-cerebellar networks during response inhibition and anterior cingulate during error-related processes. *Human Brain Mapping, 28*(11), 1163–1177.

Ruff, C. C., Ugazio, G., & Fehr, E. (2013). Changing social norm compliance with noninvasive brain stimulation. *Science, 342*(6157), 482–484.

Schmidt, M. F. H., & Sommerville, J. A. (2011). Fairness expectations and altruistic sharing in 15-month-old human infants. *PLoS One, 6*(10).

Silver, I. M., & Shaw, A. (2018). Pint-sized public relations: The development of reputation management. *Trends in Cognitive Sciences, 22*(4), 277–279.

Shaw, A., Montinari, N., Piovesan, M., Olson, K. R., Gino, F., & Norton, M. I. (2014). Children develop a veil of fairness. *Journal of Experimental Psychology: General, 143*(1), 363–375.

Sheskin, M., Chevallier, C., Lambert, S., & Baumard, N. (2014). Life-history theory explains childhood moral development. *Trends in Cognitive Sciences, 18*(12), 613–615.

Smith, C. E., Blake, P. R., & Harris, P. L. (2013). I should but I won't: Why young children endorse norms of fair sharing but do not follow them. *PLoS One, 8*(3), e59510.

Sommerville, J. A., Schmidt, M. F., Yun, J. E., & Burns, M. (2013). The development of fairness expectations and prosocial behavior in the second year of life. *Infancy, 18*(1), 40–66.

Spitzer, M., Fischbacher, U., Herrnberger, B., Gron, G., & Fehr, E. (2007). The neural signature of social norm compliance. *Neuron, 56*(1), 185–196.

Steinbeis, N. (2018). Taxing behavioral control diminishes sharing and costly punishment in childhood. *Developmental Science, 21*(1), e12492.

Steinbeis, N., Bernhardt, B. C., & Singer, T. (2012). Impulse control and underlying functions of the left DLPFC mediate age-related and age-independent individual differences in strategic social behavior. *Neuron, 73*(5), 1040–1051.

Steinbeis, N., Bernhardt, B. C., & Singer, T. (2014). Age-related differences in function and structure of rSMG and reduced functional connectivity with DLPFC explains heightened emotional egocentricity bias in childhood. *Social Cognitive and Affective Neuroscience, 10*(2), 302–310.

Steinbeis, N., & Crone, E. A. (2016). The link between cognitive control and decision-making across child and adolescent development. *Current Opinion in Behavioral Sciences, 10*, 28–32.

Steinbeis, N., Haushofer, J., Fehr, E., & Singer, T. (2016). Development of behavioral control and associated vmPFC-dlPFC connectivity explains children's increased resistance to temptation in intertemporal choice. *Cerebral Cortex, 26*(1), 32–42.

Steinbeis, N., & Over, H. (2017). Enhancing behavioral control increases sharing in children. *Journal of Experimental Child Psychology, 159,* 310–318

Starmans, C., Sheskin, M., Bloom, P., Christakis, N. A., & Brown, G. D. (2017). Why people prefer unequal societies. *Nature Human Behaviour, 1*(4), 82.

Svetlova, M., Nichols, S. R., & Brownell, C. A. (2010). Toddlers' prosocial behavior: From instrumental to empathic to altruistic helping. *Child Development, 81*(6), 1814–1827.

Takagishi, H., Kameshima, S., Schug, J., Koizumi, M., & Yamagishi, T. (2010). Theory of mind enhances preference for fairness. *Journal of Experimental Child Psychology, 105*(1–2), 130–137.

Thijssen, S., Ringoot, A. P., Wildeboer, A., Bakermans-Kranenburg, M. J., El Marroun, H., Hofman, A., … & White, T. (2015). Brain morphology of childhood aggressive behavior: A multi-informant study in school-age children. *Cognitive, Affective, and Behavioral Neuroscience, 15*(3), 564–577.

Tomasello, M. (2014). The ultra-social animal. *European Journal of Social Psychology, 44*(3), 187–194.

Tomasello, M. (2016). *A natural history of human morality.* Cambridge, MA: Harvard University Press.

Tooby, J., & Cosmides, L. (1992). The psychological foundations of culture. In J. Barkow, L. Cosmides, & J. Tooby (Eds.), *The adapted mind: Evolutionary psychology and the generation of culture* (pp. 19–136). New York: Oxford University Press.

Tops, M., & Boksem, M. A. (2011). A potential role of the inferior frontal gyrus and anterior insula in cognitive control, brain rhythms, and event-related potentials. *Frontiers in Psychology, 2,* 330.

Vaish, A., Carpenter, M., & Tomasello, M. (2009). Sympathy through affective perspective-taking and its relation to prosocial behavior in toddlers. *Developmental Psychology, 45,* 534–543.

Vygotsky, L. (1978). *Mind in society.* Cambridge, MA: Harvard University Press.

Warneken, F. (2016). Insight into the biological foundation of human altruistic sentiments. *Current Opinion in Psychology, 7,* 51–56.

Warneken, F., & Tomasello, M. (2007). Helping and cooperation at 14 months of age. *Infancy, 11*(3), 271–294.

Williams, A., O'Driscoll, K., & Moore, C. (2014). The influence of empathic concern on prosocial behavior in children. *Frontiers in Psychology, 5,* 425.

Yoder, K. J., & Decety, J. (2014a). The good, the bad, and the just: Justice sensitivity predicts neural response during moral evaluation of actions performed by others. *Journal of Neuroscience, 34*(12), 4161–4166.

Yoder, K., J., & Decety, J. (2014b). Spatiotemporal neural dynamics of moral judgments: A high-density EEG/ERP study. *Neuropsychologia, 60,* 39–45.

Yoder, K. J., Harenski, C., Kiehl, K. A., & Decety, J. (2015). Neural networks underlying implicit and explicit moral evaluations in psychopathy. *Translational Psychiatry, 5,* e625.

Zahn-Waxler, C., Radke-Yarrow, M., Wagner, E., & Chapman, M. (1992). Development of concern for others. *Developmental Psychology, 28*(1), 126–136.

13   Selective Prosocial Behavior in Early Childhood

Valerie A. Kuhlmeier, Tara A. Karasewich, and Kristen A. Dunfield

Overview

Human social interactions frequently include behaviors that are generated on behalf of others. By maintaining reciprocal social systems, the resource investment inherent in this prosocial behavior will be repaid over time, solving an otherwise potential "problem" for the evolution of prosociality. The partner choice model characterizes the advantage of being *selective* with our prosocial behavior so that we can increase the probability of a return on investment. Here, we make the case for selective partner choice even in early instances of human prosocial behavior and outline three potential cognitive mechanisms supporting this selectivity that flow from developmental and comparative psychology research and theory.

Introduction

A quote from Fred Rogers—"Mr. Rogers" of children's television fame—often spreads across social media sites after tragic events. Rogers is quoted as saying that he learned from his mother that when the news covered scary incidents, he should "look for the helpers" because "you will always find people who are helping." Rogers was correct: we are a highly social species, with much of our social behavior enacted on behalf of others, and we readily identify these behaviors in those around us.

Our prosocial behaviors are flexible responses to others' needs that start to appear in the first few years of life (for reviews, see Carpendale, Hammond, & Atwood, 2012; Dunfield, 2014; Eisenberg, Fabes, & Rapp, 2006; Grossman, 2018). Others' needs may include instrumental needs (i.e., an individual cannot complete a goal-directed behavior such as retrieving an out-of-reach object, and we can intervene by *helping*), material desire (i.e., an individual does not have a desired resource, and we can intervene by *sharing*), and emotional distress (i.e., an individual is experiencing a negative emotional state, and we can intervene by *comforting*). Yet natural and social scientists working within evolutionary theoretical frameworks have long noted that widespread engagement in prosociality is, in some ways, surprising because it requires expending personal resources for the benefit of others, including individuals with whom we share no appreciable genetic relatedness (e.g., Axelrod, 1984).

This "problem" of prosociality has been addressed by seminal models that emphasize reciprocity; when reciprocity is maintained, our resource cost can be repaid over time. One such model considers the advantage of being *selective* with our prosocial behavior so that we can increase the probability of a return on investment. In this chapter, we will first present the basic characteristics of this "partner choice" model (for more detailed discussion, see Baumard, André, & Sperber, 2013; Bshary & Noë, 2003; Roberts, 1998). Then we focus on the development of prosocial behavior, making the case for selective partner choice even in early childhood. We end by proposing cognitive, decision-making processes that may support this selective prosociality.

Thus, the chapter organization flows from at least three of Tinbergen's "four questions" (i.e., levels of analysis: Tinbergen, 1963) in relation to selective prosocial behavior: (a) ultimate causes such as function and adaptiveness, (b) proximate causes such as development, and (c) proximate causes such as cognitive mechanisms. Suggestive comparisons from the study of nonhuman animal cognition and behavior will be presented throughout.

The Partner Choice Model of Reciprocity

Reciprocity is a feature of human social organization, appearing as both direct reciprocity (e.g., B helps A after A helped B; Trivers, 1971) and indirect reciprocity (e.g., B helps A after A helped C, Alexander, 1987; or B helps C after A helped B, Leimgruber, 2018). However, it is possible for our resources (broadly construed as time, energy, tangible goods, etc.) to be directed to ineffective or unwilling members of the group who do not provide a good return on the investment. The ways in which reciprocal systems can be maintained in the face of this risk have been defined in various models. One of these models—"partner control" or "partner fidelity"—is exemplified by situations in which two individuals are forced into interaction (Noë, 2006). The partner is predetermined, and interactions may occur over repeated rounds, and thus the best strategy is one of "tit-for-tat" in which prosociality is met with prosociality and undesirable behaviors are met with punishment (e.g., Axelrod & Hamilton, 1981). In the partner control model, the discouragement and prevention of cheating can maintain a reciprocally prosocial system.

The "partner choice" model, by contrast, considers situations in which individuals can be selective when entering social interactions, and the emphasis is placed on choosing prosocial partners and being chosen as one (e.g., Baumard et al., 2013; Bull & Rice, 1991; Roberts, 1998). Cleaner fish, for example, consume ectoparasites on the surfaces of client fish that, in turn, benefit from the cleaning (for review, see Bshary & Noë, 2003). When cleaner wrasse "cheat" by eating the client's mucus, the clients seemingly find this aversive and swim away to find other cleaners who were previously observed cleaning fish without conflict. Thus, in partner choice models, a preference for good partners maintains reciprocity and selects for prosocial behavior within a species.

It is currently thought that the partner choice model is more characteristic of nonhuman animal social behavior than partner control (Schino & Aureli, 2016), and in humans, evidence of partner choice behavior is found earlier in development than partner control behavior (e.g., Warneken, 2018). We will summarize this evidence of early-developing partner choice, but an important caveat must first be presented. The partner choice model, as defined above, is discussed both in terms of selecting prosocial partners and being seen by others as a potential prosocial partner. For humans, the latter is often considered at a proximate, mechanistic level in relation to the development of "reputational concern"—or caring about how you are evaluated by others—and this concern is observed later in development (five years of age and older; see Engelmann & Rapp, 2018, for a review). Here, we will specifically focus on the choice of partner: the evaluation of potential partners and subsequent selective engagement in prosocial behavior by very young children (see also Kuhlmeier, Dunfield, & O'Neill, 2014).

Partner Choice in Early Human Development

In toddlerhood, some of the earliest instances of prosocial behavior are selective in terms of recipient. Yet even earlier, during infancy, the foundations

of partner choice are present. The existence of behaviors that support partner choice during infancy may not seem adaptive at first glance: infants
can seldom "choose" a social partner, and their prosocial behaviors are limited. However, infants' *evaluation* of social interaction may have adaptive
value as preparation for later partner selection, or perhaps as part of an
attachment mechanism during infancy (i.e., an "ontogenetic adaptation,"
Bjorklund & Pellegrini, 2000; Wynn, 2009). In this section, we start by discussing the current evidence for social evaluation during infancy. Then we
present findings from studies of toddlers and preschool-age children suggesting these social evaluations are often followed by selective engagement
of prosocial behavior.

Social Evaluation in Infancy

Partner choice behavior in humans requires an evaluative system that
both distinguishes between positive and negative interactions and encourages approach or other affiliative behaviors directed toward individuals
involved in positive interactions. Premack and Premack (1997) provided
one of the first demonstrations that by twelve months of age infants distinguish between simple helping and hindering behaviors; importantly, this
distinction is based on the underlying valence of the actions. Specifically,
in a looking time habituation/dishabituation paradigm, infants appeared
to group helping and caressing interactions together, and group hindering
and hitting interactions together, even though the surface characteristics
of the helping and hindering scenes were arguably more similar to each
other than either scene was to its similarly valenced event. That is, infants
appeared to categorize the events by valence, such that helping and caressing share a sense of positivity, and hindering and hitting share negativity.

Infants' social evaluation also influences their approach (e.g., reaching)
behavior. After witnessing puppets that either helped or hindered others,
infants as young as eight to ten months selectively reached for the helping
puppet (Hamlin, Wynn, & Bloom, 2007; see Hamlin, 2013b, for a review).
By eight months, infants appear to integrate information about the intentionality of the helping or hindering actions in their approach behavior,
favoring, for example, individuals who try but fail to aid over successful
hinderers (Hamlin, 2013a). Further, infants appear to have similar expectations regarding others' preferences: after observing helping and hindering events, nine- and twelve-month-old infants expected the recipients to

approach the helpers over hinderers (e.g., Fawcett & Liszkowski, 2012; see Kuhlmeier, 2013, for a review).

Comparable results have been obtained after infants observe other actions, such as harming (i.e., reaching for victims over harmful agents; Kanakogi et al., 2013) and sharing. In relation to the latter, infants appear to expect goods to be distributed equally (Schmidt & Sommerville, 2011; Sommerville, Schmidt, Yun, & Burns, 2013), and by ten months they expect agents to approach "fair" over "unfair" distributors (e.g., Geraci & Surian, 2011; Meristo & Surian, 2013). Together, these studies suggest that the evaluative processes—and accompanying approach behavior—that would be necessary for selective prosociality develop within the first year of life. We acknowledge, of course, that social evaluation becomes more sophisticated with age as new means of evaluation develop and as cultural norms regarding "good" and "bad" behavior become more strongly represented (e.g., Dahl, Schuck, & Campos, 2013; Lee, Xu, Fu, Cameron, & Chen, 2010), but development in infancy may create initial working models upon which later evaluation is elaborated.

Selective Prosocial Behavior

Starting within the second year of life, social evaluation appears to influence the production of prosocial behavior. Laboratory-based procedures have manipulated the interactions that young children witness, varying the behavioral and physical characteristics of the actors. Children's subsequent engagement in selective prosocial behavior, typically helping or sharing, toward these individuals is then measured. The interactions children witness have included helping, hindering, harming, and sharing, as well as other behaviors and physical characteristics that are not as explicitly pro- or antisocial. Experimental paradigms are designed with either the child as a member of the interaction (i.e., direct reciprocity) or with the child as an observer of one actor's behavior toward another actor (i.e., indirect reciprocity).

For example, in a study in which children were the direct recipients of either helpful or unhelpful actions, three-year-olds were presented with a puppet who was willing to communicate the solution to a puzzle and one who declared that he knew but was "not telling" (Dunfield, Kuhlmeier, & Murphy, 2013). Children applied the label "helpful" to the puppet that was willing to provide the answer, and this evaluation was consistent with their selective helping behavior. In one condition, children selectively delivered a dropped object to the informative puppet more frequently than to the unwilling puppet; in another, children selectively provided information to the informative puppet. (For a similar finding of selective helping after cooperative interactions, see Allen, Perry, & Kaufman, 2018.)

Young children also selectively help individuals who have shown the intention to provide resources to them. In Dunfield and Kuhlmeier (2010), twenty-one-month-old children selectively picked up a dropped object for an individual who, in a previous interaction, intended—but failed—to provide them with a desired toy over one who intentionally did not provide a toy. Another experimental condition in that study indicated that the children were selective even when both actors' earlier actions resulted in the same outcome (i.e., the children received the toy), yet only one of the actors showed the overt intention to provide (i.e., the other actor's actions were accidental).

To date, there is more experimental research on young children's selective prosocial behavior in situations in which they are third-party witnesses of social interactions than when they are direct participants. Young children, for example, selectively share resources with individuals who were observed helping others over individuals who have hindered. Preschool children in Kenward and Dahl (2011) witnessed events in which a puppet was trying to climb a ladder or trying to dig a hole and was helped by one puppet but hindered by another. Four-year-old children distributed more "biscuits" to the helper than the hinderer. These children also justified the uneven distribution in relation to the recipients' previous actions. Important to note, however, was that when there were many biscuits to distribute (e.g., eight or nine biscuits), children did give equal amounts to each actor. Thus, it is possible that factors such as an "equality bias" may trump the selective distribution of goods when resources are plentiful, yet selectivity based on recipients' history of "good" or "bad" behavior occurs when resources are scarce. These findings from Kenward and Dahl were consistent with a slightly earlier study that found that three-year-old children selectively helped an actor who previously did *not* harm another actor over one who showed the intention to harm (Vaish, Carpenter, & Tomasello, 2010).

Further evidence for selective prosocial behavior after third-party observation of interactions comes from studies in which actors were seen to share (or not share) goods with others. Dahl and colleagues (2013) found that two-year-old children were more likely to help an individual who had recently returned a desired object to another individual than one who had not returned the object. Consistent with these findings, Surian and Franchin (2017) found that eighteen- to thirty-three-month old children selectively helped an individual who had previously distributed resources equally over one who had distributed unequally, and Olson and Spelke (2008) found that three-year-olds directed a doll to give more resources to a doll that had previously given to others. Notably, though, in another condition, Olson and Spelke found that children directed the doll to give more to someone who gave directly to the doll than to someone who gave to other dolls, suggesting that early selective sharing behavior is constrained by a nuanced evaluation of the previously witnessed interaction and the individuals involved.

The studies presented thus far have measured children's selective prosocial behavior immediately after they have been directly involved in, or have witnessed, social interactions. Previous interactions, in the form of friendships, may also influence young children's selective prosociality. Moore (2009), for example, found that four- to six-year-old children shared stickers more often with their friends than other familiar peers and strangers, but treated friends and strangers similarly when there was no personal cost to providing stickers. Friends were also favored in Olson and Spelke (2008), yet here the past history of prosocial interactions may have been inferred: three-year-olds directed a doll to give more items to her doll friends. Similar results have been found by Paulus and Moore (2014) and Paulus (2016), in which preschool-age children gave more resources to friends and expected others to share more with friends as well.

Social evaluation and subsequent selective prosociality in young children, though, does not appear to require the observation of, or possibly even the inference of, others' explicitly pro- or antisocial behavior in the past. For example, young children engage in selective helping behavior based on defined group membership and similarity to the self, even without previous observation of social interactions. In O'Neill and Kuhlmeier (2014), two-year-old children selectively helped a puppet that was previously described as being "on their team" (group membership) or was wearing the same color shirt as the child (similarity to self) over non-team members and dissimilar puppets. Consistent findings are reported with slightly older children; Sparks and colleagues (2017) found that four- to six-year-olds gave more to in-group members and those who shared their interests, and Engelmann and colleagues (2013) observed that five-year-olds shared more with in-group members (see also Dunham, Baron, &

Carey, 2011). Selective prosociality also appears to follow direct experience of interpersonal synchrony; fourteen-month-olds selectively helped individuals who had moved synchronously with them over those who moved asynchronously, and extended this partner choice to members of the synchronous partner's social group (see Cirelli, 2018, for review).

In summary, laboratory research over the past decade demonstrates that young children are often selective in terms of the recipient of their helping and sharing behaviors. It is difficult, however, to pinpoint the age at which selective prosociality appears in development. Helping behavior, such as picking up a dropped object, is observed at fourteen months of age (Warneken and Tomasello, 2007), and informing is found at twelve months (Liszkowski, 2005), yet by our reading there have been few experimental attempts to examine selective helping in toddlers at this age except for the work of Cirelli and colleagues using synchronous movement (e.g., 2018). Some existing studies, though, do find age differences within their sample, with two- or three-year-old children showing less frequent selectivity than slightly older children (e.g., Kenward & Dahl, 2011; Dahl et al., 2013).

Of course, a study may not find evidence for selective prosocial behavior in early development if children did not understand the social interaction that occurred well enough to form an evaluation on which to base their selectivity. For example, if young children cannot infer the intention that underlies an action, and thus do not recognize that a negative outcome is accidental, their evaluation of the actor may differ from that of older participants (see Hilton & Kuhlmeier, 2018, for review). It is also important to note that even if selective helping or sharing is found in young children's earliest instances of prosocial behavior, the cognitive processes underlying that selectivity may differ across development and situation.

Cognitive Routes to Selective Prosocial Behavior

Though the interdisciplinary study of reciprocity has provided partner choice models that detail the ultimate, adaptive function of being selective in the engagement of prosocial behavior, the more proximate, cognitive mechanisms underlying this selectivity—particularly in early human development—have not been determined. It is tempting to assume that the decision-making process involved is based on consideration of the likelihood that a potential partner will provide benefits that repay or exceed the investment made (i.e., an expectation of reciprocation). As some have pointed out, though, this is an erroneous assumption that conflates ultimate function and proximate causation (e.g., Barrett, Henzi, & Rendall, 2007; Schino & Aureli, 2009, 2016).

Under the partner choice model, selective prosocial behavior would be selected for because it would result in higher chances of reciprocity (and thus benefits) in the long run, but the proximate decision-making process need not consider and plan for the future. Indeed, the existence of selective social behavior in both humans and a wide phylogenetic range of nonhuman animals suggests both that selectivity is a fundamental factor in the maintenance of reciprocity and that the proximate mechanisms that support it may range from highly constrained innate predispositions to more flexible individual and social learning processes and rational inference (e.g., Bshary & Noë, 2003; Warneken, 2013). With this in mind, we propose three possible cognitive routes to selective prosocial behavior. These three routes are not assumed to be exhaustive, but they have been chosen and elaborated based on existing research.

Each of the three routes begins with the observation of others' behavior or other characteristics (e.g., similarity to the self). The first two routes propose that the affective valence associated with the behavioral event or characteristic becomes associated with the individual involved—for instance, an individual who enables another's goals is, by association, a *positive person*. This process of evaluation is consistent with recent proposals for why children (and adults under cognitive load) prefer lucky individuals (e.g., "affective tagging"; Olson, Dunham, Dweck, Spelke, & Banaji, 2008), why children prefer individuals who share similar food preferences (Field, 2006), and how nonhuman primates engage in prosocial partner choice (e.g., Schino & Aureli, 2010). In the first route, the observer then directs his or her similarly valenced actions to the individual—for instance, they selectively help a helper or avoid a hinderer (see also Hamlin, 2014, for a similar proposal regarding infant selective reaching behavior).

It is possible that in some situations, though, even young children's selective prosociality stems from a behavioral prediction or an expectation of further prosocial behavior by the partner in the future. Infants predict individuals to continue to act on certain objects after initial observation of object-directed actions (e.g., Cannon & Woodward, 2012), and at two years of age, children predictively gaze to the likely location of another person's search behavior (e.g., Southgate, Senju, & Csibra, 2007). Particularly relevant to the present discussion, when faced with a person in distress, two-year-olds predictively look toward that person's friend in expectation of intervention, potentially because of expectations built from past prosocial interactions (Beier, Carpenter, & Tomasello, 2010). Additional support comes from a task in which a social partner was fixed (i.e., a task associated with partner *control* models). Warneken and Tomasello (2013) found that three-year-old children based their sharing behavior on the sharing behavior of a fixed partner over repeated encounters (i.e., showing "contingent reciprocity"). Thus, in our second route the initial valence association engenders a behavioral prediction such that subsequent behavior by this individual is predicted to have the same valence (see Brosseau-Liard & Birch, 2010, and Fusaro, Corriveau, & Harris, 2011, for similar proposals regarding how children choose good informants). This prediction is followed by the engagement of similarly valenced actions by the child observer (e.g., selective helping).

In the third cognitive route, the observation of a behavioral event or an individual's characteristic is followed by a disposition attribution, such that an individual's behavior or appearance is explained by an internal aspect of that individual (e.g., Kuhlmeier, 2013). For adults, preexisting beliefs and the observation of behavior give rise to inferences about others' dispositions (e.g., Kelley, 1973; Molden, Plaks, & Dweck, 2006). It is possible that, for young children, the observation of helping behavior may lead to an attribution that a person is a *nice* or *good-intentioned* person, which in turn allows for the prediction that future behavior will be consistent with this disposition, and the child observer directs his or her own prosocial behavior toward the individual. Note that we differentiate the second and third route by the specificity of the child's behavioral prediction. In the second route, the prediction is more general and simply matched in valence to the previous action or characteristic (e.g., someone who is wearing the same color t-shirt might also engage in future sharing: a "halo effect"). In the third route, disposition attribution allows for a more specific prediction that is more tightly connected with the observed behavior: someone who has helped is nice and therefore will likely be prosocial again.

Importantly we do not suggest that once a child is capable of making disposition attributions such as these that the child never uses either of the other two routes; both developmental and situational factors would determine which route is taken. Even adults may rely on the first or second routes when limited information is available, when under cognitive load, or when quick decision-making is required. (For example, see Apperly & Butterfill, 2009, and Penn & Povinelli, 2013, for similar proposals in relation to the cognitive mechanisms underlying the human ability to attribute beliefs to others.) Indeed, we hope that this chapter stimulates further systematic study of the cognitive and emotional proximate causes of selective prosociality, both across human cultures and across species.

Conclusion

We suggest that in early development, the prosocial behaviors produced by young children fit what would be predicted by partner choice models of reciprocity. Even in infancy, our social brain appears to evaluate interactions, ascribing positive or negative valence to them. These social evaluations become part of a decision-making process that leads to helping and sharing behaviors that are selective in terms of recipient. The details of the cognitive processes and the accompanying neurological processes underlying selective prosociality have yet to be fully elaborated, although as noted here testable predictions can be created from existing research and theory. In sum, the application of partner choice models to the study of childhood prosocial development sheds light on the very factors that encourage (or discourage) prosocial behavior in our social interactions.

References

Alexander, R. D. (1987). *The biology of moral systems*. Piscataway, NJ: Transaction.

Allen, M., Perry, C., & Kaufman, J. (2018). Toddlers prefer to help familiar people. *Journal of Experimental Child Psychology, 174*, 90–102.

Apperly, I. A., & Butterfill, S. A. (2009). Do humans have two systems to track beliefs and belief-like states? *Psychological Review, 116*, 953–970.

Axelrod, R. (1984). *The evolution of cooperation*. New York: Basic Books.

Axelrod, R., & Hamilton, W. D. (1981). The evolution of cooperation. *Science, 211*(4489), 1390–1396.

Barrett, L., Henzi, P., & Rendall, D. (2007). Social brains, simple minds: Does social complexity really require cognitive complexity? *Philosophical Transactions of the Royal Society of London B, 362*, 561–575.

Baumard, N., André, J. B., & Sperber, D. (2013). A mutualistic approach to morality: The evolution of fairness by partner choice. *Behavioral and Brain Sciences, 36*(01), 59–78.

Beier, J. S., Carpenter, M., & Tomasello, M. (2010, April). Young children's understanding of third-party social relationships. Poster presented at the XVIIth Biennial Conference on Infant Studies, Baltimore.

Bjorklund, D. F., and Pellegrini, A. D. (2000). Child development and evolutionary psychology. *Child Development, 71,* 1687–1708.

Bull, J. J., & Rice, W. R. (1991). Distinguishing mechanisms for the evolution of co-operation. *Journal of Theoretical Biology, 149*(1), 63–74.

Brosseau-Liard, P. E., & Birch, S. A. (2010). "I bet you know more and are nicer too!": What children infer from others' accuracy. *Developmental Science, 13,* 772–778.

Bshary, R., & Noë, R. (2003). Biological markets: The ubiquitous influence of partner choice on the dynamics of cleaner fish–client reef fish interactions. In P. Hammerstein (Ed.), *Genetic and cultural evolution of cooperation* (pp. 167–184). Cambridge, MA: MIT Press.

Cannon, E. N., & Woodward, A. L. (2012). Infants generate goal-based action predictions. *Developmental Science, 15,* 292–298.

Carpendale, J. I., Hammond, S. I., & Atwood, S. (2012). A relational developmental systems approach to moral development. *Advances in Child Development and Behavior, 45,* 125–153.

Cirelli, L. K. (2018). How interpersonal synchrony facilitates early prosocial behavior. *Current Opinion in Psychology, 20,* 35–39.

Dahl, A., Schuck, R. K., & Campos, J. J. (2013). Do young toddlers act on their social preferences? *Developmental Psychology, 49,* 1964–1970.

Dunfield, K. A. (2014). A construct divided: Prosocial behavior as helping, sharing, and comforting subtypes. *Frontiers in Psychology, 5,* 958.

Dunfield, K. A., & Kuhlmeier, V. A. (2010). Intention-mediated selective helping in infancy. *Psychological Science, 21,* 523–527.

Dunfield, K. A., Kuhlmeier, V. A., & Murphy, L. (2013). Children's use of communicative intent in the selection of cooperative partners. *PLoS One, 8,* e61804.

Dunham, Y., Baron, A. S., & Carey, S. (2011). Consequences of "minimal" group affiliations in children. *Child Development, 82,* 793–811.

Eisenberg, N., Fabes, R. A., & Spinrad, T. L. (2006). Prosocial development. In N. Eisenberg, W. Damon, & R. M. Lerner (Eds.), *Handbook of child psychology: Vol. 3, Social, emotional, and personality development* (6th ed., pp. 646–718). New York: John Wiley & Sons.

Engelmann, J. M., Over, H., Herrmann, E., & Tomasello, M. (2013). Young children care more about their reputation with ingroup members and potential reciprocators. *Developmental Science, 16,* 952–958.

Engelmann, J. M., & Rapp, D. J. (2018). The influence of reputational concerns on children's prosociality. *Current Opinion in Psychology, 20,* 92–95.

Fawcett, C., & Liszkowski, U. (2012). Infants anticipate others' social preferences. *Infant and Child Development, 21,* 239–249.

Field, A. P. (2006). I don't like it because it eats sprouts: Conditioning preferences in children. *Behavior Research and Therapy, 44,* 439–455.

Fusaro, M., Corriveau, K. H., & Harris, P. (2011). The good, the strong, and the accurate: Preschoolers' evaluations of informant attributes. *Journal of Experimental Child Psychology, 110,* 561–574.

Geraci, A., & Surian, L. (2011). The developmental roots of fairness: Infants' reactions to equal and unequal distributions of resources. *Developmental Science, 14,* 1012–1020.

Grossmann, T. (2018). How to build a helpful baby: A look at the roots of prosociality in infancy. *Current Opinion in Psychology, 20,* 21–24.

Hamlin, J. K. (2013a). Failed attempts to help and harm: Intention versus outcome in preverbal infants' social evaluations. *Cognition, 128,* 451–474.

Hamlin, J. K. (2013b). Moral judgment and action in preverbal infants and toddlers: Evidence for an innate moral core. *Current Directions in Psychological Science, 22,* 186–193.

Hamlin, J. K. (2014). The origins of human morality: Complex socio-moral evaluations by preverbal infants. In J. Decety and Y. Christen (Eds.), *New Frontiers in Social Neuroscience.* New York: Springer.

Hamlin, J. K., Wynn, K., & Bloom, P. (2007). Social evaluation by preverbal infants. *Nature, 450,* 557–559.

Hilton, B. C., & Kuhlmeier, V. A. (2018). Intention attribution and the development of moral evaluation. *Frontiers in Psychology, 9,* 2663.

Kanakogi, Y., Okumura, Y., Inoue, Y., Kitazaki, M., & Itakura, S. (2013). Rudimentary sympathy in preverbal infants: Preference for others in distress. *PLoS One, 8,* e65292.

Kelley, H. H. (1973). The process of causal attribution. *American Psychologist, 28,* 107–128.

Kenward, B., & Dahl, M. (2011). Preschoolers distribute scarce resources according to the moral valence of recipients' previous actions. *Developmental Psychology, 47*(4), 1054–1064.

Kuhlmeier, V. A. (2013). The social perception of helping and hindering. In M. D. Rutherford & V. A. Kuhlmeier (Eds.), *Social Perception*. Cambridge, MA: MIT/Bradford Press.

Kuhlmeier, V. A., Dunfield, K. A., & O'Neill, A. C. (2014). Selectivity in early prosocial behavior. *Frontiers in Psychology, 5,* 1–6.

Lee, K., Xu, F., Fu, G., Cameron, C. A., & Chen, S. (2010). Taiwan and Mainland Chinese and Canadian children's categorization and evaluation of lie- and truth-telling: A modesty effect. *British Journal of Developmental Psychology, 19,* 525–542.

Leimgruber, K. L. (2018). The developmental emergence of direct reciprocity and its influence on prosocial behavior. *Current Opinion in Psychology, 20,* 122–126.

Liszkowski, U. (2005). Human twelve-month-olds point cooperatively to share interest with and helpfully provide information for a communicative partner. *Gesture, 5,* 135–154.

Meristo, M., & Surian, L. (2013). Do infants detect indirect reciprocity? *Cognition, 129,* 102–113.

Molden, D. C., Plaks, J. E., & Dweck, C. S. (2006). "Meaningful" social inferences: Effects of implicit theories on inferential processes. *Journal of Experimental Social Psychology, 42,* 738–752.

Moore, C. (2009). Fairness in children's resource allocation depends on the recipient. *Psychological Science, 20,* 944–948.

Noë, R. (2006). Cooperation experiments: Coordination through communication versus acting apart together. *Animal Behaviour, 71,* 1–18.

Olson, K. R., Dunham, Y., Dweck, C. S., Spelke, E. S., & Banaji, M. R. (2008). Judgments of the lucky across development and culture. *Journal of Personality and Social Psychology, 94,* 757–776.

Olson, K. R., & Spelke, E. S. (2008). Foundations of cooperation in young children. *Cognition, 108,* 222–231.

O'Neill, A. C., & Kuhlmeier, V. A. (2014). Similarity and group membership: Influences on perception and behavior in toddlers. Paper presented at the 2014 meeting of the International Conference on Infant Studies, Berlin.

Paulus, M. (2016). Friendship trumps neediness: The impact of social relations and others' wealth on preschool children's sharing. *Journal of Experimental Child Psychology, 146,* 106–120.

Paulus, M., & Moore, C. (2014). The development of recipient-dependent sharing behavior and sharing expectations in preschool children. *Developmental Psychology, 50,* 914–921.

Penn, D. C., & Povinelli, D. J. (2013). The comparative delusion: The "behaviorisitic/ mentalistic" dichotomy in comparative theory of mind research. In J. Metcalfe & H. S. Terrace (Eds.), *Agency and joint attention* (pp. 61–81). Oxford: Oxford University Press.

Premack, D., & Premack, A. J. (1997). Infants attribute value± to the goal- directed actions of self-propelled objects. *Journal of Cognitive Neuroscience, 9,* 848–856.

Roberts, G. (1998). Competitive altruism: From reciprocity to the handicap principle. *Proceedings of the Royal Society B: Biological Sciences, 265*(1394), 427–431.

Schino, G., & Aureli, F. (2009). Reciprocal altruism in primates: Partner choice, cognition, and emotions. *Advances in the Study of Behavior, 39,* 45–69.

Schino, G., & Aureli, F. (2010). Primate reciprocity and its cognitive requirements. *Evolutionary Anthropology, 19,* 130–135.

Schino, G., & Aureli, F. (2016). Reciprocity in group-living animals: Partner control versus partner choice. *Biological Reviews, 92,* 665–672.

Schmidt, M. F., & Sommerville, J. A. (2011). Fairness expectations and altruistic sharing in 15-month-old human infants. *PLoS One, 6,* e23223.

Sommerville, J. A., Schmidt, M. F., Yun, J. E., & Burns, M. (2013). The development of fairness expectations and prosocial behavior in the second year of life. *Infancy, 18,* 40–66.

Southgate, V., Senju, A., & Csibra, G. (2007). Action anticipation through attribution of false beliefs in 2-year-olds. *Psychological Science, 18,* 587–592.

Sparks, E., Schinkel, M. G., & Moore, C. (2017). Affiliation affects generosity in young children: The roles of minimal group membership and shared interests. *Journal of Experimental Child Psychology, 159,* 242–262.

Surian, L., & Franchin, L. (2017). Toddlers selectively help fair agents. *Frontiers in Psychology, 8,* 944.

Tinbergen, N. (1963). On aims and methods in ethology. *Zeitschrift für Tierpsychologie, 20,* 410–433.

Trivers, R. L. (1971). The evolution of reciprocal altruism. *Quarterly Review of Biology, 46*(1), 35–57.

Vaish, A., Carpenter, M., & Tomasello, M. (2010). Young children selectively avoid helping people with harmful intentions. *Child Development, 81,* 1661–1669.

Warneken, F. (2013). From partner choice to equity–and beyond? *Behavioral and Brain Sciences, 36,* 102.

Warneken, F. (2018). How children solve the two challenges of cooperation. *Annual Review of Psychology, 69,* 205–229.

Warneken, F., & Tomasello, M. (2007). Helping and cooperation at 14 months of age. *Infancy, 11,* 271–294.

Warneken, F., & Tomasello, M. (2013). The emergence of contingent reciprocity in young children. *Journal of Experimental Child Psychology, 116,* 338–350.

Wynn, K. (2009). Constraints on natural altruism. *British Journal of Psychology, 100,* 481–485.

14 What Do We (Not) Know about the Genetics of Empathy?

Lior Abramson, Florina Uzefovsky, and Ariel Knafo-Noam

Overview

Individual differences in empathy are evolutionarily adaptive and thus are expected to relate to genetic individual variation. In this chapter we review the current literature on genetic variation related to empathy, focusing specifically on findings from twin and genetic association studies. This literature shows that empathy is related to genetic variation across individuals, but also that this relationship varies through development and is specific to different aspects of empathy (i.e., cognitive and emotional). It also shows that theoretically meaningful biological systems (i.e., systems of oxytocin, vasopressin, dopamine, and endorphins) account for the genetic variability of empathy by affecting attention, reactivity, motivation, and regulation processes directed toward social stimuli. Finally, we present some of the challenges and the directions the field is heading toward, emphasizing the need for direct comparisons between different empathy components and between genes from multiple biological systems. Such an approach should help in unfolding the complexity of empathy as part of the social mind.

Introduction

One of the most fascinating phenomena of our social mind is our ability to empathize with others, an ability that enables humans and other species to form social groups (De Waal & Preston, 2017). This importance may explain our fascination with the origins of individual differences in empathy. For example, when a British team reported a genome-wide analysis accounting for about 11 percent of the variance in empathy (Warrier et al., 2018a), it generated great media attention. Interestingly, the media responses to the exact same findings ranged from "if you're empathetic, it might be genetic" in *Smithsonian* magazine (Katz, 2018), to "empathetic people are made, not born…only 10% of the variation…was down to genes," in the *Telegraph* (Knapton, 2018). This chapter provides a scientific angle to the debate: What is empathy? And what is the role of genes in this key social phenomenon? We will review what is known (and point to what is yet to be discovered) on the genetic origins of individual differences in human empathy as examined in twin and genetic-association studies. First, we will briefly review the phenotypical differences in this multifaceted construct.

The Components and Processes Involved in Empathy

Empathy is defined here as the ability to recognize the emotions of others and to share those emotions, while maintaining other-oriented focus and self-other distinction (Uzefovsky & Knafo-Noam, 2016; Walter, 2012). Thus, empathy is, by definition, a complex construct, comprising an affective and a cognitive component, and requiring motivational and regulatory processes (Decety, 2011; Tousignant, Eugène, & Jackson, 2017). These aspects are not independent from one another. Instead, they influence each other and interact to create various empathic reactions. Importantly, individuals vary in all these elements, which may explain the broad scope of individual differences in empathy.

The *affective component of empathy* refers to our ability to share others' emotions—to experience an emotion similar that of another, even though the event that caused the emotion did not directly happen to us. It is generated mainly through bottom-up, automatic processes of affective arousal (Decety, 2011), such as emotional mimicry (e.g., synchronization of facial expressions or postures) and emotion contagion—feeling the emotion of another by mere association (Walter, 2012). These processes are supported by shared neuronal activation during emotion-observation and emotion-experience (Decety, 2011; Tousignant et al., 2017). The affective component of empathy is present very early in development. For example, newborns and infants show emotional contagion in the form of contagious crying (Geangu, Benga, Stahl, & Striano, 2010). Also, brain areas that support shared self-other neuronal representations are activated already in infancy (Tousignant et al., 2017). Although present very early, affective empathy may increase (e.g., Kienbaum, 2014) or change along development, perhaps due to influences of other empathy-related processes, as we will describe.

The *cognitive component of empathy,* conceptually related to the construct of affective theory of mind (ToM) or affective perspective taking (Uzefovsky & Knafo-Noam, 2016), refers to recognition and understanding of others' emotions. It is supported by cognitive ToM (Decety 2011) and emotional knowledge such as emotional face recognition (Kosonogov, Titova, & Vorobyeva, 2015) and understanding causal relations between events and emotions. It is affected by top-down processes and executive abilities (e.g., Vetter, Altgassen, Phillips, Mahy & Kliegel, 2013), which may increase in importance as the social situation becomes more complicated, abstract, or detailed. The brain areas related to that component support mentalizing, projecting, and planning abilities, and are activated already in childhood. However, they continue to develop, and their activation becomes more efficient with age (Decety, 2011; Tousignant et al., 2017). Emotional knowledge (e.g., better differentiation between emotions; Widen, 2013) also increases and improves during childhood, and more refined aspects of emotional intelligence continue to improve during adulthood (decreasing again in old age) (Cabello, Sorrel, Fernández-Pinto, Extremera, & Fernández-Berrocal, 2016). These behavioral findings further support the prolonged development of cognitive empathy.

Empathy does not take place toward every other person and in any situation (Decety, 2011). It is important, therefore, to consider the motivational aspects of empathy. Specifically, two motivational processes underlie empathy. First, a motivation to engage and interact with social stimuli is often needed for the individual to attend and concentrate on others. A second type of motivation refers to the desire to care and improve the other's well-being. This motivation is not necessary for empathy (although it can help by increasing attention to others' emotions). Instead, it stands in the basis of what the literature calls sympathy (Eisenberg, 2010) or empathic concern (Decety, 2011)—warm feelings and concern toward another person, without necessarily resonating with that person's emotions. Both motivations are supported by brain areas that process reward-related behavior (Decety, 2011), and are present very early in development. This is evident by studies showing that infants prefer social stimuli over nonsocial stimuli (Simion, Regolin, & Bulf, 2008) and prosocial figures over antisocial figures (Van de Vondervoort & Hamlin, 2018), suggesting intrinsic social and prosocial motivations from early on in development.

Another process required for empathy is regulation—intrinsic processes aimed at adjusting mental and physiological states adaptively to context (Nigg, 2017). Specifically, regulation is necessary for optimal levels of emotional arousal. It enables decreasing of strong vicarious emotions that may arise in empathic situations (Decety 2011; Eisenberg, 2010), and by that prevent self-distress—a negative emotion that, like empathy, is elicited by others' affective states, but unlike empathy is a self-centered emotion (Walter, 2012). On the other hand, if one's motivation is to experience higher empathy (due to prosocial, normative, or other reasons), regulation can enhance the emotional reaction to the other's mental state (Eisenberg, 2010). Regulation of empathic resonance is supported by the same, mainly anterior and prefrontal brain areas that support other forms of regulation and cognitive control (Decety & Michalska, 2010; Tousignant et al., 2017). These areas develop slowly and continue their development until adulthood (Decety & Michalska, 2010). Thus, with age empathy becomes more refined, regulated, and flexible.

To conclude, empathy includes both an affective and a cognitive component, and it depends on motivational and regulatory processes. As people vary in all these components and processes, it is important to describe the phenotypical (i.e., observed) individual differences in the components of empathy.

Phenotypical Individual Differences in the Components of Empathy

Empathy shows stable, trait-like individual differences already in infancy and early childhood (Knafo, Zahn-Waxler, Van Hulle, Robinson, & Rhee, 2008). Although it can be seen as a separate temperament dimension (Knafo & Israel, 2012), individual differences in empathy have been linked to differences in broader temperament and personality traits. Specifically, three temperament dimensions have been associated with empathic responses: emotional reactivity, regulation, and approach.

Emotional reactivity, the tendency to react with emotional intensity, is relevant mainly to affective empathy, as it should apply also when thinking about events that occur to others. Most studies have focused on negative emotional reactivity, probably because it is considered a key dimension of individual differences (Shiner et al., 2012) and because empathy is usually investigated in the context of negative events. However, there is also evidence for a relation between positive emotionality and empathy (Volbrecht, Lemery-Chalfant, Aksan, Zahn-Waxler, & Goldsmith, 2007). The relation between empathy and emotionality is complex, as the literature provides evidence for both negative and positive relations with empathy (Eisenberg, 2010). This inconsistency may be explained by the idea that different negative emotionality tendencies (e.g., fear, anger, and sadness) relate differently to empathic behaviors (Abramson, Paz, & Knafo-Noam, 2019). Moreover, the tendency to experience negative emotions may have different effects on individuals based on their regulation abilities. If individuals can modulate their emotions as needed, their negative emotionality should contribute to, or at least not impair, their empathy. In contrast, if individuals are low in regulating their negative emotions, they may show lower empathy due to being overwhelmed and prone to self-distress (Eisenberg, 2010).

Another trait that may be important to individual differences in emotional empathy is, therefore, *regulation*. Indeed, associations between regulation and empathy, as well as empathic concern and prosocial behavior, were found through most of the life span, from toddlerhood to adulthood (Eisenberg, 2010). Regulation involves aspects of executive functions (Nigg, 2017), which in turn predict emotion recognition (Vetter et al., 2013). Moreover, regulation deficits as expressed in attention-deficit and hyperactivity disorder (Nigg, 2017) were associated with difficulty to recognize emotions (Bora & Pantelis, 2016). Thus, regulation is also important to the cognitive component of empathy.

Approach tendencies are the least investigated in the context of individual differences in empathy. However, theory suggests that the continuum of approach versus withdrawal should relate to empathy through the motivation to engage and interact with social stimuli (Decety, 2011). Indeed, studies showed that infants and toddlers with a tendency to avoid novel toys exhibited less empathy toward an experimenter (e.g., Liew et al., 2011). Studies with adults (e.g., DeYoung, Quilty, Peterson, & Gray, 2014) suggest a relation between empathy and openness, a personality trait that is preceded by childhood approach behaviors like curiosity and active explorations (Caspi & Shiner, 2007).

While investigating the associations between empathy and temperament traits reveal phenotypical differences in individuals' empathy, genetic investigations uncover the sources that create these individual differences.

We will now turn to reviewing the literature on the genetic underpinning of human empathy.

Genetic Underpinnings of Individual Differences in Empathy

Empathy has ancient evolutionary roots (De Waal & Preston, 2017). However, evolutionarily speaking, individual differences in empathy should also be present because empathy has both advantages and disadvantages for survival fitness, depending on the individual's environment (Nettle, 2006). From this perspective, variability in empathy should stem, to some extent, from genetic variation between individuals. This genetic variation has been primarily investigated from two approaches: twin studies and gene-associations studies.

Heritability of Empathy as Examined in Twin Studies

One of the most common ways to estimate genetic influences on a trait is by examining monozygotic and dizygotic twins who were reared together. In short, the resemblance of twins in a specific trait is compared between monozygotic twins, who share almost 100 percent of their genes with each other, and between dizygotic twins, who share on average 50 percent of the genetic variation. The twin model assumes that monozygotic twins share environmental experiences to the same degree as dizygotic twins. Thus, if monozygotic twins resemble each other more than dizygotic twins, the difference in resemblance is attributed to genetic influences. If, however, monozygotic twins resemble each other to the same extent as dizygotic twins, one could say that shared environmental influences make them behave similarly. The extent to which twins behave differently from one another, despite their shared genetic and environmental background, is attributed to nonshared environmental experiences and to measurement error.

Twin research has shown that empathy is moderately heritable. In a meta-analysis of eight studies, Knafo and Uzefovsky (2013) found that 35 percent of the variance in empathy was accounted for by genetic factors. Interestingly, genetic factors were more evident as participants' age increased, indicating that genetic influences on empathy increase with age, similar to the patterns found for other traits such as intelligence (Plomin, DeFries, Knopik, & Neiderhiser, 2016) and prosocial behavior (Knafo & Plomin, 2006).

Several processes may account for the age-increase in genetic influences. One process is *innovation*—the appearance of new genetic influences that were not evident before due to changes in brain structures or functions (Plomin et al., 2016). For example, as regulatory abilities develop during childhood, they become more relevant to the expression of empathy. Thus, as individuals grow up, genetic influences on empathy regulation may be more evident (Knafo & Uzefovsky, 2013; Melchers, Montag, Reuter, Spinath, & Hahn, 2016).

A second relevant process could be *genetic amplification* or *gene-environment correlations* (Plomin et al., 2016). As children grow up, they select, modify, and create their own environment, which suits more of their initial genetic tendencies. This environment, in turn, affects the developing individuals by amplifying their initial tendencies, including the expression of the genes at their basis. A third process is that the environment itself changes with age (Knafo & Plomin, 2006). The social situations that individuals encounter become more complex as they grow up (e.g., larger peer groups), which in turn change the relevance of genetic differences to children's empathic reactions. This process can be termed *gene-environment relevance*. An important conclusion is, therefore, that genetic and environmental effects both change with age. Thus, although (as this chapter's opening attests) genetic effects are often interpreted as suggesting that individuals are "born" with empathy, both genetic and environmental effects contribute to development of empathy as a dynamic process.

Although it is tempting to think only of these theoretical reasons, a possible methodological explanation also may account for the finding of an age increase in the heritability of empathy. Studies on children have used mainly observational methods to assess empathy, such as the child's reactions to a crying experimenter (e.g., Knafo et al., 2008). Observational methods capture one situation, which is affected not only by the individual's traits but also by the specific context in which the situation occurs and which may be attributed to shared or nonshared environmental influences. Studies on adults, on the other hand, have used mainly self-reports to assess empathy, attempting to capture an individual's average behavior beyond a specific situation. Thus, differences in heritability estimations may be attributed to the methodological differences between age-varying samples.

In line with that possibility, a study that examined maternal reports on seven-year-old twins (Knafo-Noam, Uzefovsky, Israel, Davidov, &

Zahn-Waxler, 2015) found that empathic concern showed high heritability estimates (76 percent). This suggests that, when using questionnaires, reports on children's emotional and motivational aspects of empathy are also highly heritable. The only way to disentangle the methodological and theoretical explanations for the age increase in the heritability of empathy is by performing longitudinal studies, or studies that use the same method with participants of different ages. For example, Knafo et al. (2008) examined twin toddlers who were observed in empathic responding at four time points, from the ages of fourteen to thirty-six months. They showed that, indeed, genetic influences increased and shared-environmental effects decreased with age. More studies with broader age ranges and multiple methods are required for a better understanding of this effect.

As we claim throughout this chapter, to understand the mechanisms underlying empathy it is important to consider its different components. Indeed, Knafo and Uzefovsky (2013) found in their meta-analysis that 30 percent of the variation in emotional empathy was due to genetic factors, with no evidence for shared-environmental influences. By contrast, individual differences in cognitive empathy were accounted for by both genetic (26 percent of the variance) and shared environmental (17 percent of the variance) effects. Since the meta-analysis was published, additional studies found patterns that support this conclusion (although for different results, which also differed by participants' gender, see Toccaceli et al., 2018). Melchers et al. (2016) found that self-reported aspects of emotional empathy (empathic concern and personal distress) exhibited 54 percent to 57 percent heritability. By contrast, self-reported perspective-taking, reflecting cognitive empathy, showed weaker heritability (27 percent of the variance) and a small shared environmental effect (9 percent).

In the study of Melchers and colleagues (2016), performance in the "Reading the Mind in the Eyes" test, a task that examines emotion recognition from the eyes' expressions, reflecting a certain aspect of cognitive empathy (Baron-Cohen, Wheelwright, Hill, Raste, & Plumb, 2001), also showed a relatively weaker genetic component (30 percent of the variance). Another study that used the same test (Warrier et al., 2018b) found similar estimations of genetic effects (28 percent of the variance). The best fitting model included only genetic and non-shared environmental influences. However, the next best model was a model of shared and nonshared environment, which allocated 22 percent of the variance to shared environmental effects.

As a third example, in a study on seven-year-old twins, facial and vocal emotion recognition (aspects of cognitive empathy) were explained mainly by the shared environment (44 percent of the variance) (Schapira, Elfenbein, Amichay-Setter, Zahn-Waxler, & Knafo-Noam, 2019).

A possible reason for the difference in emotional and cognitive empathy heritability estimates is that, compared with cognitive empathy, emotional empathy may be more strongly associated with the temperament traits of emotionality, regulation, and approach, which are highly heritable (Shiner et al., 2012). By contrast, cognitive empathy may rely more on learning experiences (Melchers et al., 2016) and be affected by socioemotional and cultural nuances. As noted, emotional empathy develops substantially in infancy, whereas cognitive empathy shows longer development. The development of cognitive empathy in parallel to increasing acquisition of social experiences allows for more input from the environment and fits with the notion that it is more influenced by it.

Further evidence that genes play a role in empathy comes from studies on the low end of the empathy spectrum, namely callous unemotional traits— defined as lack of empathy, lack of guilt, and superficial affect (Henry et al., 2018). Callous unemotional traits are especially linked with low emotional empathy (Jones, Happé, Gilbert, Burnett, & Viding, 2010). In a longitudinal study, Henry and colleagues (2018) investigated twin children's callous unemotional traits longitudinally, at four time points from seven to twelve years of age. Heritability estimates in the four time points ranged from 41 percent to 61 percent. Interestingly, the authors found that genetic factors contributed to stability in callous unemotional traits, but also that new genetic factors, which had not been present at the earliest age, emerged and contributed to changes in callous unemotional traits along development. Importantly, genetic influences on callous unemotional traits of adolescent twins overlap with genetic influences on agreeableness (Mann, Briley, Tucker-drob, & Harden, 2015), a broad personality trait that includes also empathic concern (Caspi & Shiner, 2007). This finding suggests that conclusions regarding the genetics of callous unemotional traits may apply also to genetic variation in the emotional and motivational aspects of empathy in the normal population.

Genetics of Empathy as Examined in Genetic Association Studies

Previously we reviewed twin-design studies on the heritability of empathy. Studies that directly examined people's genes also have suggested
that empathy has a genetic component. Most candidate gene studies have
focused on genes that regulate the activity of the neuropeptide and hormone oxytocin, which is related to many interpersonal behaviors such
as trust, generosity, and cooperation (for a review, see Bartz, Zaki, Bolger &
Ochsner, 2011). Three main explanations were proposed regarding the
mechanism through which oxytocin influences social behaviors (Bartz et
al., 2011). The first is social anxiety reduction, according to which oxytocin
modulates stress-related brain (e.g., the amygdala) and hormonal systems
(e.g., cortisol release), which in turn increase prosocial motivations. The
second is an increase in social salience, according to which oxytocin alters
the perceptual sensitivity to social cues. The third is that oxytocin increases
affiliative motivation, for example by operating on reward-dependent
dopaminergic activation (Chang et al., 2014).

Regardless of the specific mechanism through which oxytocin operates,
research suggests that variations in oxytocin genes play a role in individual
differences in empathy. Most studies focused on the oxytocin receptor gene
(*OXTR*), as oxytocin is broadly distributed in the brain and exerts its effects
through this receptor (Jurek and Neumann, 2018). The genetic sequence
of the receptor includes several single nucleotide polymorphisms (SNPs;
common changes in a single letter of the gene sequence) that have been
investigated in relation to empathy.

For example, in a recent meta-analysis on thirteen studies, Gong and colleagues (2017) showed that adults' empathy was related to the rs53576 SNP
in the *OXTR*. As the meta-analysis examined both cognitive and emotional
empathy together, it could not compare the role of oxytocin in specific
emotional and cognitive empathy components. Such a comparison may
not only inform us about the etiology of cognitive and emotional empathy, but also uncover information about the mechanism through which
oxytocin operates. For example, a stronger relation with cognitive empathy
may support the social saliency mechanism, whereas a stronger relation
with emotional empathy may support the mechanisms of affiliative motivation and/or anxiety reduction. There is evidence that oxytocinergic genes
relate to cognitive (Rodrigues, Saslow, Garcia, John, & Keltner, 2009) and
emotional (Uzefovsky et al., 2015) empathy, but genetic studies with larger

sample sizes that directly compare between these components are needed to reach firmer conclusions.

Another set of genes that may account for variability in empathy are those that regulate the activity of vasopressin, a neuropeptide and hormone that plays a role in the social and affiliative behaviors of many species (Uzefovsky, Shalev, Israel, Knafo, & Ebstein, 2012). Research has suggested that vasopressin relates to increased aggression as well as to autonomic responses and attention to negative social stimuli (Skuse & Gallagher, 2009), especially in men who respond to other men (Thompson, George, Walton, Orr, & Benson, 2006; Uzefovsky et al., 2012). Interestingly, vasopressin in women was found to have an opposite effect: it increases friendliness ratings of other women (Thompson et al., 2006). The authors concluded that vasopressin increases attention and biases perception of threatening stimuli. In men it activates a "fight or flight approach" that makes them view other men as more threatening; in women it activates a "tend and befriend" approach (Taylor et al., 2000) that makes them view other women as friendlier. This pattern fits the idea that vasopressin operates through altering attentional sociocognitive processes, which might suggest that it explains individual differences in cognitive empathy more than emotional empathy. Indeed, variations in the gene *AVPR1a*, which modulates vasopressin activation, explained individual differences in cognitive but not emotional empathy (Pearce, Wlodarski, Machin, & Dunbar, 2017; Uzefovsky et al., 2015). However, another study found the opposite pattern (Wu, Shang, & Su, 2015). Obviously, more research is needed to understand the role of vasopressin in cognitive and emotional empathy.

A third group of genes that may contribute to empathy are those that regulate the activity of dopaminergic systems. Dopamine is a pervasive neurotransmitter in the brain, responsible for the activity of various systems. Probably most relevant to empathy is the mesolimbic dopaminergic pathway, which sends dopaminergic input from the ventral tegmental area to the nucleus accumbens and plays a crucial role in reward-dependent behaviors (Decety, 2011; Gregory, Cheng, Rupp, Sengelaub, & Heiman, 2015). This system's activation increases in response to viewing salient social stimuli (e.g., an infants' cry; Gregory et al., 2015). Moreover, individual differences in dopaminergic activation in the nucleus accumbens in response to one's own infant explain individual differences in maternal behavior (Atzil et al., 2017).

These findings suggest that individual differences in the mesolimbic system should relate to variations in empathy through the motivation to engage with other people. Apart from the mesolimbic system, dopaminergic activity in the prefrontal cortex should also relate to social behaviors and to empathy in particular (Skuse & Gallagher, 2008). Indeed, a few candidate gene studies have shown relations between dopaminergic genes and self-reported empathic concern (Pearce et al., 2017; Ru et al., 2017). In addition, two studies found relations between the dopamine receptor gene D4 and cognitive empathy (Uzefovsky et al., 2014; Ben-Israel, Uzefovsky, Ebstein, & Knafo-Noam, 2015). However, in these two studies the effects were moderated in opposite directions by the participants' sex, which highlights the need of further research.

Most studies have examined the oxytocin, vasopressin, and dopamine systems separately. However, these systems operate simultaneously, interact, and influence each other (Atzil et al., 2017; Chang et al., 2014). Thus, a full understanding of the relation between empathy and genetic variations in these systems would be accomplished only if they are investigated together (Machin & Dunbar, 2011). Pearce et al. (2017) examined the relationship between what they called "disposition sociality"—which included mostly empathy measures (i.e., Reading the Mind in the Eyes test, self-reports on empathy, but also self-reports on attachment styles)—to SNPs that regulate the activation of receptors for these three systems and other important systems. Importantly, because they examined the relations of all these SNPs at once, they were able to examine the relative contribution of each SNP group, controlling for other SNPs. The authors found that dopamine and vasopressin SNPs accounted for part of the variance in empathy, whereas oxytocin SNPs were less related to dispositional empathy and more related to specific measures of dyadic relationships. This finding may indicate that the influence of oxytocin on empathy is context specific, or that it is mediated by other biological systems. For example, as oxytocin amplifies the activity of striatal dopamine (Chang et al., 2014), variations in oxytocinergic genes influence dopaminergic activity. Once dopaminergic or other genes are considered, the influence of the oxytocin system, although an important part of the biological process, might be less pronounced in behavioral differences.

Interestingly, in Pearce and colleagues' study (2017) the SNPs that explained variation in empathy the most were those from the endorphin system, which were relatively less examined in the context of empathy.

Endorphins play a crucial role in reactivity and regulation of physical and emotional pain, as well as of social stimuli (e.g., touch and synchrony). They are found in high density in the mesolimbic system, and they contribute to reward-dependent behavior (Machin & Dunbar, 2011). Thus, they may contribute to empathy by increasing motivation and reactivity to social interactions and by regulating an individual's emotions and decreasing self-distress. Indeed, a recent study showed that blocking the activation of the opioid system (which includes endorphins) decreased the participants' responsivity to both firsthand pain and empathy to others' pain (Rütgen et al., 2018).

The advantage of candidate gene studies is that they are theory driven, are often based on animal models, and thus provide meaningful explanations of the mechanisms behind behavioral differences. At the same time, a main criticism of these studies, as was demonstrated by the findings of Pearce and colleagues (2017), is that they ignore many genes that were not considered for methodological or historical reasons (e.g., the endorphin system) and that may have stronger associations with psychological traits (Psychiatric GWAS Consortium Coordinating Committee, 2009). Also, these studies may lack the power to detect meaningful, though small, associations. In recent years, large-scale genome-wide association studies (GWAS) have been increasingly used to overcome these problems. These studies examine, in very large samples, hundreds of thousands of SNPs simultaneously that represent most of human common alleles, thus identifying relevant genes for behavioral variations in a more robust manner (Psychiatric GWAS Consortium Coordinating Committee, 2009).

Using this method, two studies found evidence that empathy is heritable. Warrier et al. (2018a) found that around 11 percent of the variance in self-reported empathy was accounted for by variations in people's SNPs. This estimation is lower than the estimation found in Knafo and Uzefovsky's meta-analysis (2013) on twin studies, probably because the GWAS method, which examines only SNP variability, does not capture some of the genetic variance captured by twin studies. In another study with an overlapping sample (Warrier et al., 2018b), the participants' SNPs variation explained around 6 percent of the variance in the Reading the Mind in the Eyes test. Age variability, which was very high, was controlled for in the analysis. The authors suggested that since heritability of empathy increases with age (Knafo & Uzefovsky, 2013), controlling for age probably resulted in an underestimation of empathy heritability. Notably, the GWAS method does not rely on the twin design assumption of similarly equal environments for monozygotic and dizygotic twins. Thus, it provides an independent replication of the role of genes in individual differences in empathy.

Interplay with the Environment

This chapter has focused on the "main effects" of genes, but a key pathway for future research is to investigate the interplay of environmental and genetic factors in the development of empathy. A handful of studies have addressed gene–environment interactions, showing, for example, that parenting interacted with the dopamine receptor D4 gene in predicting children's empathy (Knafo & Uzefovsky, 2013), or that childhood maltreatment moderated the *OXTR* SNP rs53576 influence on emotional empathy (Flasbeck, Moser, Kumsta, & Brüne, 2018). Also important are gene–environment correlations, and specifically the evocative gene–environment correlations in which an individual's genetically affected behavior elicits an environmental reaction that can then affect the developing person. For example, a common finding linking sensitive parenting to child empathy (Stern & Cassidy, 2018) could reflect a process in which an empathic child elicits more sensitive parenting.

Future Directions: Genetic Relations between Empathy and Other Phenotypes

A better understanding of the genetics of empathy may be accomplished by examining its genetic overlap with other phenotypical traits (e.g., temperament traits). Identifying a genetic overlap between traits, for example, by multivariate genetic studies, indicates that psychologically different traits have nevertheless common genetic architecture (Plomin et al., 2016). In the context of empathy, identifying which phenotypical traits genetically overlap with the different components of empathy can help identify the processes—emotional, cognitive, motivational, or regulatory—that account for genetic differences.

Going from the psychological to the physiological level, because both temperament (Shiner et al., 2012) and empathy are partially heritable, we propose that some genetic factors influence both phenomena by operating on the same neurophysiological and hormonal systems. For instance, genetic factors that increase physiological arousal were proposed to increase both tendencies for high emotionality and tendencies for affective empathy (Davis, Luce, & Kraus, 1994). However, multivariate genetic analysis providing empirical evidence for this process is still lacking.

Evidence from genetic association studies has suggested a possible genetic overlap also between empathy and approach, perhaps through the dopamine or endorphin systems. For example, in the study by Pearce et al. (2017), dopamine and endorphin were related not only to empathy but also to broader aspects of social approach and affiliation—namely, quality of dyadic relations and size of social network. The authors suggested that because these systems influence reward-related behavior and because social stimuli are intrinsically rewarding, genetic variations in these systems regulate a wide range of social behaviors (including empathy and approach). Using the GWAS method, Warrier and colleagues (2018b) found a genetic overlap between performance in the Reading the Mind in the Eyes test and the personality trait of openness—which, as mentioned, reflects aspects of approach behavior (Caspi & Shiner, 2007). Further, they found a marginal genetic overlap between performance on the test and brain dopaminergic circuits activation, although not from the reward mesolimbic system.

Taken together, the theories and preliminary evidence suggest that empathy may have a shared genetic background with the traits emotional reactivity and approach, and future twin studies could examine this hypothesis directly.

Conclusions

People show meaningful variation in empathic responses to others' emotions, which in turn influences many aspects of their interpersonal interactions and well-being (Stern & Cassidy, 2018). This variation is grounded in the evolutionary origins of our species (Nettle, 2006) and thus should manifest in genetic variation across people. In this chapter, we presented the most recent findings regarding phenotypic and genetic individual differences in empathy. The literature shows that different aspects of empathy are influenced by individuals' genetics and that theoretically meaningful biological systems explain this genetic variability.

We have also stressed the aspects that must be further investigated. Specifically, to truly understand the genetics of empathy, researchers must advance into simultaneous examination of different empathy components, while controlling for genes from multiple biological systems. In addition, examination of the genetic overlap between empathy and other relevant phenotypical traits (e.g., emotional reactivity, regulation, and approach) may further shed light on the etiology of empathy question.

By understanding the genetic variations of empathy, researchers could advance intervention programs for clinical disorders characterized by empathy deficits. Such knowledge would help in finding the most effective drugs that operate on the most relevant biological systems (Pearce et al., 2017) and in tailoring specific programs according to the individual's genetic and biological baseline (Bartz et al., 2011). Although we did not refer to this aspect enough in this chapter, numerous environmental factors have a crucial role in shaping empathy, both directly and through moderating genetic influences (Knafo & Uzefovsky, 2013). Future research that provides a strong theoretical understanding of the genetic basis of empathy, combined with characterization of the environmental factors that influence it, could tell the story of empathy as an important part of the social mind.

References

Abramson, L., Paz, Y., & Knafo-Noam, A. (2019). From negative reactivity to empathic responding: Infants high in negative reactivity express more empathy later in development, with the help of regulation. *Developmental Science, 22*(3), e12766. doi: 10.1111/desc.12766

Atzil, S., Touroutoglou, A., Rudy, T., Salcedo, S., Feldman, R., Hooker, J. M.,…Barrett, L. F. (2017). Dopamine in the medial amygdala network mediates human bonding. *Proceedings of the National Academy of Sciences, 114*(9), 2361–2366. doi: 10.1073/pnas.1612233114

Baron-Cohen, S., Wheelwright, S., Hill, J., Raste, Y., & Plumb, I. The 'Reading the Mind in the Eyes' test revised version: A study with normal adults, and adults with Asperger syndrome or high-functioning autism. *Journal of Child Psychology and Psychiatry, 42*, 241–251.

Bartz, J. A., Zaki, J., Bolger, N., & Ochsner, K. N. (2011). Social effects of oxytocin in humans: Context and person matter. *Trends in Cognitive Sciences, 15*(7), 301–309. doi: 10.1016/j.tics.2011.05.002

Ben-Israel, S., Uzefovsky, F., Ebstein, R. P., & Knafo-Noam, A. (2015). Dopamine D4 receptor polymorphism and sex interact to predict children's affective knowledge. *Frontiers in Psychology, 6*, 1–12. doi: 10.3389/fpsyg.2015.00846

Bora, E., & Pantelis, C. (2016). Meta-analysis of social cognition in attention-deficit/hyperactivity disorder (ADHD): Comparison with healthy controls and autistic spectrum disorder. *Psychological Medicine, 46*(4), 699–716. doi: 10.1017/S0033291715002573

Cabello, R., Sorrel, M. A., Fernández-Pinto, I., Extremera, N., & Fernández-Berrocal, P. (2016). Age and gender differences in ability emotional intelligence in adults: A cross-sectional study. *Developmental Psychology, 52*(9), 1486–1492. doi: 10.1037/dev0000191

Caspi, A., & Shiner, R. L. (2007). Personality development. In W. Damon, & R. Lerner (Series Eds.) & N. Eisenberg (Vol. Eds.), *Handbook of child psychology: Vol.3. Social, emotional, and personality development* (6th ed.) (pp. 300-365). New York: Wiley.

Chang, W. H., Lee, I. H., Chen, K. C., Chi, M. H., Chiu, N. T., Yao, W. J.,...Chen, P. S. (2014). Oxytocin receptor gene rs53576 polymorphism modulates oxytocin-dopamine interaction and neuroticism traits-A SPECT study. *Psychoneuroendocrinology, 47*, 212–220. doi: 10.1016/j.psyneuen.2014.05.020

Davis, M. H., Luce, C. & Kraus, S. J. (1994), The heritability of characteristics associated with dispositional empathy. *Journal of Personality, 62*, 369–391. doi: 10.1111/j.1467-6494.1994.tb00302.x

Decety, J. (2011). The neuroevolution of empathy. *Annals of the New York Academy of Sciences of the United States of America, 1231*(1), 35–45. doi: 10.1111/j.1749-6632.2011.06027.x

Decety, J., & Michalska, K. J. (2010). Neurodevelopmental changes in the circuits underlying empathy and sympathy from childhood to adulthood. *Developmental Science, 13*(6), 886–899. doi: 10.1111/j.1467-7687.2009.00940.x

De Waal, F. B. M., & Preston, S. D. (2017). Mammalian empathy: Behavioural manifestations and neural basis. *Nature Reviews Neuroscience, 18*(8), 498–509. doi: 10.1038/nrn.2017.72

DeYoung, C. G., Quilty, L. C., Peterson, J. B., & Gray, J. R. (2014). Openness to experience, intellect, and cognitive ability. *Journal of Personality Assessment, 96*(1), 46–52. doi: 10.1080/00223891.2013.806327

Eisenberg, N. (2010). Empathy-related responding: Links with self-regulation, moral judgment, and moral behavior. In M. Mikulincer and P. R. Shaver (Eds.), *Prosocial motives, emotions, and behavior: The better angels of our nature* (pp. 129–148). Washington, DC: American Psychological Association.

Flasbeck, V., Moser, D., Kumsta, R., & Brüne, M. (2018). The OXTR single-nucleotide polymorphism rs53576 moderates the impact of childhood maltreatment on empathy for social pain in female participants: Evidence for differential susceptibility. *Frontiers in Psychiatry, 9*, 1–9. doi: 10.3389/fpsyt.2018.00359

Geangu, E., Benga, O., Stahl, D., & Striano, T. (2010). Contagious crying beyond the first days of life. *Infant Behavior and Development, 33*(3), 279–288. doi: 10.1016/j .infbeh.2010.03.004

Gong, P., Fan, H., Liu, J., Yang, X., Zhang, K., & Zhou, X. (2017). Revisiting the impact of OXTR rs53576 on empathy: A population-based study and a meta-analysis. *Psychoneuroendocrinology, 80,* 131–136. doi: 10.1016/j.psyneuen.2017.03.005

Gregory, R., Cheng, H., Rupp, H. A., Sengelaub, D. R., & Heiman, J. R. (2015). Oxytocin increases VTA activation to infant and sexual stimuli in nulliparous and postpartum women. *Hormones and Behavior, 69,* 82–88. doi: 10.1016/j.yhbeh.2014.12.009

Henry, J., Dionne, G., Viding, E., Petitclerc, A., Feng, B., Vitaro, F.,…Boivin, M. (2018). A longitudinal twin study of callous-unemotional traits during childhood. *Journal of Abnormal Psychology, 127*(4), 374–384. doi: 10.1037/abn0000349

Jones, A. P., Happé, F. G. E., Gilbert, F., Burnett, S., & Viding, E. (2010). Feeling, caring, knowing: Different types of empathy deficit in boys with psychopathic tendencies and autism spectrum disorder. *Journal of Child Psychology and Psychiatry, and Allied Disciplines, 51*(11), 1188–1197. doi: 10.1111/j.1469-7610.2010.02280.x

Jurek, B., & Neumann, I. D. (2018). The oxytocin receptor: From intracellular signaling to behavior. *Physiological Reviews, 98*(3), 1805–1908. doi: 10.1152/physrev.00031.2017

Katz, B. (2018, March 13). If you're empathetic, it might be genetic. *Smithsonian .com.* https://www.smithsonianmag.com/smart-news/if-youre-empathetic-it-might -be-genetic-180968466/

Kienbaum, J. (2014). The development of sympathy from 5 to 7 years: Increase, decline, or stability? A longitudinal study. *Frontiers in Psychology, 5,* 1–10. doi: 10.3389/ fpsyg.2014.00468

Knafo, A. & Israel, S. (2012). Empathy, prosociality, and other aspects of kindness. In M. Zentner & R. Shiner, (Eds.). *The handbook of temperament: Theory and research* (pp. 168–179). New York: Guilford Press.

Knafo, A., & Plomin, R. (2006). Prosocial behavior from early to middle childhood: Genetic and environmental influences on stability and change. *Developmental Psychology, 42*(5), 771–786. doi: 10.1037/0012-1649.42.5.771

Knafo, A., & Uzefovsky, F. (2013). Variation in empathy: The interplay of genetic and environmental factors. In M. Legerstee, D. W. Haley, & M. H. Bornstein (Eds.), *The infant mind: Origins of the social brain* (pp. 97–120). New York: Guilford Press.

Knafo, A., Zahn-Waxler, C., Van Hulle, C., Robinson, J. L., & Rhee, S. H. (2008). The developmental origins of a disposition toward empathy: Genetic and environmental contributions. *Emotion, 8*(6), 737–752. doi: 10.1037/a0014179

Knafo-Noam, A., Uzefovsky, F., Israel, S., Davidov, M., & Zahn-Waxler, C. (2015). The prosocial personality and its facets: Genetic and environmental architecture of mother-reported behavior of 7-year-old twins. *Frontiers in Psychology, 6,* 1–9. doi: 10.3389/fpsyg.2015.00112

Knapton, S. (2018, March 12). Empathetic people are made, not born, new research suggests. *The Telegraph.* https://www.telegraph.co.uk/science/2018/03/12/empathetic -people-made-not-born-new-research-suggests/

Kosonogov, V., Titova, A., & Vorobyeva, E. (2015). Empathy, but not mimicry restriction, influences the recognition of change in emotional facial expressions. *Quarterly Journal of Experimental Psychology, 68*(10), 2106–2115. doi: 10.1080/17470218.2015 .1009476

Liew, J., Eisenberg, N., Spinrad, T. L., Eggum, N. D., Haugen, R. G., Kupfer, A., …Baham, M. E. (2011). Physiological regulation and fearfulness as predictors of young children's empathy-related reactions. *Social Development, 20*(1), 111–134. doi: 10.1111/j.1467-9507.2010.00575.x

Machin, A. J., & Dunbar, R. I. M. (2011). The brain opioid theory of social attachment: A review of the evidence. *Behaviour, 148*(9–10), 985–1025. doi: 10.1163 /000579511X596624

Mann, F. D., Briley, D. A., Tucker-drob, E. M., & Harden, K. P. (2015). A behavioral genetic analysis of callous-unemotional traits and big five personality in adolescence. *Journal of Abnormal Psychology, 124*(4), 982–993. doi: 10.1037/abn0000099

Melchers, M., Montag, C., Reuter, M., Spinath, F. M., & Hahn, E. (2016). How heritable is empathy? Differential effects of measurement and subcomponents. *Motivation and Emotion, 40*(5), 720–730. doi: 10.1007/s11031-016-9573-7

Nettle, D. (2006). The evolution of personality variation in humans and other animals. *American Psychologist, 61*(6), 622–631. doi: 10.1037/0003-066X.61.6.622

Nigg, J. T. (2017). Annual research review: On the relations among self-regulation, self-control, executive functioning, effortful control, cognitive control, impulsivity, risk-taking, and inhibition for developmental psychopathology. *Journal of Child Psychology and Psychiatry and Allied Disciplines, 58*(4), 361–383. doi: 10.1111/jcpp.12675

Pearce, E., Wlodarski, R., Machin, A., & Dunbar, R. I. M. (2017). Variation in the β-endorphin, oxytocin, and dopamine receptor genes is associated with different dimensions of human sociality. *Proceedings of the National Academy of Sciences of the United States of America, 114*(24), E4898–E4898. doi: 10.1073/pnas.1708178114

Plomin, R., DeFries, J. C., Knopik, V. S., & Neiderhiser, J. M. (2016). Top 10 replicated findings from behavioral genetics. *Perspectives on Psychological Science, 11*(1), 3–23. doi: 10.1177/1745691615617439

Psychiatric GWAS Consortium Coordinating Committee (2009). Genomewide association studies: History, rationale, and prospects for psychiatric disorders. *American
Journal of Psychiatry, 166*(5), 540–556. doi: 10.1176/appi.ajp.2008.08091354

Rodrigues, S. M., Saslow, L. R., Garcia, N., John, O. P., & Keltner, D. (2009). Oxytocin receptor genetic variation relates to empathy and stress reactivity in humans.
Proceedings of the National Academy of Sciences of the United States of America, 106(50),
21437–21441. doi: 10.1073/pnas.0909579106

Ru, W., Fang, P., Wang, B., Yang, X., Zhu, X., Xue, M., … Gong, P. (2017). The impacts
of Val158Met in Catechol-O-methyltransferase (COMT) gene on moral permissibility
and empathic concern. *Personality and Individual Differences, 106,* 52–56. doi: 10.1016/j
.paid.2016.10.041

Rütgen, M., Seidel, E. M., Pletti, C., Riečanský, I., Gartus, A., Eisenegger, C., & Lamm,
C. (2018). Psychopharmacological modulation of event-related potentials suggests
that first-hand pain and empathy for pain rely on similar opioidergic processes. *Neuropsychologia, 116,* 5–14. doi: 10.1016/j.neuropsychologia.2017.04.023

Schapira, R., Elfenbein, H. A., Amichay-Setter, M., Zahn-Waxler, C., & Knafo-Noam,
A. (2019). Shared environment effects on children's emotion recognition. *Frontiers
in Psychiatry, 10,* 1–6. doi:10.3389/fpsyt.2019.00215

Shiner, R. L., Buss, K. A., Mcclowry, S. G., Putnam, S. P., Saudino, K. J., & Zentner, M.
(2012). What is temperament now? Assessing progress in temperament research on
the twenty-fifth anniversary of Goldsmith et al. (1987). *Child Development Perspectives, 6*(4), 436–444. doi: 10.1111/j.1750-8606.2012.00254.x

Simion, F., Regolin, L., & Bulf, H. (2008). A predisposition for biological motion in
the newborn baby. *Proceedings of the National Academy of Sciences of the United States
of America, 105*(2), 809–813. doi: 10.1073/pnas.0707021105

Skuse, D. H., & Gallagher, L. (2009). Dopaminergic-neuropeptide interactions in the
social brain. *Trends in Cognitive Sciences, 13*(1), 27–35. doi: 10.1016/j.tics.2008.09.007

Stern, J. A., & Cassidy, J. (2018). Empathy from infancy to adolescence: An attachment perspective on the development of individual differences. *Developmental
Review, 47,* 1–22. doi: 10.1016/j.dr.2017.09.002

Taylor, S. E., Klein, L. C., Lewis, B. P., Gruenewald, T. L., Gurung, R. A. R., &
Updegraff, J. A. (2000). Biobehavioral responses to stress in females: Tend-and-
befriend, not fight-or-flight. *Psychological Review, 107*(3), 411–429. doi: 10.1037//
0033-295X.107.3.411

Thompson, R. R., George, K., Walton, J. C., Orr, S. P., & Benson, J. (2006). Sex-
specific influences of vasopressin on human social communication. *Proceedings of
the National Academy of Sciences of the United States of America, 103*(20), 7889–7894.
doi: 10.1073/pnas.0600406103

Toccaceli, V., Fagnani, C., Eisenberg, N., Alessandri, G., Vitale, A., & Stazi, M. A. (2018). Adult empathy: Possible gender differences in gene-environment architecture for cognitive and emotional components in a large Italian twin sample. *Twin Research and Human Genetics, 21*(3), 214–226. doi: 10.1017/thg.2018.19

Tousignant, B. Eugène, F., & Jackson, P. L. (2017). A developmental perspective on the neural bases of human empathy. *Infant Behavior and Development, 48*, 5–12. doi: 10.1016/j.infbeh.2015.11.006

Uzefovsky, F., & Knafo-Noam, A. (2016). Empathy development throughout the lifespan. In J. A. Sommerville & J. Decety (Eds.), *Social cognition: Frontiers in developmental science series* (pp. 71–97). New York: Psychology Press.

Uzefovsky, F., Shalev, I., Israel, S., Edelman, S., Raz, Y., Mankuta, D., ... Ebstein, R. P. (2015). Oxytocin receptor and vasopressin receptor 1a genes are respectively associated with emotional and cognitive empathy. *Hormones and Behavior, 67*, 60–65. doi: 10.1016/j.yhbeh.2014.11.007

Uzefovsky, F., Shalev, I., Israel, S., Edelman, S., Raz, Y., Perach-Barzilay, N., ... Ebstein, R. P. (2014). The dopamine D4 receptor gene shows a gender-sensitive association with cognitive empathy: Evidence from two independent samples. *Emotion, 14*(4), 712–721. doi: 10.1037/a0036555

Uzefovsky, F., Shalev, I., Israel, S., Knafo, A., & Ebstein, R. P. (2012). Vasopressin selectively impairs emotion recognition in men. *Psychoneuroendocrinology, 37*(4), 576–580. doi: 10.1016/j.psyneuen.2011.07.018

Van de Vondervoort, J. W., & Hamlin, J. K. (2018). The early emergence of sociomoral evaluation: Infants prefer prosocial others. *Current Opinion in Psychology, 20*, 77–81. doi: 10.1016/j.copsyc.2017.08.014

Vetter, N. C., Altgassen, M., Phillips, L., Mahy, C. E. V, & Kliegel, M. (2013). Development of affective theory of mind across adolescence: Disentangling the role of executive functions. *Developmental Neuropsychology, 38*(2), 114–25. doi: 10.1080/87565641.2012.733786

Volbrecht, M. M., Lemery-Chalfant, K., Aksan, N., Zahn-Waxler, C., & Goldsmith, H. H. (2007). Examining the familial link between positive affect and empathy development in the second year. *Journal of Genetic Psychology, 168*(2), 105–130. doi: 10.3200/GNTP.168.2.105-130

Walter, H. (2012). Social cognitive neuroscience of empathy: Concepts, circuits, and genes. *Emotion Review, 4*(1), 9–17. doi: 10.1177/1754073911421379

Warrier, V., Grasby, K. L., Uzefovsky, F., Toro, R., Smith, P., Chakrabarti, B., ... Baron-Cohen, S. (2018a). Genome-wide meta-analysis of cognitive empathy: Heritability, and correlates with sex, neuropsychiatric conditions and cognition. *Molecular Psychiatry, 23*(6), 1402–1409. doi: 10.1038/mp.2017.122

Warrier, V., Toro, R., Chakrabarti, B., iPSYCH-Broad Autism Group, Børglum, A. D., Grove, J....Baron-Cohen, S. (2018b). Genome-wide analyses of self-reported empathy: Correlations with autism, schizophrenia, and anorexia nervosa. *Translational Psychiatry, 8*(1), 1-10. doi: 10.1038/s41398-017-0082-6

Widen, S. C. (2013). Children' s interpretation of facial expressions: The long path from valence-based to specific discrete categories. *Emotion Review, 5*(1), 72–77. doi: 10.1177/1754073912451492

Wu, N., Shang, S., & Su, Y. (2015). The arginine vasopressin V1b receptor gene and prosociality: Mediation role of emotional empathy. *PsyCh Journal, 4*(3), 160–165. doi: 10.1002/pchj.102

15 The Development of Children's Sharing Behavior: Recipients' and Givers' Characteristics

Hagit Sabato and Tehila Kogut

Overview

This chapter examines the development of children's sharing behavior while addressing two types of factors that shape sharing decisions at different developmental stages. First, the recipient's characteristics, such as the recipients' level of neediness, their group affiliation, and their identifiability and scope. Second, the prospective giver's characteristics, such as the child's level of theory of mind, subjective well-being, popularity among peers, and cultural background. Finally, we discuss how the interaction between the giver's and receiver's characteristics may affect children's sharing decisions at different developmental stages.

Introduction

The willingness to help others is one of the most important facets of human social behavior. Over the past few decades, a great body of research has examined the development of children's prosociality, emphasizing both the genetic and the environmental factors that play a role in this process from early ages to adolescence. One main robust finding is the substantial increase in prosocial behavior during childhood (e.g., Eisenberg and Fabes 1998; Thompson, Barresi, & Moore, 1997). Although rudimentary prosocial behavior appears even at the early age of eighteen months, when asked to share valuable resources with another at a cost to themselves, young children between the ages of three and six are relatively selfish, and the proportion of children who allocate any resources at all increases with age (e.g., Blake & Rand, 2010; Kogut, 2012;). For example, employing the dictator-game paradigm, Fehr, Bernhard, and Rockenbach (2008) found that at the ages of three to four years, the overwhelming majority of children behave selfishly, whereas most children at the ages of seven to eight prefer resource allocations that promote equity.

Along with the general increase in their willingness to share with others, with age, children become increasingly sensitive to different situational factors as well as to normative cues in their sharing decisions—for example, rewarding the recipient according to work productivity and to charity norms (Sigelman & Waitzman, 1991), paying attention to the recipient's (disadvantaged) group status (Elenbaas & Kilen, 2016), and becoming concerned with one's own reputation and the need to appear fair to others (e.g. Shaw et al., 2014).

In this chapter, we discuss these developmental processes, while addressing two types of factors that shape children's sharing decisions at different developmental stages: first, the recipient characteristics, such as the recipients' level of neediness, their group affiliation, and their identifiability and scope; and second, the prospective giver's characteristics (i.e., the child who makes the allocation decision), such as the child's level of theory of mind (ToM), subjective well-being, popularity among peers, and cultural background. Finally, we discuss how the interaction between the giver's and receiver's characteristics may affect children's sharing decisions at different developmental stages. Although many studies examine each type of characteristic separately, we argue that new insights may be obtained by looking at the interaction between the two types of characteristics, as we demonstrate later in this chapter.

Characteristics of the Recipient

Specific versus General Recipients
Sensitivity to prominent characteristics of the recipient increasingly play a role in children's sharing decisions as they grow. Children from four and a half to six years begin to react differently toward particular individuals, by sharing more with friends than with strangers (e.g. Moore, 2009). Around the age of six, children begin to exhibit the *identifiability effect,* meaning that adding identifying information about the recipient (e.g., a name or a picture) increases their willingness to share, even when the identifying information conveys no meaningful information (Kogut, Slovic, & Vastfjall, 2016). Research on the identifiable-victim effect among adult subjects

suggests identifiable targets arouse in the perceiver an intense emotional reaction that, in turn, increases sharing (e.g., Kogut, 2011; Kogut & Ritov, 2005a, 2005b; Small & Loewenstein, 2003).

Interestingly, the recipient's identifiability interacts with the prospective giver's level of ToM, an important variable in children's cognitive development (Kogut et al. 2016), such that children who have acquired higher levels of ToM share more with identifiable recipients than with unidentifiable ones, beyond age. Similarly, the development of higher levels of ToM was found to be related to the development of the *singularity effect:* a greater willingness to share with a single, identified recipient than with a group of similar recipients (whether identified or not; Kogut & Ritov, 2005a, 2005b). Although younger children are overall less willing to share with others, they give more of their endowment to a group of recipients than to a single recipient. However, this tendency reverses for older children and children with higher levels of ToM, who exhibit the singularity effect by giving more of their endowment to a single, identified target (Kogut & Slovic, 2016). Here, too, the presentation of a specific, single recipient to whom one can relate increases sharing only for children in whom the ability for perspective-taking (ToM) has developed to an advanced stage.

Consequently, the results of Kogut and Slovic (2016) suggest a developmental reversal, such that younger children give more weight to the number of recipients than older children, who, like adults, exhibit a spontaneous reaction to a specific identified recipient and tend to share more with single identified recipients (see figure 15.1). This developmental reversal, like other reversals described in the literature (false-memory, framing, and conjunction-fallacy effects; e.g., Reyna, Chick, Corbin, & Hsia, 2014), are largely explained by the fuzzy-trace theory, which suggests decision-making becomes more intuitive and categorical with age (Reyna & Brainerd, 2011). However, both the identifiability and the singularity effects were found only when the recipient was described as needy (a poor child in the studies with children, and a sick or poor person in the studies with adults). This finding underscores the importance of the recipient's neediness in prosocial decisions in general and in sharing decisions in particular.

The Recipient's Neediness

The results of several studies have demonstrated that the extent to which children consider the relative needs of the recipient in their sharing

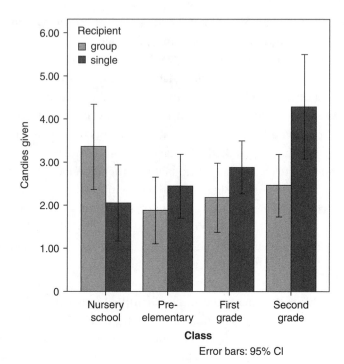

Figure 15.1
Number of candies given to a poor child (children) as a function of age group and type of recipient (single or group). Error bars: 95% confidence interval. Adapted from Kogut and Slovic (2016).

decisions develops during childhood. At younger ages children (preschoolers) demonstrate a preference for sharing more with a richer (rather than poorer) recipient (e.g., Paulus, 2014); however, from the age of five onward, awareness of the neediness of the recipient increases (Malti et al., 2016; Rizzo, Elenbaas, Cooley, & Killen, 2016; Schmidt, Svetlova, Johe, & Tomasello, 2016). Specifically, when asked to share candy with another child, only after age six did children exhibit a significant increase in sharing when confronted with a needy recipient (Kogut et al., 2016).

Similarly, Elenbaas and Killen (2016) found that older children (ages ten to eleven but not younger children, ages five to six) allocated significantly more societal resources to disadvantaged recipients than to recipients who were not described as disadvantaged. Examining children's perceptions of the obligation to help others as a function of the recipient's level of need,

Weller and Lagattuta (2013) found that five- to thirteen-year-olds predicted people would more often self-sacrifice to help as well as feel more obligated to help in high-need situations (seeing an injured person) versus low-need ones (seeing someone who wants to have a turn with a toy). However, only children aged nine and older made need-level distinctions when judging whether to self-sacrifice to help, expecting more self-sacrificing to help in high- versus low-need situations (Weller & Lagattuta, 2013).

The increase with age in sharing with needy versus less needy recipients may be attributed to both cognitive and social developmental processes. As described previously, in the study by Kogut, Slovic, and Västfjäll (2016), children with higher levels of ToM, who were better able to take the perspective of the needy recipient, exhibited a greater preference to share with a needy recipient than with a less needy one, suggesting a cognitive account for this developmental pattern. In addition, the increasing importance that children between the ages of five and thirteen place on the neediness of the recipient reflects their internalization of the common social norm to help the needy—a norm that is common in most societies (e.g., Sabato & Kogut, 2018).

Although internalizing the norms of their society increases children's sensitivity to the needs of others, it may also increase sensitivity to other social norms and group stereotypes, including the common knowledge that "charity begins at home" and the preference for in-group sharing, which may lead to out-group discrimination.

Group Affiliation

Indeed, research suggests that along with the increase in prosociality and children's sensitivity to specific characteristics of the recipient, children also develop parochialism with age, an increasing preference for members of their own social group.

The origins of group divisions begin early in development. In the first years of life, children already show a preference for in-group members in terms of race (Dunham, Baron, & Banaji, 2008) and gender (e.g., Shutts, 2015). In a meta-analysis of 113 studies, Raabe and Beelman (2011) show children express a bias toward their ethnic in-group (over ethnic out-group) as young as age three, which increases until the age of seven. This preference also translates into action, when children (three years and older) are asked to distribute resources between members of different groups, the self not included (e.g., Renno & Shutts, 2015). Examining costly sharing with in-group and out-group recipients, Fehr, Bernhard, and Rockenbach (2008) found three- to eight-year-old children are more likely to share their resources with members of their own school than with children from a different school.

However, an intergroup context does not automatically lead to in-group bias. Several studies found no in-group bias in children's prosociality (e.g., Kinzler & Spelke, 2011, in a sharing task), and some studies have even reported out-group favoritism (e.g. Sierksma, Lansu, Karremans, & Bijlstra 2018). These contradictory findings may suggest children's preference for the in-group depends on a variety of situational factors. Such factors may include perceptions of the in-group's behavior (Wilks & Nielsen, 2018), the group's status (e.g. Dunham, Newheiser, Hoosain, Merrill, & Olson, 2014), or the specific content of the out-group's stereotype (Sierksma et al., 2018).

In this section, we have discussed different characteristics of the recipient that may affect the development of children's sharing decisions. However, variations in sharing behavior and in reactions to different recipients are largely affected by the prospective giver's characteristics, including cultural background, social surroundings, and personality. We next discuss the role of such characteristics in the development of children's sharing behavior and how they may interact with the recipient characteristics we have reviewed.

Giver's Characteristics

Culture and Religiousness

Although prosocial behaviors are common and valued in most cultures, the motivation for such behaviors and the situations that trigger them may vary from one culture to the next, leading to varying levels of cooperation across societies and cultures (e.g., Henrich et al., 2005; Levine, Norenzayan, & Philbrick, 2001). Society's prosocial values are transmitted to children at a very young age (House et al., 2013) and increasingly influence their behavior (Abramson, Daniel, & Knafo-Noam, 2018). Thus, the development of the willingness to share seems to follow a similar pattern at early ages of children across different cultures.

Yet population differences in prosocial behavior emerge during middle childhood, as children internalize the norms of their specific culture. In a study of children of six different societies, House and colleagues (2013) demonstrated a substantial degree of environmental and cultural variation in children's costly prosocial behavior during middle childhood. For example, Western children become substantially more averse to inequity after the age of seven or eight (Blake & McAuliffe, 2011; Fehr et al., 2008; Kogut, 2012), but this increase is not consistently apparent in non-Western populations (e.g., House et al., 2013; Rochat et al., 2009). Similarly, in middle childhood, children are influenced by features pertaining to their respective households. For example, Miller, Kahle, and Hastings (2015) found that children from less wealthy families behaved more altruistically than children from wealthier families—a phenomenon echoed among adults (e.g., Piff, Kraus, Côté, Cheng, & Keltner 2010).

According to this literature, middle childhood (beginning around the age of seven) appears to be an important phase in which children become increasingly involved and integrated in their cultural and social community (Lancy & Grove, 2011). This period is therefore particularly important for the development of conformity to local social norms and to internalization of the values of one's own society—including values related to one's religious environment.

Research on the relationship of religiosity and sharing behavior among children is rare. Among adults, research points to a modest but consistent link between prosocial inclinations and religiosity (e.g., Saroglou, 2013). Moreover, this link may depend on the attributes of the helping target such that the help proffered by religious people may be restricted to in-group recipients (e.g., Batson, Schoenrade, & Ventis, 1993). They are less likely to help causes or targets that are not commensurate with their group's values and norms (e.g., Pichon & Saroglou, 2009) because their prosociality may be driven by the wish to create a positive impression (Galen, 2012).

In a recent study, Decety et al. (2015) demonstrated a negative association between religiosity and sharing behavior among children between ages five and twelve from six different countries, challenging the view that religiosity facilitates prosocial behavior in children (but see also Shariff, Willard, Muthukrishna, Kramer, & Henrich, 2016).

Considering possible variables that may reconcile the inconsistent findings regarding the relationship of religion and prosociality in children, Sabato and Kogut (2018) recently suggested religious children's sharing may be dependent on the neediness of the recipient because religion promotes sharing with the poor and needy (*charitable giving*) rather than sharing per se. For example, in Christian theology, charity is the greatest of the three theological virtues. In Judaism, giving to charity is a commandment, not voluntary. Similarly, giving to charity is one of the five pillars of Islam. Examining second and fifth graders' actual sharing decisions, the authors found the recipient's neediness moderated the relation between children's household religiosity and prosocial behavior: religious children shared more than nonreligious children only when the recipient was described as needy, not when a non-needy recipient was considered (see figure 15.2). This finding was replicated with a sample from a minority population of Christian Arab children in Israel from religious and nonreligious households (Sabato & Kogut, 2020). This finding highlights the importance of the examination of the interaction between characteristics of prospective givers (e.g., religiosity) and the recipient (e.g., neediness) in predicting children's sharing behavior.

Subjective Well-Being

In addition to the effect of external variables such as religion and culture, personal characteristics such as the prospective giver's subjective well-being (SWB) may also be related to individual differences in children's willingness

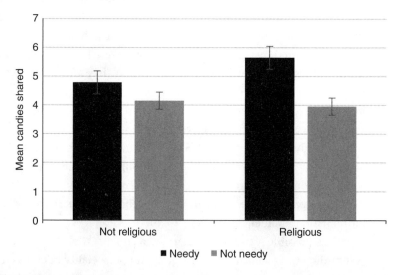

Figure 15.2
The number of pieces of candy shared by the children as a function of religiosity and the recipient's neediness. Adapted from Sabato & Kogut (2018).

to share with others. Research into the link between SWB and prosocial behavior among adults suggests happy people are more likely to engage in prosocial behavior (e.g., Aknin, Dunn, & Norton, 2012). For example, Aknin, Dunn, and Norton (2013) examined the correlation between charitable giving and happiness in 136 countries. In 120 of these, the authors found a positive correlation between giving and happiness—even when controlling for income and other demographic variables.

Many researchers describe the associations between SWB and prosociality as a virtuous circle: Happy people engage more in prosocial behaviors, and doing good deeds enhances their well-being. For example, Meier and Stutzer (2008) and Plagnol and Huppert (2010) found such an association in experimental and correlational studies (respectively) in the context of volunteering behavior. Aknin and colleagues (2012) found it, too, in an experimental context of monetary spending—namely, prosocial spending increased happiness, which in turn encouraged future prosocial spending. Although this literature offers convincing evidence of a bilateral link between SWB and prosociality among adults in various behaviors, the question of whether this association also exists among children (and if so, the age at which it emerges) has not yet been examined, despite extensive research into the development of prosocial behavior in children.

To date, only a handful of studies have examined the developmental pattern of the link between giving to others and the satisfaction derived from such behavior. Paulus and Moore (2017) show preschool children (aged three to six years) anticipated experiencing positive emotions when they shared with others and negative emotions when they did not. Moreover, these individual differences in emotional understanding predicted the children's generosity in a sharing task. Understanding the emotional reward of giving appears to grow and strengthen with age. For example, Weller and Lagattuta (2013) found that when they asked children to predict the emotions of other characters in prosocial moral dilemmas, the older children (aged seven to thirteen years) attributed more intensely positive emotions to self-sacrificing helpers than younger children did (at the ages of five or six years).

However, in these studies, the children had been asked to judge hypothetical sharing behavior (by themselves or by others) rather than reporting actual satisfaction from sharing. In a recent study, Wu, Zhang, Guo, and Gros-Louis (2017) examined the positive emotions of preschoolers (ages three to five) immediately after sharing (either autonomous or obligated). They found that in the autonomous-sharing condition (although not in the obligated one), children expressed greater happiness when they shared the items in question (stickers) with their prospective recipient than when they kept them for themselves. Notably, in that study, emotional responses were analyzed by an observer rather than stated by the children themselves.

Asking the children about the satisfaction they derived from actual giving immediately after having the opportunity to share candy with another child in a dictator game, Kogut (2012) found that only at the ages of eight to ten years were children significantly happier when sharing (especially when sharing half of their endowment), compared with other children of their age who chose not to share at all or who had shared less than half their endowment (younger children in particular—under the age of six—tended to behave selfishly and keep all or most of their candy). Moreover, satisfaction from sharing occurred mostly when the sharing felt voluntary (i.e., in the dictator game) and less so when it felt obligatory or necessary (i.e., in the ultimatum game, in which sharing is driven by greater strategic considerations).

Kogut's (2012) findings suggest the link between actual costly giving and conscious satisfaction (as reported by the children) increases with age and is less likely to form before the age of eight. Moreover, this association is apparent mostly with regard to voluntary sharing, and less so when sharing is driven by other strategic motivations.

To the best of our knowledge, the inverse association—that SWB (life satisfaction) encourages giving—has not been directly examined among children. However, initial evidence shows an association between manipulated positive emotions (as opposed to general, more stable SWB) and giving. For example, enhanced feelings of success (through a bowling game in which the child earned higher scores in most of the game's trials) or a positive mood (induced by asking the children to recall events that made them happy) increased prosocial behavior among third and fourth graders (e.g., Isen, Horn, & Rosenhan, 1973).

In a recent study, we examined the association between children's SWB and their sharing behavior (Sabato & Kogut, 2019a). Schoolchildren (second or fifth graders) were interviewed in private, and they rated their subjective general happiness in life (using emojis) and had an opportunity to share sweets with a recipient under one of two between-subject conditions: *perceived-high obligation* (a recipient in poverty) and *perceived-low obligation* (a temporarily needy recipient). The results have provided initial evidence of an increased association between SWB and sharing decisions with age. Whereas SWB was not significantly correlated with the incidence of sharing by younger children, it was a positive predictor of sharing behavior among fifth graders. Manipulating the perceived obligation to share (by emphasizing the causes beyond the recipient's need), we found that higher levels of SWB were linked to sharing only in the perceived-low-obligation condition. The children with lower SWB behaved as expected by the norm and shared to a similar degree as the children with higher SWB when sharing felt obligatory. However, when sharing was less obligatory, higher levels of SWB were linked to higher levels of sharing.

According to the *warm-glow-giving* theory of Andreoni (1990), people gain utility from the act of giving in the form of positive emotional feelings that they experience from helping others. During their lifetime, people learn from experience that helping others makes them feel better, and such positive emotions serve as a positive reinforcing feedback that can shape future behavior through the anticipation of a similar emotional response to similar behaviors. This understanding enhances prosocial behavior, which in turn increases SWB; therefore, the mutual association between helping others and SWB is likely to develop with experience and to increase with age. Specifically, one may expect this association to emerge when satisfaction is derived from the act of giving and sharing with others—namely, around the age of eight (Kogut, 2012)—and to gradually increase as the child grows older. Sabato and Kogut (2019a) provide another example of how the prospective giver's characteristics may interact with situational factors (perceived obligation to help) in predicting children's sharing decisions at different developmental stages.

Whereas the research we have discussed examined children's general subjective well-being, children's well-being in specific life domains may also be related to their sharing behavior. Specifically, in middle childhood, social interactions and peer acceptance become major factors in children's daily lives, affecting their behavior in various contexts, such as their motivation and achievements (e.g. Buhs, Ladd, & Herald, 2006), including their prosocial behavior.

Social Status and Peer Acceptance

Research among adult participants has suggested that being rejected decreases prosociality (Twenge, Baumeister, DeWall, Ciarocco, & Bartels,

2007). Social acceptance and perceived support from one's peer group have been found to be highly, positively correlated with prosocial and moral inclinations also in early adolescence (Wentzel & McNamara, 1999). Studies of these correlations in developmental and educational research suggest children who score high in sociometric popularity measures[1] (by peers, as well as in teacher's reports) are perceived by their friends and teachers as being more helpful, cooperative, kind, and more often engaged in prosocial interaction than less popular and rejected children (Berg, Lansu, & Cillessen, 2015; Rodkin, Ryan, Jamison, & Wilson, 2013).

These studies offer convincing evidence of a solid association among popularity, peer acceptance, and children's prosociality; however, they gauge the children's prosocial behaviors based on evaluations by peers and teachers of the children's general tendencies toward kindness, helpfulness, and cooperation, rather than on measures of actual behavior. In addition, these general evaluations of the children's prosociality do not consider the characteristics of the prosocial situation in question, such as the recipient (i.e., the person the child is interacting with). In our view, when determining whether less socially accepted children are indeed less prosocial in general, it is important to examine the possible interactions between the children's social status and the recipient's characteristics, and whether their behavior depends on situational factors—namely, whether they behave in a less prosocial manner only when they do not feel accepted by fellow members of their group.

A study by Sabato and Kogut (2019b) examined the link between fourth and fifth graders' (ages ten to twelve) social status among their peer group and their tendency to behave in a prosocial manner. We used sociometric nominations to determine the children's level of popularity among their peers and examined actual sharing behavior with another child who was either a fellow in-group member (a child from their class) or an outgroup member (a child from a different class). The results replicated earlier findings that children's social acceptance within their class and their sharing behavior are correlated; however, this correlation was significant only in the in-group condition, and not when the children shared with out-group recipients.

---

1. Sociometric popularity measures are based on friends' actual preferences and are distinct from "perceived popularity," which is based on peers' global assessment. For a broad discussion regarding the differences between the two measurements, see Cillessen and Marks (2011).

These findings highlight the idea that the link between social acceptance and prosociality may largely depend on situational factors. Specifically, children who are less socially accepted may act in a less prosocial manner toward fellow in-group members because they are surrounded by children who are not their friends, rather than because of inherent factors such as a lack of social skills. The finding that less socially accepted children share with out-group members as much as popular children do suggests less socially accepted children may act in a more prosocial manner when the beneficiary is not a part of the group where they feel shunned or less accepted.

Because previous research was based mainly on teachers' reports of the children's prosociality, these teachers likely based their assessments on the children's behavior within their class (where the teacher is likely to observe the children). The above findings highlight the importance of gauging children's prosociality in other environments and contexts in addition to their class group when examining the link between social status and prosociality.

Conclusion

Our discussion of the features that shape the development of children's prosocial behavior demonstrates that with age, the characteristics of the recipient—such as identifiability, scope, neediness, and group affiliation—as well as characteristics of the prospective giver—such as culture, religiosity, subjective well-being, and peer acceptance—increasingly play a role in sharing decisions.

Most importantly, the research that we reviewed highlights the importance of the interaction between the prospective helper's and the recipient's (or the situation's) characteristics in our understanding of the development of prosociality in children. For example, children's household religiosity interacts with the neediness of the recipient in predicting sharing, increasing the sharing with needy (not with non-needy) recipients. SWB interacts with perceived obligation to help (increasing sharing only when children feel less obligated to help). Finally, social status (popularity) among peers interacts with the recipient's group affiliation, predicting sharing with in-group but not outgroup recipients.

Future research may continue the investigation of the role such interactions play in explaining children's sharing and other prosocial decisions, promoting our understanding of the development of such behaviors in children. This investigation may include the interaction between different cultural aspects of the children's environment and the recipient's characteristics. For example, studies show the singularity effect is more dominant among people from individualistic cultures, as compared with collectivist cultures (Kogut, Slovic, & Vastfjall, 2015). Future research may examine how children's culture (collectivism versus individualism) interacts with the singularity effect (one versus several recipients) in predicting sharing behavior.

References

Abramson, L., Daniel, E., & Knafo-Noam, A. (2018). The role of personal values in children's costly sharing and non-costly giving. *Journal of Experimental Child Psychology, 165,* 117–134.

Aknin, L. B., Barrington-Leigh, C. P., Dunn, E. W., Helliwell, J. F., Burns, J., Biswas-Diener, R.,…Norton, M. I. (2013). Prosocial spending and well-being: Cross-cultural evidence for a psychological universal. *Journal of Personality and Social Psychology, 104,* 635–652.

Aknin, L. B., Dunn E. W., & Norton, M. I. (2012). Happiness runs in a circular motion: Evidence for a positive feedback loop between prosocial spending and happiness. *Journal of Happiness Studies, 11,* 347–355.

Andreoni, J. (1990). Impure altruism and donations to public goods: A theory of warm-glow giving. *Economic Journal, 100,* 464–477.

Batson, C. D., Schoenrade, P., & Ventis, W. L. (1993). *Religion and the individual: A social-psychological perspective.* Oxford: Oxford University Press.

Berg, Y. H., Lansu, T. A., & Cillessen, A. H. (2015). Measuring social status and social behavior with peer and teacher nomination methods. *Social Development, 24,* 815–832.

Blake, P. R., & McAuliffe, K. (2011). "I had so much it didn't seem fair": Eight-year-olds reject two forms of inequity. *Cognition, 120,* 215–224.

Blake, P. R., & Rand, D. G. (2010). Currency value moderates equity preference among young children. *Evolution and Human Behavior, 31,* 210–218.

Buhs, E. S., Ladd, G. W., & Herald, S. L. (2006). Peer exclusion and victimization: Processes that mediate the relation between peer group rejection and children's classroom engagement and achievement? *Journal of Educational Psychology, 98,* 1–13.

Cillessen, A. H. N., & Marks, P. E. L. (2011) Conceptualizing and measuring popularity. In Cillessen, A. H., Schwartz, D., & Mayeux, L. (Eds.), *Popularity in the peer system* (pp. 25–55). New York: Guilford Press.

Decety, J., Cowell, J. M., Lee, K., Mahasneh, R., Malcolm-Smith, S., Selcuk, B., and Zhou, X. (2015). The negative association between religiousness and children's altruism across the world. *Current Biology, 25,* 2951–2955.

Dunham, Y., Baron, A. S., & Banaji, M. R. (2008). The development of implicit intergroup cognition. *Trends in Cognitive Sciences, 12,* 248–253.

Dunham, Y., Newheiser, A. K., Hoosain, L., Merrill, A., & Olson, K. R. (2014). From a different vantage: Intergroup attitudes among children from low-and intermediate-status racial groups. *Social Cognition, 32,* 1–21.

Eisenberg, N., & Fabes, R. (1998). Prosocial development. In W. Damon & N. Eisenberg (Eds.), *Handbook of child psychology: Vol. 3. Social, emotional, and personality development* (5th ed., pp. 701–778). New York: Wiley.

Elenbaas, L., & Killen, M. (2016). Children rectify inequalities for disadvantaged groups. *Developmental Psychology, 52,* 1318–1329.

Fehr, E., Bernhard, H., & Rockenbach, B. (2008). Egalitarianism in young children. *Nature, 454,* 1079–1083.

Galen, L. W. (2012). Does religious belief promote prosociality? A critical examination. *Psychological Bulletin, 138,* 876–882.

Henrich, J., Boyd, R., Bowles, S., Camerer, C., Fehr, E., Gintis, H.,…Henrich, N. S. (2005). "Economic man" in cross-cultural perspective: Behavioral experiments in 15 small-scale societies. *Behavioral and Brain Sciences, 28,* 795–815.

House, B. R., Silk, J. B., Henrich, J., Barrett, H. C., Scelza, B. A., Boyette, A. H.,…Laurence, S. (2013). Ontogeny of prosocial behavior across diverse societies. *Proceedings of the National Academy of Sciences of the United States of America, 110,* 14586–14591.

Isen, A. M., Horn, N., & Rosenhan, D. L. (1973). Effects of success and failure on children's generosity. *Journal of Personality and Social Psychology, 27,* 239–247.

Kinzler, K. D., & Spelke, E. S. (2011). Do infants show social preferences for people differing in race? *Cognition, 119,* 1–9.

Kogut, T. (2011). Someone to blame: When identifying a victim decreases helping. *Journal of Experimental Social Psychology, 47,* 748–755.

Kogut, T. (2012). Knowing what I should, doing what I want: From selfishness to inequity aversion in young children's sharing behavior. *Journal of Economic Psychology, 33,* 226–236.

Kogut, T., & Ritov, I. (2005a). The "identified victim" effect: An identified group, or just a single individual? *Journal of Behavioral Decision Making, 18,* 157–167.

Kogut, T., & Ritov, I. (2005b). The singularity of identified victims in separate and joint evaluations. *Organizational Behavior and Human Decision Processes, 97,* 106–116.

Kogut, T., & Slovic, P. (2016). The development of scope insensitivity in sharing behavior. *The Journal of Experimental Psychology: Learning, Memory, and Cognition, 42,* 1972–1981.

Kogut, T., Slovic, P., & Västfjäll, D. (2015). Scope insensitivity in helping decisions: Is it a matter of culture and values? *Journal of Experimental Psychology: General, 144,* 1042–1052.

Kogut, T., Slovic, P., & Västfjäll, D. (2016). The effect of recipient identifiability and neediness on children's sharing behavior. *Journal of Behavioral Decision Making, 29,* 353–362.

Lancy, D. F., & Grove, M. A. (2011). Getting noticed. *Human Nature, 22,* 281–302.

Levine, R. V., Norenzayan, A., & Philbrick, K. (2001). Cross-cultural differences in helping strangers. *Journal of Cross-Cultural Psychology, 32,* 543–560.

Malti, T., Gummerum, M., Ongley, S., Chaparro, M., Nola, M., & Bae, N. Y. (2016). "Who is worthy of my generosity?" Recipient characteristics and the development of children's sharing. *International Journal of Behavioral Development, 40,* 31–40.

Meier, S., & Stutzer, A. (2008). Is volunteering rewarding in itself? *Economica, 75,* 39–59.

Miller, J. G., Kahle, S., & Hastings, P. D. (2015). Roots and benefits of costly giving: Children who are more altruistic have greater autonomic flexibility and less family wealth. *Psychological Science, 26,* 1038–1045.

Moore, C. (2009). Fairness in children's resource allocation depends on the recipient. *Psychological Science, 20,* 944–948.

Paulus, M. (2014). The early origins of human charity: developmental changes in preschoolers' sharing with poor and wealthy individuals. *Frontiers in Psychology, 5,* 344 (1–9).

Paulus, M., & Moore, C. (2017). Preschoolers' generosity increases with understanding of the affective benefits of sharing. *Developmental Science, 20*(3), e12417.

Pichon, I., & Saroglou, V. (2009). Religion and helping: Impact of target thinking styles and just-world beliefs. *Archive for the Psychology of Religion, 31,* 215–236.

Piff, P. K., Kraus, M. W., Côté, S., Cheng, B. H., & Keltner, D. (2010). Having less, giving more: the influence of social class on prosocial behavior. *Journal of Personality and Social Psychology, 99,* 771–184.

Plagnol, A. C., & Huppert, F. A. (2010). Happy to help? Exploring the factors associated with variations in rates of volunteering across Europe. *Social Indicators Research, 97,* 157–176

Raabe, T., & Beelman, A. (2011). Development of ethnic, racial, and national prejudice in childhood and adolescence: A multinational meta-analysis of age differences. *Child Development, 82,* 1715–1737.

Renno, M. P., & Shutts, K. (2015). Children's social category-based giving and its correlates: Expectations and preferences. *Developmental Psychology, 51,* 533–543.

Reyna, V. F., & Brainerd, C. J. (2011). Dual processes in decision making and developmental neuroscience: A fuzzy-trace model. *Developmental Review, 31*(2–3), 180–206.

Reyna, V. F., Chick, C. F., Corbin, J. C., & Hsia, A. N. (2014). Developmental reversals in risky decision making: Intelligence agents show larger decision biases than college students. *Psychological Science, 25,* 76–84.

Rizzo, M. T., Elenbaas, L., Cooley, S., & Killen, M. (2016). Children's recognition of fairness and others' welfare in a resource allocation task: Age related changes. *Developmental Psychology, 52,* 1307–1317.

Rochat, P., Dias, M. D., Liping, G., Broesch, T., Passos-Ferreira, C., Winning, A., & Berg, B. (2009). Fairness in distributive justice by 3-and 5-year-olds across seven cultures. *Journal of Cross-Cultural Psychology, 40,* 416–442.

Rodkin, P. C., Ryan, A. M., Jamison, R., & Wilson, T. (2013). Social goals, social behavior, and social status in middle childhood. *Developmental Psychology, 49,* 1139–1150.

Sabato, H., & Kogut, T. (2018). The association between religiousness and children's altruism: The role of the recipient's neediness. *Developmental Psychology, 54,* 1363–1371.

Sabato, H., & Kogut, T. (2019a). Feel good, do good? Subjective well-being and sharing behavior among children. *Journal of Experimental Child Psychology, 177,* 335–350.

Sabato, H., & Kogut, T. (2019b). Sharing and belonging: Social status and children's sharing behavior with ingroup and outgroup members. Manuscript under review.

Sabato, H., & Kogut, T. (2020). The development of prosociality among Christian Arab children in Israel: The role of children's household religiosity and of the recipient's neediness. Manuscript under review.

Saroglou, V. (2013). Religion, spirituality, and altruism. In K. I. Pargament, J. J. Exline, & J. W. Jones (Eds.), *APA handbooks in psychology. APA handbook of psychology, religion, and spirituality: Vol. 1. Context, theory, and research* (pp. 439–457). Washington, DC: American Psychological Association.

Schmidt, M. F., Svetlova, M., Johe, J., & Tomasello, M. (2016). Children's developing understanding of legitimate reasons for allocating resources unequally. *Cognitive Development, 37,* 42–52.

Shariff, A. F., Willard, A. K., Muthukrishna, M., Kramer, S. R., & Henrich, J. (2016). What is the association between religious affiliation and children's altruism? *Current Biology, 26,* R699–R700.

Shaw, A., Montinari, N., Piovesan, M., Olson, K. R., Gino, F., & Norton, M. I. (2014). Children develop a veil of fairness. *Journal of Experimental Psychology: General, 143,* 363–373.

Shutts, K. (2015). Young children's preferences: Gender, race, and social status. *Child Development Perspectives, 9*, 262–266.

Sierksma, J., Lansu, T. A. M., Karremans, J. C., & Bijlstra, G. (2018). Children's helping behavior in an ethnic intergroup context: Evidence for outgroup helping. *Developmental Psychology, 54*, 916–928.

Sigelmann, C. K., & Waitzmann, K. A. (1991). The development of distributive justice orientations: Contextual influences on children's resource allocations. *Child Development, 62*, 1367–1378.

Small, D. A., & Loewenstein, G. (2003). Helping a victim or helping the victim: Altruism and identifiability. *Journal of Risk and Uncertainty, 26*, 5–16.

Thompson, C., Barresi, J., & Moore, C. (1997). The development of future-oriented prudence and altruism in preschoolers. *Cognitive Development, 12*, 199–212.

Twenge, J. M., Baumeister, R. F., DeWall, C. N., Ciarocco, N. J., & Bartels, J. M. (2007). Social exclusion decreases prosocial behavior. *Journal of Personality and Social Psychology, 92*, 56–66.

Weller, D., & Lagattuta, K. H. (2013). Helping the in-group feels better: Children's judgments and emotion attributions in response to prosocial dilemmas. *Child Development, 84*, 253–268.

Wentzel, K. R., & McNamara, C. C. (1999). Interpersonal relationships, emotional distress, and prosocial behavior in middle school. *Journal of Early Adolescence, 19*, 114–125.

Wilks, M., & Nielsen, M. (2018). Children disassociate from antisocial in-group members. *Journal of Experimental Child Psychology, 165*, 37–50.

Wu, Z., Zhang, Z., Guo, R., & Gros-Louis, J. (2017). Motivation counts: Autonomous but not obligated sharing promotes happiness in preschoolers. *Frontiers in Psychology, 8*, 867.

IV Social Categorization

16   The Role of Essentialism in Children's Social Judgments

Susan A. Gelman and Rachel D. Fine

Overview

Psychological essentialism is an intuitive belief that certain categories, such as those regarding gender, race, ethnicity, ability, or traits, reflect not just observable features, but also a deeper, nonobvious reality. Research with adults links psychological essentialism to stereotyping, prejudice, and inequality. Moreover, extensive prior research documents that children readily construe some categories in essentialist ways, viewing them as discrete, immutable, biologically based, and having rich inductive potential. At the same time, there is considerable cultural and developmental variation in which social categories are essentialized. This chapter provides an up-to-date overview of recent research on psychological essentialism in children's social categories. We discuss how essentialism can distort children's social judgments of others, and the implications of essentialism for important social consequences, including stereotyping and intergroup bias. We end by discussing factors that influence when and why children essentialize, and possible interventions that may reduce essentialism.

Introduction

Social judgments are inextricably tied to categories. When evaluating and interacting with others, we do not simply consider them as unique individuals, but also as members of the groups to which they belong. Barack Obama is not just an individual with a distinctive personal history (born in Hawaii and served as the forty-fourth president of the United States); he also belongs to numerous overlapping sets: men, African Americans, fathers, husbands, English-speakers, Democrats, attorneys, Christians. Furthermore—and most important—social judgments follow from categories such as these. We daily make many decisions about how to interact with others—closeness versus distance, cooperation versus competition, friendship versus enmity, trust versus suspicion, us versus them; all are informed by the categories we have formed and activate with ease.

This is equally true of children. The human child is an avid systematizer, who detects similarities and differences, and uses these regularities to extend knowledge and draw new inferences. The past twenty years have revealed surprising sensitivity on the part of children to the structures and patterns in their environment (Saffran, Aslin, & Newport, 1996). This sensitivity extends to children's social world. Even preverbal infants sort people into categories, making use of sometimes subtle cues to identify an individual's gender, race, and age (Quinn, Lee, & Pascalis, 2019).

This early drive to categorize, coupled with the link between categories and social judgments, leads to a puzzle that motivates and frames this chapter: Why do some categories have powerful social consequences, whereas others are relatively ignored? Hirschfeld (1996) gave an example from Sartre (1948/1995) of a woman who claimed that her dislike of Jews was due to a negative experience with Jewish furriers. As Sartre mused, Why did she choose to hate Jews rather than furriers? From a logical perspective, either inference could follow from the (perceived) evidence, yet categories differ from one another in their link to prejudice and intolerance. This problem is especially acute when one considers the astonishing variety of categories people use to organize their social world (gender, race, ethnicity, traits, nationality, occupation, caste, appearance, social status…). Which categories form the basis of people's social reasoning, why these categories, and how do these linkages emerge in development?

In this chapter we focus on one important factor, psychological essentialism (Allport, 1954; Gelman, 2003; James, 1890; Keil, 1989; Medin & Ortony, 1989). Psychological essentialism is the belief that certain categories share not just superficial appearances but also a deeper reality. For example, birds share not just obvious features (feathers, two legs) but also a wealth of nonobvious commonalities. Upon knowing that something is a bird, one can make inferences about its behavior (breathes air, lays eggs), internal features (hollow bones, warm-blooded), and enough more to keep ornithologists busy for generations to come. These hidden features are more important than outward appearances: an animal may look like a bird but not be one (bat, pterodactyl), or may be a bird but not look like one (penguin, dodo). Furthermore, both the surface and deeper commonalities that category members share are assumed to have an internal, inherent cause (also known as the "essence"). This essence may be a known biological feature such as genes or blood, or it may be an unspecified "placeholder"—a belief that there is some underlying cause, even if it has not yet been specified or discovered.

Psychological essentialism is supported by adults' judgments about several interrelated components (Haslam, Rothschild, & Ernst, 2000), four of which are reviewed later (see table 16.1). Adults judge that category members are alike in many nonobvious ways (inductive potential), that category identity and category-linked features cannot change (immutability), that commonalities within a category are due to nature and thus are inborn (biological basis), and that category boundaries are objective and absolute (discreteness). When applied to the category of "birds," these assumptions seem for the most part unproblematic, even insightful (but see Gelman & Rhodes, 2012; Leslie, 2013). Birds are indeed alike in nonobvious ways (as with their hollow bones and ZW chromosomal system), have a fixed identity (a bird cannot transform into another species), are biologically based

Table 16.1
Components of Essentialism and Consequences for Distortions of Social Others

Component of Essentialism	Description	Distortions of Social Others	Sample Task with Children
Informativeness	Category members are alike in many nonobvious ways	Treating superficial differences between people as predictive of deeper differences	Inductive inferences
Immutability	Category identity and category-linked features cannot change	Underestimating the possibility of environmental influence or change over time	Transformations
Biological basis	Commonalities within a category are due to nature (inborn)	Assuming that causes inhere in individuals, not in larger structures or systems	Innate potential
Discreteness	Category boundaries are objective and absolute	Denying overlapping categories and fuzzy boundaries	Boundaries

(birds have characteristic genetic structures), and have discrete category boundaries (a typical bird such as a robin and an atypical bird such as a penguin are equally, fully birds).

However, when these components are applied to categorizing people, they lead to systematic distortions (table 16.1). The component of inductive potential leads to treating superficial differences between people as predictive of deeper differences, as when racial differences are falsely assumed to predict differences in intelligence or athleticism (Pauker, Ambady, & Apfelbaum, 2010). The component of immutability entails underestimating the possibility of environmental influence or change over time, as when gender imbalances in various professions (e.g., science, nursing) are assumed to be the natural order of things. The component of biological basis leads to viewing features as inherent in individuals and underestimating the importance of structural causes, as when the prominence of Jews in banking is seen as evidence of an inherent greediness rather than historical restrictions on land ownership and other forms of economic activities (Penslar, 2001). The component of discreteness denies the existence of overlap and fuzzy boundaries, as when intersex individuals are surgically transformed as infants to "fit" into a single gender category (Hughes, Houk, Ahmed, & Lee, 2006), or when transgender individuals are forced to use the bathroom corresponding to the gender listed on their birth certificate.

Given these distortions, it is not surprising that essentialism predicts a host of social consequences in adults, including stereotyping and prejudice (Bastian & Haslam, 2006; Keller, 2005; Meyer & Gelman, 2016; Williams & Eberhardt, 2008), acceptance of inequality (Hussak & Cimpian, 2019), punitive judgments (Kraus & Keltner, 2013), lack of agency (Dar-Nimrod & Heine, 2011), biased categorizations of multiracial individuals (Gaither et al., 2014; Ho, Roberts, & Gelman, 2015), and gender imbalances in academia (Leslie, Cimpian, Meyer, & Freeland, 2015). Essentialism can extend as well to political attitudes (Roberts, Ho, Rhodes, & Gelman, 2017) and perceived connection to national events. For example, when surveying people in Latvia and Hong Kong, nations that have been victims of atrocities, Zagefka, Pehrson, Mole, and Chan (2010) found that the more people essentialized nationality as biological, the more they felt anger and reluctance to forgive. This was partially due to a sense of connection to their compatriots who were victims. On the other side, for ethnic Russians in Latvia, higher levels of biological essentialism among those with ties to a country that committed atrocities related to feeling more guilt for these atrocities.

Childhood Essentialism

Children show an early-emerging tendency to essentialize the natural world (Gelman, 2004) across many different cultural contexts (e.g., Moya, Boyd, & Henrich, 2015; Waxman, Medin, & Ross, 2007). They also show a pervasive but considerably more variable tendency to essentialize kinds of people (Diesendruck, 2013). In this section we first discuss how to measure essentialism, and then review a few key findings demonstrating childhood essentialism, using animal and gender categories as illustrative cases.

Measuring essentialism is key to determining when, why, and with what consequences people's social categories direct their social judgments, but it is also a challenge. The concept at the heart of this construct—a category's essence—is typically only a placeholder rather than a well-articulated, specified representation. For example, the belief that there is some underlying, inherent property that causes someone to be shy may be quite vague on the details (e.g., is it something in the brain? in the blood? in the DNA?).

Consequently, psychological essentialism is best measured not by asking what the essence is, but rather by measuring its consequences. Essentialism questionnaires administered to adults thus include items such as the following, for which participants can rate their level of agreement or disagreement on each of a range of different categories:

- "Some categories have sharper boundaries than others. For some, membership is clear-cut, definite, and of an 'either/or' variety; people either belong to the category or they do not. For others, membership is more 'fuzzy'; people belong to the category in varying degrees" (Haslam, et al., 2000).

- "Some categories are natural categories that exist in the real world and someone has to discover what they are. Other categories are invented by a culture and decided upon by experts" (Ahn, Flanagan, Marsh, & Sanislow, 2006).

Measuring essentialism in childhood poses yet a further challenge, because young children cannot respond to questions of this level of complexity. However, the basic components of essentialism can be assessed via more engaging methods that ask children to make simple predictions, as we will describe later (see also Gelman, 2003; Rhodes & Mandalaywala, 2017).

Inductive inferences. One of the most potent consequences of essentialism is informativeness—the novel inferences that people draw on the basis of category membership. Category informativeness has been measured in children by teaching them new facts about a set of individuals and testing how participants generalize those facts to other individuals (Gelman & Markman, 1986). For example, a child might learn that a "bird" (flamingo) feeds its baby mashed-up food, whereas a "bat" (black bat in flight) feeds its baby milk. Then they are presented with another "bird" (blackbird in flight) and are asked whether it feeds its baby mashed-up food or milk. Children thus have a choice of generalizing the new information on the basis of category membership (as indexed by the basic-level name [bird]) or on the basis of perceptual similarity (as indicated by the outward appearance—bat and blackbird are alike in shape and color). Studies from multiple laboratories have indicated that preschool children typically draw inferences about animal kinds based on category membership (Booth, 2014; Gelman & Davidson, 2013; Jaswal, Lima, & Small, 2009; Tarlowski, 2018), thus suggesting that they possess an essentialist belief that category members are alike in nonobvious ways. This same method has been used with social categories. For example, in one trial children might see a picture of a boy and learn that he has "andro" in his blood, and a picture of a girl and learn that she has "estro" in her blood, and then see a child with long hair who is labeled as a "boy" and are asked if he has andro or estro in his blood. Again, preschool children typically answer in accordance with the individual's gender category (Gelman, Collman, & Maccoby, 1986).

Transformations. Essentialized concepts are thought to be immutable: no matter what you do to a raccoon, you cannot transform it into a skunk, because its internal essence remains untouched. Keil (1989) developed a classic task to assess this belief in children, presenting hypothetical scenarios in which an item undergoes a series of human-initiated modifications that radically change its appearance and features. For example, a raccoon had its fur painted, some belly fat removed, and a smelly sac inserted, resulting in an animal that looks and acts like a skunk. Following these sorts of operations, second-grade children report that the animal's original identity is maintained (in this case, that the animal is still a raccoon), and even preschoolers report maintenance of identity for transformation involving costumes. A similar belief about gender is in place by about six to seven years of age or even earlier; children report that gender cannot change if a person's external features change (e.g., a girl getting her hair cut, a boy wearing a purse; this is the classic "gender constancy" task; Ruble et al., 2007). However, studies generally have not asked children to consider transgender individuals, whose gender is viewed as changeable by many adults (Elischberger, Glazier, Hill, & Verduzco-Baker, 2016). This remains an important topic for future research.

Innate potential.   Although there is no meaningful answer to the question "nature or nurture?" young children seem to have robust intuitions that certain qualities are inborn and resistant to environmental influence. A "switched-at-birth" task asks participants to predict the properties of individuals whose birth and adoptive parents differ in key respects. For example, children may learn about a newborn rabbit raised by monkeys, and be asked if the animal would have long or short ears, and whether it would eat carrots or bananas. Four-year-olds consistently judge that category-typical properties will emerge regardless of rearing environment (e.g., the rabbit raised by monkeys will nonetheless have long ears and eat carrots; Gelman & Wellman, 1991). A modified version has been used with gender, in which a newborn infant is described as living on an island with individuals who are all either the same sex as the infant (e.g., boy raised entirely with/by males) or different from the infant (e.g., boy raised entirely with/ by females) (Taylor, Rhodes, & Gelman, 2009). On this version of the task, young children (four to five years) predict that the child's birth gender will determine a host of behaviors, abilities, and preferences (e.g., a boy raised with females will prefer to play football rather than sew). They have little biological knowledge, but view gender-linked behaviors as inborn, inflexible, and biologically based.

Boundaries.   Do children view category boundaries as natural and objective (essentializing), or as conventionalized and subjective (nonessentializing)? One way this question has been addressed is by asking children to judge unconventional pairings—that is, pairs of individuals that cross a categorical divide. For example, in one set of studies (Rhodes & Gelman, 2009), children were presented with two animals from two different categories (e.g., a dog and a cat). They then learned that a group of people from a far-away community said that these two items were the same kind, and the child's task was to say whether or not these other people might be right. Five-year-olds said no, these other people could not be right; a dog and a cat are not the same kind, regardless of what others might think. In other words, they treated animal boundaries as natural and objective. They were also asked to make judgments about artifact kinds (e.g., the far-away people said that a car and a train were the same kind). Here children felt this was fine; the boundary between cars and trains is more subjective and conventional, and thus could vary by context. What about when children were presented with items differing in gender (e.g., a smiling little boy and a smiling little girl of about the same age, matched for race)? If the far-away people say they are the same kind of thing, might that be right? Here again, children rejected the consensus, reporting that it is wrong to treat the boy and the girl as the same kind. In effect, they viewed the boundary between male and female as akin to the boundary between cat and dog. This is particularly striking, as the boy and girl shared other category memberships (e.g., race, age, emotional expression).

Cultural and Developmental Variation in Childhood Essentialism

As noted previously, despite children's firm foundation of essentialism regarding animal kinds, there is substantial variation in children's essentialist reasoning regarding social kinds. This section briefly summarizes examples of that variation, organized around three conclusions. (a) Social essentialism in children exhibits a complex interaction of content, age, and cultural context. (b) Cultural influences on children's social essentialism can range in scale. (c) Different components of essentialism are separable early in development.

Social essentialism in children exhibits a complex interaction of content, cultural context, and age.   Children's social essentialism varies as a function of category content (gender, race, etc.), cultural context (i.e., the community in which the child lives), and child age—as well as interactions among these factors. The importance of considering all three factors simultaneously is illustrated by a study of essentialist reasoning in two different communities in southeast Michigan, only seventy-five miles apart: a liberal, ethnically diverse, midsized city and a conservative, rural, ethnically homogeneous town (Rhodes & Gelman, 2009). Children ranging in age from five to eighteen were tested on their essentialist beliefs about race, gender, animals, and artifacts. Children in the two communities did not differ in their essentialism of animals (consistently high) or artifacts (consistently low). However, differences emerged when reasoning about gender and race. Children in both communities started out high in gender essentialism, with responses in the two communities diverging over time: those in the midsized city showed decreasing gender essentialism with age, whereas those in the rural community maintained high levels throughout the age range. In contrast, children in both communities started out low in race essentialism, but diverged over time: those in the midsized city maintained low levels throughout the age range, whereas those in the rural community increased in racial essentialism with age.

More generally, a number of laboratories have found that for the category of gender, essentialism emerges early and in robust fashion across a range of societies, whereas for other categories, including race, ethnicity, nationality, socioeconomic status, or team membership, essentialism arises inconsistently, and often only at a later age if at all (Astuti et al., 2004; Davoodi, Soley, Harris, & Blake, 2019; Diesendruck, Birnbaum, Deeb, & Segall, 2013; Mandalaywala et al., 2018; Roberts & Gelman, 2017). It is notable that gender and race are treated differently from one another, given that both have visible correlates that children detect from an early age.

Relatedly, essentialism can either increase or decrease with age. In some studies conducted in the United States, for example, essentialism has been found to decrease over the elementary school years for gender (Taylor, 1996), personality traits (Gelman, Heyman, & Legare, 2007; Heyman & Gelman, 2000), religion (in some cases; Chalik, Leslie, & Rhodes, 2017), nationality (Hussak & Cimpian, 2019), and language (Hirschfeld & Gelman, 1997; Kinzler & Dautel, 2012). By contrast, children's essentialism of race, ethnicity, and religion often increase with age, across different cultural contexts (Diesendruck, Birnbaum, Deeb, Segall, & Ben-Eliyahu, 2010; Rhodes, Leslie, & Tworek, 2012; Smyth, Feeney, Eidson, & Coley, 2017). These findings suggest that young children have access to multiple causal models to explain human difference (Vasilyeva, Gopnik, & Lombrozo, 2018) and may be on the lookout for environmental cues to ramp up or tamp down essentialism, a point to which we return in the following section.

Cultural influences on children's social essentialism can range in scale. To date, relatively few studies have provided direct comparisons of childhood essentialism across cultural contexts, but those that exist suggest that cultural differences can have a profound effect (though see Davoodi et al., 2019, for an exception). One way this can occur is by introducing categorical distinctions that may not even exist outside that community (e.g., Catholic/Protestant; Dalit/Brahmin; royalty/commoner). Another way this can occur is by variations in cultural salience. As just one example, in a study of children in the United States and Israel, children's essentialism of race (White/Black) and ethnicity (Arab/Jewish) showed distinct developmental trends from ages five to ten. In the U.S. sample, race and ethnicity essentialism both increased with age, whereas in the Israeli sample, race essentialism decreased with age, and ethnicity essentialism was high at age five and remained stable over time (Diesendruck, Goldfein-Elbaz, Rhodes, Gelman, & Neumark, 2013).

Cultural influences can range in scale, from large scale (e.g., India versus the United States) to smaller scale (e.g., different regions of the United States) to what we might consider microscale, such as neighboring communities (Rhodes & Gelman, 2009), different schools within a community (Smyth et al., 2017), or different families within a community (Diesendruck & Haber, 2009). Even within a cultural community, there are subcultures that can have surprisingly large effects on children's reasoning. For example, Pauker, Xu, Williams, and Biddle (2016) found that children growing up in Massachusetts, a racially homogeneous state, showed an increase with age in essentialism and stereotyping of racial groups, whereas those in Hawaii, a racially diverse state, did not show this pattern.

More research is needed to understand the mechanisms underlying these differences. It would also be informative to test whether contexts that modify stereotyping also modify essentialism. Take, for example, children in Sweden who were enrolled in a gender-neutral preschool (Shutts, Kenward, Falk, Ivegran, & Fawcett, 2017). Children in the gender-neutral preschool were more like to choose playmates of another gender and gender stereotype less than their peers in conventional preschools. It was not the case that children in gender-neutral preschools noticed gender less; rather, they did not use the category as readily in their decisions. This task does not directly measure essentialism, but the findings are consistent with the "informativeness" component of essentialism. It would be fascinating in the future to examine gender essentialism in these children.

Different components of essentialism are separable early in development. As noted earlier, essentialism has multiple components. Any given social category may exhibit just a subset of these components (Rhodes & Mandalaywala, 2017). For example, people may treat age categories (e.g., teenager, old person) as deeply informative and having a biological reality, even though they also recognize that age is not immutable (teens grow into adults) and that the boundaries between age groups are fuzzy. Conversely, a given social category may be immutable (Daughter of the American Revolution, lefty) but not deeply informative. For adults, these various components tend to be mutually informative and intercorrelated. Adults show positive correlations among various essentialist judgments about different kinds of people. Moreover, upon learning that a new social category has one essentialist component, adults are likely to infer that it has other essentialist components as well (Gelman et al., 2007).

However, younger children appear to be less prone to assume interconnectedness among essentialist components than older children and adults (Gelman et al., 2007). For example, first-grade children who learned that a character was born with a given characteristic (she was born "very gogi") were no more likely to report that this feature was immutable or not under environmental influence, whereas third, fifth, and seventh graders did make this inference, as well as adults (Gelman et al., 2007). This can lead to different patterns of findings across tasks. Consider, for example, a study of ethnicity in Israel (Diesendruck, Birnbaum, et al., 2013). On a switched-at-birth task, the youngest children (kindergarteners) treated ethnic categories as biological, a belief that was revised over the next few years, resulting in lower essentialism by sixth grade. At the same time, on a task requiring children to assess the relative biological status of different social categories, they increasingly fine-tuned their expectations, resulting in increasing levels of ethnic essentializing over time on a task requiring children to judge the relative stability of ethnicity compared to occupation or body build (Diesendruck, Birnbaum, et al., 2013).

Environmental Effects

The variability we have discussed indicates that experiences play an important role in when and how children essentialize. These experiences do not create childhood essentialism (see Gelman, 2003, for discussion); rather, in the words of Rhodes and Mandalaywala (2017), children's essentialism of social kinds is due to "the interplay between children's conceptual biases and the environment they encounter." Identifying factors that influence children's essentialism of social kinds takes us one step closer to developing effective interventions to reduce essentialist thinking, with the goal of lessening the negative downstream consequences reviewed earlier (e.g., stereotyping, prejudice, and feeling a lack of control or agency to make positive changes). Here we briefly mention three factors that play a role in essentialist reasoning: children's contact with others, their own identity, and the language that they hear.

Contact. An extensive literature in social psychology indicates that whom children and youth come in contact with in school, at home, or in the broader community relates to how they view social categories (Pettigrew & Tropp, 2006). This also seems to be the case for essentialist reasoning. For example, White undergraduates who go to a more racially diverse college than their hometown tend to show less race essentialism after spending time at college (Pauker, Carpinella, Meyers, Young, & Sanchez, 2018). Similarly, the finding that children in Hawaii show lower levels of essentialism than children in Massachusetts may be due to the greater opportunities for cross-race contact in Hawaii (Pauker et al., 2016). Children in segregated schools also show greater levels of essentialism when reasoning about religious categories in Northern Ireland (Smyth et al., 2017), ethnic categories in Israel (Deeb, Segall, Birnbaum, Ben-Eliyahu, & Diesendruck, 2011), or racial categories in the United States (Hirschfeld, 1995). Moreover, Black-White multiracial children were more likely to impose discrete category boundaries on race (i.e., categorizing other Black-White individuals as more Black than White) when they had experienced greater levels of contact with White monoracial individuals (Roberts & Gelman, 2017).

There are other examples of contact effects on social concepts that have not yet been studied with regard to essentialism and thus are ripe for future study. For example, siblings of gender nonconforming and transgender children have direct, daily contact with an individual whose preferences and behaviors run counter to traditional gender stereotypes. These individuals, like their siblings, are less likely to endorse gender stereotype, and more accepting of others who violate gender stereotypes, compared with cisgender children without such extensive exposure to transgender people (Olson & Enright, 2018). These results provide a more nuanced view of how children's environment can influence how they conceptualize gender, and would be valuable to study with regard to essentialism.

Identity. Identity also plays an important role in how children essentialize categories. For example, in a study of children in a multicultural

London community, children viewed their own ethnicity as more stable than other ethnicities (Woods, 2017). This difference was hypothesized to be due to children's personal investment in their own ethnic identity, with its implications for family belonging and unity. Sometimes having a stronger connection with an identity, such as religion, corresponds to more essentializing of the category to which the individual belongs (Chalik et al., 2017; Smyth et al., 2017). However, identity can also predict lower rates of essentialism. Such is the case with multiracial children and adults who view race as less discrete (Roberts & Gelman, 2017). Political beliefs also relate to how much a person essentializes a particular category, such as race (Rhodes & Gelman, 2009) and nationality (Hussak & Cimpian, 2019).

The mechanisms underlying these different directions of influence are not well understood but are likely to reflect not just identity per se but also the nature of the experiences afforded by that identity (e.g., Morton & Postmes, 2009). For example, consider the case of essentializing language differences, such as believing that whether one speaks English or Portuguese is passed down from one's birth parents or is immutable over the course of development—beliefs that have been documented in monolingual children (Hirschfeld & Gelman, 1997; Kinzler & Dautel, 2012). Bilingual children show different rates of language essentialism, depending on whether they are simultaneously bilingual (exposed to and learned two languages from birth) or sequentially bilingual (first learned one language and then, after a few years, learned a second language) (Byers-Heinlein & Garcia, 2015). Children who had the experience of acquiring a second language at a later age were less essentialist about language (on a switched-at-birth task, viewing a character as more likely to speak the language of their adoptive parents than of their birth parents), compared with simultaneous bilinguals or monolinguals. This result seems to reflect sequential bilingual children's own firsthand experience with the effects of environmental input on their language skills. Surprisingly, this same result of lower levels of essentialism extended to their judgments about animal vocalizations and physical traits (Byers-Heinlein & Garcia, 2015).

The location of one's identity within the larger social structure (e.g., as a minority member or as belonging to a category with less power) also has consequences for essentialism. For example, in India, adults of a higher-ranking caste were more likely to view caste as determined by birth than those of a lower-ranking caste (Mahalingam, 2007). Analogously, when reasoning about poverty and wealth, wealthier Chilean children felt wealth was more inheritable and stable whereas less wealthy children felt it was less inheritable and more easily changed (del Río & Strasser, 2011).

The effect of a school environment as more or less integrated, noted earlier, is attenuated by the child's own location within the social structure. Children who are in the ethnic majority in an ethnically integrated school environment are more aware of others' ethnicity but also show a decrease in ethnic essentialism sooner than ethnic majority children in nonintegrated schools and children in the ethnic minority (Deeb et al., 2011). In general, children in the ethnic majority who come into more contact with people in the ethnic minority show less ethnic essentialism (Deeb et al., 2011).

Language.   When essentialist or antiessentialist messages are explicitly provided, they can shift attitudes. For example, undergraduates in a Psychology of Women course became more constructionist and less essentialist in their views on gender after taking the course (Yoder, Fischer, Kahn, & Groden, 2007). Conversely, when six-year-old children heard a story with overtly essentialist messaging (e.g., biological basis, immutability, discreteness), they were more likely to exhibit interethnic biases (Diesendruck & Menahem, 2015). However, children rarely receive overtly essentialist messages from their parents—for example, they rarely are told that certain properties are innate or caused by an internal essence (Gelman et al., 1998; Gelman, Taylor, & Nguyen, 2004).

In contrast, children hear a variety of linguistic devices that subtly imply that a category is natural, coherent, and richly informative (Gelman & Roberts, 2017). For example, lexicalizing a concept (stating it in the form of a noun label) implies that it is more stable across time and contexts (e.g., children treated someone who is "a carrot-eater" as more likely to eat carrots than one who "eats carrots whenever she can"; Gelman & Heyman, 1999; see also Diesendruck et al., 2010; Roberts et al., 2017; and Waxman, 2010). Children and adults also engage in higher levels of essentialism when a novel category is characterized with a series of generic statements (e.g., "Zarpies hate ice cream") than specific statements ("This Zarpie hates ice cream"), and this is true whether "Zarpies" are animal kinds (Gelman et al., 2010) or types of people (Rhodes et al., 2012). Furthermore, statements with subject-complement structure (e.g., "Girls are as good as boys at math") suggest to young children the essentialist assumption that the complement category (in this example, boys) is naturally better than the subject category (in this case, girls)—despite the overtly egalitarian message that is expressed (Chestnut & Markman, 2018). Parents' own language use corresponds to their children's level of essentialism, such that when parents used more generic statements and categories labels while reading to their children, children showed more essentialism (Segall, Birnbaum, Deeb, & Diesendruck, 2015). How messages are relayed to children and the person who relays the message may be key points to address when trying to reduce a child's essentialist beliefs.

Conclusions

Essentialism is a cognitive bias that distorts information by treating certain categories as biological, immutable, richly informative, and sharply distinct from other categories. From the preschool age, children display a foundational tendency to essentialize not just biological kinds (birds, cows) but also certain social kinds (girls, boys). At the same time, there is considerable cultural and developmental variation in essentialist beliefs. Not all categories are essentialized, and which social categories are essentialized is especially variable. Category content, cultural context, and child age all play a powerful role in the developmental trajectory of social essentialism, often interacting in combination. These different patterns reflect a variety of environmental factors, including a child's own identity, their contact with others, and both explicit and implicit messages provided in language.

An important direction for the future is to develop and test interventions for reducing essentialism. While essentialist beliefs have been found to have negative downstream consequences for adults (for stereotyping, prejudice, and discrimination), little is known regarding the effects on children (Rhodes, Leslie, Saunders, Dunham, & Cimpian, 2018), nor have effective interventions been developed, especially those that can be "scaled-up" for use in the real world. A further point to consider for the future is the timing of interventions. Determining when to intervene (i.e., at what age, in what context, and by whom) may be critical. For example, would anti-essentialism messages backfire if introduced at an age when children are entrenched in essentialist views of the category in question? Or is earlier always better in order to inoculate children to messages that they may be hearing in the future? These and other questions await further research.

References

Ahn, W. K., Flanagan, E. H., Marsh, J. K., & Sanislow, C. A. (2006). Beliefs about essences and the reality of mental disorders. *Psychological Science, 17*(9), 759–766.

Allport, G. W. (1954). *The nature of prejudice.* Cambridge, MA: Addison-Wesley.

Astuti, R., Solomon, G. E. A., & Carey, S. (2004). Constraints on conceptual development: A case study of the acquisition of folkbiological and folksociological knowledge in Madagascar. In *Monographs of the Society for Research in Child Development* (pp. 1–135). Boston: Blackwell.

Bastian, B., & Haslam, N. (2006). Psychological essentialism and stereotype endorsement. *Journal of Experimental Social Psychology, 42*(2), 228–235.

Booth, A. (2014). Conceptually coherent categories support name-based inductive inference in preschoolers. *Journal of Experimental Child Psychology, 123,* 1–14.

Byers-Heinlein, K., & Garcia, B. (2015). Bilingualism changes children's beliefs about what is innate. *Developmental Science, 18*(2), 344–350.

Chalik, L., Leslie, S., & Rhodes, M. (2017). Developmental psychology religion cultural context shapes essentialist beliefs about religion. *Developmental Psychology, 53*(6), 1178–1187.

Chestnut, E. K., & Markman, E. M. (2018). "Girls are as good as boys at math" implies that boys are probably better: A study of expressions of gender equality. *Cognitive Science, 42,* 2229–2249.

Dar-Nimrod, I., & Heine, S. J. (2011). Genetic essentialism: On the deceptive determinism of DNA. *Psychological Bulletin, 137*(5), 800–818.

Davoodi, T., Soley, G., Harris, P. L., & Blake, P. R. (2019). Essentialization of social categories across development in two cultures. *Child Development.* Advance online publication. doi: 10.1111/cdev.13209

Deeb, I., Segall, G., Birnbaum, D., Ben-Eliyahu, A., & Diesendruck, G. (2011). Seeing isn't believing: The effect of intergroup exposure on children's essentialist beliefs about ethnic categories. *Journal of Personality and Social Psychology, 101*(6), 1139–1156.

Del Río, M. F., & Strasser, K. (2011). Chilean children's essentialist reasoning about poverty. *British Journal of Developmental Psychology, 29*(4), 722–743.

Diesendruck, G. (2013). Essentialism: The development of a simple, but potentially dangerous, idea. In M. Banaji & S. Gelman (Eds.), *Navigating the social world: The early years* (pp. 263–268). New York: Oxford University Press.

Diesendruck, G., Birnbaum, D., Deeb, I., & Segall, G. (2013). Learning what is essential: Relative and absolute changes in children's beliefs about the heritability of ethnicity. *Journal of Cognition and Development, 14,* 546–560.

Diesendruck, G., Birnbaum, D., Deeb, I., Segall, G., & Ben-Eliyahu, A. (2010). The development of social essentialism: The case of Israeli children's inferences about Jews and Arabs. *Child Development, 81*(3), 757–777.

Diesendruck, G., Goldfein-Elbaz, R., Rhodes, M., Gelman, S., & Neumark, N. (2013). Cross-cultural differences in children's beliefs about the objectivity of social categories. *Child Development, 84*(6), 1906–1917.

Diesendruck, G., & Haber, L. (2009). God's categories: The effect of religiosity on children's teleological and essentialist beliefs about categories. *Cognition, 110*(1), 100–114.

Diesendruck, G., & Menahem, R. (2015). Essentialism promotes children's interethnic bias. *Frontiers in Psychology, 6,* 1180.

Elischberger, H. B., Glazier, J. J., Hill, E. D., & Verduzco-Baker, L. (2016). "Boys don't cry"—or do they? Adult attitudes toward and beliefs about transgender youth. *Sex Roles, 75*(5–6), 197–214.

Gaither, S. E., Schultz, J. R., Pauker, K., Sommers, S. R., Maddox, K. B., & Ambady, N. (2014). Essentialist thinking predicts decrements in children's memory for racially ambiguous faces. *Developmental Psychology, 50*(2), 482–488.

Gelman, S. A. (2003). *The essential child: Origins of essentialism in everyday thought.* New York: Oxford University Press.

Gelman, S. A. (2004). Psychological essentialism in children. *Trends in Cognitive Sciences, 8*(9), 404–409.

Gelman, S. A., Coley, J. D., Rosengren, K. S., Hartman, E., Pappas, A., & Keil, F. C. (1998). Beyond labeling: The role of maternal input in the acquisition of richly structured categories. *Monographs of the Society for Research in Child Development, 63*(1).

Gelman, S. A., Collman, P., & Maccoby, E. E. (1986). Inferring properties from categories versus inferring categories from properties: The case of gender. *Child Development, 57*(2), 396–404.

Gelman, S. A., & Davidson, N. S. (2013). Conceptual influences on category-based induction. *Cognitive Psychology, 66*(3), 327–353.

Gelman, S. A., & Heyman, G. D. (1999). The effects of lexicalization on children's inferences about. *Psychological Science, 10*(6), 489–493.

Gelman, S. A., Heyman, G. D., & Legare, C. H. (2007). Developmental changes in the coherence of essentialist beliefs about psychological characteristics. *Child Development, 78*(3), 757–774.

Gelman, S. A., & Markman, E. M. (1986). Categories and induction in young children. *Cognition, 23*(3), 183–209.

Gelman, S. A., & Rhodes, M. (2012). "Two-thousand years of stasis": How psychological essentialism impedes evolutionary understanding. In K. S. Rosengren, S.

Brem, E. M. Evans, & G. Sinatra (Eds.), *Evolution challenges: Integrating research and practice in teaching and learning about evolution.* New York: Oxford University Press.

Gelman, S. A., & Roberts, S. O. (2017). How language shapes the cultural inheritance of categories. *Proceedings of the National Academy of Sciences of the United States of America, 114*(30), 7900–7907.

Gelman, S. A., Taylor, M G., & Nguyen, S. (2004). Mother-child conversations about gender: Understanding the acquisition of essentialist beliefs. *Monographs of the Society for Research in Child Development, 69*(1).

Gelman, S. A., Ware, E. A., & Kleinberg, F. (2010). Effects of generic language on category content and structure. *Cognitive Psychology, 61*(3), 273–301.

Gelman, S. A., & Wellman, H. M. (1991). Insides and essences: Early understandings of the non-obvious. *Cognition, 38*(3), 213–244.

Haslam, N., Rothschild, L., & Ernst, D. (2000). Essentialist beliefs about social categories. *British Journal of Social Psychology, 39*(1), 113–127.

Heyman, G. D., & Gelman, S. A. (2000). Preschool children's use of trait labels to make inductive inferences. *Journal of Experimental Child Psychology, 77*(1), 1–19.

Hirschfeld, L. A. (1995). Do children have a theory of race? *Cognition, 54*(2), 209–252.

Hirschfeld, L. A. (1996). *Race in the making: Cognition, culture, and the child's construction of human kinds.* Cambridge, MA: MIT Press.

Hirschfeld, L. A., & Gelman, S. A. (1997). What young children think about the relationship between language variation and social difference. *Cognitive Development, 12,* 213–238.

Ho, A. K., Roberts, S. O., & Gelman, S. A. (2015). Essentialism and racial bias jointly contribute to the categorization of multiracial individuals. *Psychological Science, 26*(10), 1639–1645.

Hughes, I. A., Houk, C., Ahmed, S. F., & Lee, P. A. (2006). Consensus statement on management of intersex disorders. *Journal of Pediatric Urology, 2*(3), 148–162.

Hussak, L. J., & Cimpian, A. (2019). "It feels like it's in your body": How children in the United States think about nationality. *Journal of Experimental Psychology: General, 148*(7), 1153–1168.

James, W. (1890). *The principles of psychology.* New York: Holt.

Jaswal, V. K., Lima, O. K., & Small, J. E. (2009). Compliance, conversion, and category induction. *Journal of Experimental Child Psychology, 102*(2), 182–195.

Keil, F. C. (1989). *Concepts, kinds, and cognitive development.* Cambridge, MA: MIT.

Keller, J. (2005). In genes we trust: The biological component of psychological essentialism and its relationship to mechanisms of motivated social cognition. *Journal of Personality and Social Psychology, 88*(4), 686–702.

Kinzler, K. D., & Dautel, J. B. (2012). Children's essentialist reasoning about language and race. *Developmental Science, 15*(1), 131–138.

Kraus, M. W., & Keltner, D. (2013). Social class rank, essentialism, and punitive judgment. *Journal of Personality and Social Psychology, 105*(2), 247–261.

Leslie, S. J. (2013). Essence and natural kinds: When science meets preschooler intuition. *Oxford Studies in Epistemology, 4,* 108–165.

Leslie, S. J., Cimpian, A., Meyer, M., & Freeland, E. (2015). Expectations of brilliance underlie gender distributions across academic disciplines. *Science, 347*(6219), 262–265.

Mahalingam, R. (2007). Essentialism, power, and the representation of social categories: A folk sociology perspective. *Human Development, 50*(6), 300–319.

Mandalaywala, T. M., Ranger-Murdock, G., Amodio, D. M., & Rhodes, M. (2018). The nature and consequences of essentialist beliefs about race in early childhood. *Child Development, 90*(4), e437–e453.

Medin, D., & Ortony, A. (1989). Psychological essentialism. In S. Vosniadou and A. Ortony (Eds.), *Similarity and analogical reasoning* (pp. 179–195). New York: Cambridge University Press.

Meyer, M., & Gelman, S. A. (2016). Gender essentialism in children and parents: Implications for the development of gender stereotyping and gender-typed preferences. *Sex Roles, 75*(9–10), 409–421.

Morton, T. A., & Postmes, T. (2009). When differences become essential: Minority essentialism in response to majority treatment. *Personality and Social Psychology Bulletin, 35*(5), 656–668.

Moya, C., Boyd, R., & Henrich, J. (2015). Reasoning about cultural and genetic transmission: Developmental and cross-cultural evidence From Peru, Fiji, and the United States on how people make inferences about trait transmission. *Topics in Cognitive Science, 7*(4), 595–610.

Olson, K. R., & Enright, E. A. (2018). Do transgender children (gender) stereotype less than their peers and siblings? *Developmental Science, 21*(4), e12606.

Olson, K. R., Key, A. C., & Eaton, N. R. (2015). Gender cognition in transgender children. *Psychological Science, 26*(4), 467–474.

Pauker, K., Ambady, N., & Apfelbaum, E. P. (2010). Race salience and essentialist thinking in racial stereotype development. *Child Development, 81*(6), 1799–1813.

Pauker, K., Carpinella, C., Meyers, C., Young, D. M., & Sanchez, D. T. (2018). The role of diversity exposure in whites' reduction in race essentialism over time. *Social Psychological and Personality Science, 9*(8), 944–952.

Pauker, K., Xu, Y., Williams, A., & Biddle, A. M. (2016). Race essentialism and social contextual differences in children's racial stereotyping. *Child Development, 87*(5), 1409–1422.

Penslar, D. (2001). *Shylock's children: Economics and Jewish identity in modern Europe*. Berkeley: University of California Press.

Pettigrew, T. F., & Tropp, L. R. (2006). A meta-analytic test of intergroup contact theory. *Journal of Personality and Social Psychology, 90*(5), 751–783.

Quinn, P. C., Lee, K., & Pascalis, O. (2019). Face processing in infancy and beyond: The case of social categories. *Annual Review of Psychology, 70,* 165–189.

Rhodes, M., & Gelman, S. A. (2009). A developmental examination of the conceptual structure of animal, artifact, and human social categories across two cultural contexts. *Cognitive Psychology, 59*(3), 244–274.

Rhodes, M., Leslie, S. J., Saunders, K., Dunham, Y., & Cimpian, A. (2018). How does social essentialism affect the development of inter-group relations? *Developmental Science, 21*(1), e12509.

Rhodes, M., Leslie, S. J., & Tworek, C. M. (2012). Cultural transmission of social essentialism. *Proceedings of the National Academy of Sciences of the United States of America, 109*(34), 13526–13531.

Rhodes, M., & Mandalaywala, T. M. (2017). The development and developmental consequences of social essentialism. *Wiley Interdisciplinary Reviews: Cognitive Science, 8*(4), e1437.

Roberts, S. O., & Gelman, S. A. (2017). Multiracial children's and adults' categorizations of multiracial individuals. *Journal of Cognition and Development, 18*(1), 1–15.

Roberts, S. O., Ho, A. K., Rhodes, M., & Gelman, S. A. (2017). Making boundaries great again: Essentialism and support for boundary-enhancing initiatives. *Personality and Social Psychology Bulletin, 43*(12), 1643–1658.

Ruble, D. N., Taylor, L. J., Cyphers, L., Greulich, F. K., Lurye, L. E., & Shrout, P. E. (2007). The role of gender constancy in early gender development. *Child Development, 78*(4), 1121–1136.

Saffran, J. R., Aslin, R. N., & Newport, E. L. (1996). Statistical learning by 8-month-old infants. *Science, 274*(5294), 1926–1928.

Sartre, J.-P. (1995). *Anti-Semite and Jew* (G. G. Becker, Trans.). New York: Schocken. (Original work published 1948.)

Segall, G., Birnbaum, D., Deeb, I., & Diesendruck, G. (2015). The intergenerational transmission of ethnic essentialism: How parents talk counts the most. *Developmental Science, 18*(4), 543–555.

Shutts, K., Kenward, B., Falk, H., Ivegran, A., & Fawcett, C. (2017). Early preschool environments and gender: Effects of gender pedagogy in Sweden. *Journal of Experimental Child Psychology, 162,* 1–17.

Smyth, K., Feeney, A., Eidson, R. C., & Coley, J. D. (2017). Development of essentialist thinking about religion categories in Northern Ireland (and the United States). *Developmental Psychology, 53*(3), 475–496.

Tarlowski, A. (2018). Ontological constraints in children's inductive inferences: Evidence from a comparison of inferences within animals and vehicles. *Frontiers in Psychology, 9,* 520.

Taylor, M. G. (1996). The development of children's beliefs about social and biological aspects of gender differences. *Child Development, 67*(4), 1555–1571.

Taylor, M. G., Rhodes, M., & Gelman, S. A. (2009). Boys will be boys; cows will be cows: Children's essentialist reasoning about gender categories and animal species. *Child Development, 80*(2), 461–481.

Vasilyeva, N., Gopnik, A., & Lombrozo, T. (2018). The development of structural thinking about social categories. *Developmental Psychology, 54*(9), 1735–1744.

Waxman, S. R. (2010). Names will never hurt me? Naming and the development of racial and gender categories in preschool-aged children. *European Journal of Social Psychology, 40,* 593–610.

Waxman, S. R., Medin, D., & Ross, N. (2007). Folkbiological reasoning from a cross-cultural developmental perspective: early essentialist notions are shaped by cultural beliefs. *Developmental Psychology, 43*(2), 294–308.

Williams, M. J., & Eberhardt, J. L. (2008). Biological conceptions of race and the motivation to cross racial boundaries. *Journal of Personality and Social Psychology, 94*(6), 1033–1047.

Woods, R. (2017). The development of non-essentialist concepts of ethnicity among children in a multicultural London community. *British Journal of Developmental Psychology, 35*(4), 546–567.

Yoder, J. D., Fischer, A. R., Kahn, A. S., & Groden, J. (2007). Changes in students' explanations for gender differences after taking a psychology of women class: More constructionist and less essentialist. *Psychology of Women Quarterly, 31*(4), 415–425.

Zagefka, H., Pehrson, S., Mole, R. C. M., & Chan, E. (2010). The effect of essentialism in settings of historic intergroup atrocities. *European Journal of Social Psychology, 40*(5), 718–732.

17 Are Humans Born to Hate? Three Myths and Three Developmental Lessons about the Origins of Social Categorization and Intergroup Bias

Marjorie Rhodes

Overview

Are humans born to hate? The present chapter considers three common claims about the nature of social categorization—that humans are predisposed to racism, to dislike out-groups, and to think of differences between people in the same way as they think about differences between animal species—and addresses the developmental evidence on which these claims are based. Consideration of the processes underlying the emergence of these phenomena across childhood provides new insight into how to prevent the development of prejudice and improve intergroup relations.

Introduction

Racism and other forms of prejudice and discrimination have impeded the safety and survival, educational and economic opportunities, and physical and mental health, of countless people throughout the course of human history. The pervasiveness of racism around the world and throughout history and the evidence that racism and racial biases emerge early in human development have given rise to the view that racism and prejudice are the inevitable consequence of basic and inalterable features of human psychology. This view is problematic for two reasons. First, this view misrepresents what developmental science has revealed about the origins and ontogeny of racism and prejudice. Second, this view directs attention away from what we might learn from developmental science about how to prevent these negative phenomena in the next generation of children. In the present chapter, I describe three common myths about the developmental origins of racism and prejudice and three lessons from developmental science about how these phenomena emerge and might be prevented.

Myth 1: Babies Are Racist

The idea that humans are born racist, or with racial biases, originates in studies of infant looking behavior. For example, by three months, babies often look longer at faces of their own race (Bar-Haim, Ziv, Lamy, & Hodes, 2006; Kelly et al., 2005; Liu et al., 2015). By six-months, infants begin to categorize faces into groups that correspond to conventional race categories (Anzures, Quinn, Pascalis, Slater, & Lee, 2010). To illustrate this categorization behavior, after babies see a series of White faces on a screen, for example, they notice (as evidenced by increased attentive looking to the screen) if they are then shown a face of someone of a different race. This behavior suggests that the participating babies had grouped all of the White faces together as one type and then noticed a "change in type" that corresponds to race. Further, by the end of their first year, infants are often better at recognizing and remembering individual faces of people of their own race than people of other racial backgrounds (Anzures, Quinn, Pascalis, Slater, & Lee, 2013; Kelly et al., 2007). Thus, in several aspects of their visual behavior, babies appear to discriminate between members of their own race and others.

These findings are often described or interpreted as evidence that babies have racial biases, or even that humans are born racist (Kluger, 2014; Parry, 2012; Ryan, 2017). But this interpretation is wrong for two reasons. First, babies are not *born* with these behaviors. Newborn babies look equally long at people from different racial backgrounds (Kelly et al., 2005); babies develop the tendency to look longer at faces of their own race only after the first few months of life. Further, this development is entirely dependent on infants' environments. For instance, White babies growing up in Israel (who see mostly White faces) look longer at White faces; Black babies growing up in Ethiopia (who see mostly Black faces) look longer at Black faces; but Black babies growing up in Israel (who see both) look equally long at White and Black faces (Bar-Haim et al., 2006). Also, whereas babies often lose the ability to remember and recognize other-race faces by the end of the first year, babies who regularly see people of different racial backgrounds, or even who are intentionally shown diverse faces in books and media, do not (Anzures et al., 2013; Liu et al., 2015).

Babies process faces that seem familiar to them differently than faces that appear more different (Ellis, Xiao, Lee, & Oakes, 2016). Therefore, when infants grow up in environments that are racially homogeneous, they experience certain race faces as more familiar and begin to process them differently. But when babies grow up in diverse environments, they do not show these biases in their early looking behavior. All of this suggests that infant behaviors that might appear as early forms of racial biases do not reflect innate tendencies but instead are the result of early learning that takes place in particular environments.

The second reason why babies' looking behavior does not reflect early racial biases or innate racism is that even when babies grow up in homogeneous environments and *do* start to look longer at own-race faces, this longer looking does not necessarily reflect or have any implications for the development of biased attitudes, beliefs, or behaviors. It feels intuitive to assume that babies look longer at things they *like;* if this were the case, then infants' tendency to look longer at own-race faces might mean that they already prefer people with those faces. But although babies sometimes look longer at things they like (Kinzler, Doupoux, & Spelke, 2007), this is not always the case. Sometimes babies look longer at things because they are afraid of them or find them potentially threatening (LoBue & DeLoache, 2009; Rakison & Derringer, 2008). Sometimes, babies look longer at things that are familiar, surprising, or easier or harder to process (Powell & Spelke, 2018). What determines whether babies look longer at something depends on a wide variety of factors. Because there are so many reasons why babies might look longer at a particular stimulus, it is a mistake to assume that when babies look longer at faces of a particular race this pattern necessarily means they like them better.

In fact, there is considerable evidence against the idea that there is a straight developmental path from longer looking to biased attitudes. For instance, infants' tendency to look longer at faces of their own race does not even persist across the first year of life—by twelve months, babies often look longer at faces from *less* familiar racial groups (Liu et al., 2015; Singarajah et al., 2017). Also, in a direct test of infants' preferences, Kinzler and Spelke (2011) found that ten-month-old babies are just as likely to accept a toy from someone of the same or different race (even though this is a sensitive test of babies' preferences more generally).

In sum, babies are not racist, and humans are not born with racial biases. How babies begin to attend to race depends on the diversity of the environment that they grow up in. Growing up in a diverse environment, or even exposure to diverse faces via books or the media, can help infants maintain their abilities to recognize and remember individual faces (Lee, Quinn, & Pascalis, 2017). As noted previously, although there is not a straight line from infants' visual behavior to later bias in attitudes or social behavior, there is evidence that helping infants and young children retain these abilities to recognize and remember individuals from diverse racial groups can have positive consequences for intergroup relations (Lee et al., 2017). For example, providing children with targeted experiences with faces from diverse backgrounds reduces implicit race biases both in the moment and across time (Xiao et al., 2015; Qian et al., 2019). Therefore, the first lesson from developmental science on how to prevent the early emergence of racial biases is to expose infants and young children to people from diverse backgrounds.

Myth 2: People Are Predisposed to Hate Out-Groups

Another common myth about human psychology is that people are predisposed to hate out-groups. From this perspective, people cannot help but categorize into groups of "us" and "them" and to dislike people who are in the other group. If this were the case—if hating out-groups was the inevitable consequence of categorizing as "us" and "them"—then the particular stereotypes, experiences, or ideologies that people hold about certain groups are almost incidental (Dunham, 2018). From a strong version of this perspective, the simple act of categorizing is at the root of prejudice, discrimination, and intergroup conflict; the rest (e.g., group-specific stereotypes, biased ideologies and so on) is just justification.

The idea that people are predisposed to hate out-groups is often thought of as supported by the findings of social psychological studies that have tested what happens when people are placed into made-up groups for the first time during an experiment (Tajfel & Turner, 2004). This approach is useful because people do not bring any stereotypes or particular experiences with group members into the study. Researchers then vary features of the environment—whether the groups are in competition, differ in size, and so on, to see precisely what causes intergroup bias to emerge (for review, see Dunham, 2018). The striking finding from these studies is that intergroup bias emerges even in the "control" conditions: even when there is no competition, the groups are of equal size, neither group has more social power, and so on, people are still biased in favor of their own group.

As further evidence of the fundamental nature of intergroup bias, such "minimal group" effects emerge very early in childhood, long before children have extensive experience in formal social groups or teams (Dunham, Baron, & Carey, 2011). Even infants respond preferentially to people who are similar to them on arbitrary dimensions that are made salient in experimental contexts (Buttelmann, Zmvi, Daum, & Carpenter, 2012; Mahajan & Wynn, 2012).

Yet none of this indicates that out-group hate is a psychological primitive or inevitability. This is because there is a fundamental distinction between in-group love and out-group hate (Brewer, 1991). Minimal group paradigms reveal evidence of the former, but usually not of the latter. When someone in a minimal group paradigm chooses to do something nice (e.g., sharing a resource) for an in-group member rather than an out-group member, they usually do because they feel increased positive feelings for the in-group member, not because they feel more negatively toward the out-group member. Of course, this is still a form of intergroup bias and discrimination. But the most pernicious forms of intergroup bias—including prejudice, punitive treatment, dehumanization, and intergroup violence—seem uniquely motivated by out-group hate (Brewer, 1999).

In human childhood, in-group love develops before out-group hate, further reinforcing that these are not two sides of the same coin—it is possible to love one's own group and not hate the out-group at all (Nesdale, 2004). As an experimental demonstration of these effects, Buttelmann and Boehm (2014) found that already by age six children allocated more positive resources (e.g., stickers) to their in-group members than to out-group members. But, although children of this age also chose *not* to give negative resources (e.g., moldy toast) to their in-group, they did not systematically choose to give these negative things to the out-group either. They often discarded them into a neutral box. Thus, simply being placed into a minimal group did not lead six-year-old children to behave punitively toward the out-group. In this study, a tendency to behave badly to the out-group developed later in childhood. Studies on the development of racial attitudes also confirms this differentiation and age-related time lag between the emergence of in-group love and out-group hate (Nesdale, 2004).

Although feeling positively toward one's own group may indeed be a psychological inevitably (Dunham, 2018) and can itself have problematic consequences, thinking of in-group love and out-group hate as two sides of the same coin ignores the uniquely problematic nature of out-group hate and the particular circumstances that encourage it to develop. Considering the circumstances under which out-group hate develops reveals that—far from reflecting psychological primitives—the processes that give rise to out-group hate are clearly under societal control. For instance, children develop more out-group hate when they are exposed to specific derogatory stereotypes and ideologies designed to perpetuate oppressive status hierarchies (Bigler & Liben, 2007). Because of the pervasiveness of such stereotypes and ideologies, a second developmental lesson is the importance of actively preparing children to recognize and confront such stereotypes and problematic belief systems so that they will not be passive recipients of problematic messages. Whereas more research is needed on how to do this most effectively, recognizing the role of these experiential factors—instead of viewing out-group hate as the inevitable consequence of human psychology—is a step in the direction to motivate such endeavors.

Myth 3: People Inevitably Think of Differences between People as the Difference between Animal Species

The final common myth that I will address in this chapter is the idea that people inevitably think of human social categories as marking fundamentally distinct kinds of people—in the same manner as they think of tigers and sheep, for example, as fundamentally different kinds of animals (Rothbart & Taylor, 1992). This idea is pervasive in theoretical approaches to the origins of prejudice and discrimination from fields ranging from social and developmental psychology (Haslam, Rothschild, & Ernst, 2002; Hirschfeld, 1996) to philosophy (Leslie, 2017) and anthropology (Gil-White, 2001). From this perspective, racism and other forms of social prejudice are the inevitable by-product of one of the most central functions of the human cognitive system: the simple but fundamental tendency to classify individuals into categories that capture similarities across members and differences between groups.

These ideas originate in the literature on psychological essentialism. Psychological essentialism describes a set of intuitions that people hold about the structure and function of some everyday categories (Gelman, 2003). For instance, these intuitions include the beliefs that a baby animal born to tiger parents will be a tiger; that this baby tiger will inevitably grow up to have stripes, sharp teeth, and ferocious behavior, even if it is raised by a community of sheep; that this tiger will naturally have many features in common with other tigers and many differences from other animals (including those not yet observed or discovered); and that the difference between tigers and other animals is absolute, discrete, and reflects the objective and natural structure in the world (e.g., that the distinction between tigers and sheep is determined by nature and discovered by people, not the product of social construction). These ideas are referred to as psychological essentialism because they all reflect a core intuition that category identities are determined by an intrinsic category "essence" (e.g., that the baby tiger inherited a tiger essence from its parents), which determines category membership and category-related features.

In the case of a category like *tigers*, it might appear that the essentialist view of the category is roughly accurate, and even might arise from formal science education (e.g., learning about DNA in school). But neither of these things is quite right. By focusing on category identity as determined by an intrinsic entity located inside each individual animal, for example, essentialist views are inconsistent with modern understandings of species categories that emphasize population genetics (Gelman, 2003; Leslie, 2013). Also, by emphasizing stability across category members and time, essentialism leads people to neglect within-category variation, making it challenging for them to understand core scientific concepts like natural selection and speciation (Gelman & Rhodes, 2012; Shtulman & Schulz, 2008). Thus, rather than being the product of science education, essentialist intuitions impede the development of scientific understanding. Further, essentialist intuitions emerge before the onset of formal schooling—by the age of three years, children demonstrate each of the intuitions about animal categories just described (Gelman & Wellman, 1991; Waxman, Medin, & Ross, 2007). Thus, essentialist intuitions seem to reflect a flawed yet fundamental way of understanding the structure and meaning of some important everyday categories.

The myth of interest in this section is the idea that children are predisposed to think of differences between people through the same essentialist lens through which they understand animal species—that they inevitably think of differences between boys and girls, or White people and Black people, for example, in the same way as they think about the difference between tigers and sheep. Researchers from various disciplines have proposed that some social differences appear to the human mind to pattern like differences in animal species, either because they appear to be inherited from parents or to correlate with physical feature differences in a similar manner as species categories do (Davoodi, Soley, Harris, & Blake,

2019; Gil-White, 2001), or because they are labeled with the same type of noun labels that are frequently used to refer to basic-level animal species (e.g., referring to "girls" and "boys" in the same way as one might refer to "tigers" and "sheep") (Hirschfeld, 1995; Waxman, 1990). From this perspective, when people confront such differences, they cannot help but think of them as reflecting the same types of essential differences that they believe structure the biological world.

It is easy to see why such a view of the social world would be problematic and contribute to prejudice and other forms of intergroup bias. For example, an essentialist view of gender implies that it is impossible for one's gender to change over time and is associated with decreased acceptance of transgender identities and policies that support transgender rights and freedoms (Roberts, Ho, Rhodes, & Gelman, 2017). Further, essentialist views contribute to social stereotyping—leading people to assume that all members of a category should share the same features (e.g., that *all* girls should prefer pink to blue, for example; Bastian & Haslam, 2006; Taylor, Rhodes, & Gelman, 2009). Essentialist beliefs also promote problematic and inaccurate explanations for group differences—for instance, that the reason more men than women succeed in advanced mathematics is because of differences in inherent potential (Leslie, Cimpian, Meyer, & Freeland, 2015) or that the reason Black people in the United States have less wealth and social power is because they have less inherent value (Mandalaywala, Amodio, & Rhodes, 2017). Finally, because essentialism leads people to think of differences between people as if they are members of different species, essentialism can lead people to dehumanize members of social outgroups (Haslam, 2006).

Children do indeed sometimes represent some social differences as similar to species differences. For example, by age four, children often hold essentialist beliefs about gender. Children expect a baby that is born a girl will remain a girl and develop the stereotypical properties of girls (e.g., having long hair, liking tea sets, and so on), even if she grows up in an unusual environment where she is surrounded only by boys (analogous to their beliefs about a baby tiger who grows up surrounded by sheep; Taylor, 1996; Taylor et al., 2009). By age four, children also expect girls to share many features with one another even if they have very dissimilar appearances or personalities (Berndt & Heller, 1986; Gelman, Collman, & Maccoby, 1986), and they see the difference between boys and girls as reflecting the objective structure of the world instead of as socially constructed (Rhodes & Gelman, 2009).

Yet it is a myth that thinking of social differences in this way is an inevitable or inalterable consequence of human psychology. Children do not automatically think of differences between people in the same way as they think of the differences between animals. It is true they sometimes learn to think of social differences in essentialist terms, but when they do, these beliefs are the product of a protracted period of culturally embedded learning.

We know this is the case for two reasons. First, essentialism is not the product of an "on-or-off" switch in the mind that is automatically flipped "on" to understand species differences and differences between people. Essentialism includes a set of interrelated beliefs (e.g., that whether one is a tiger or not is determined by birth and is stable, that being a tiger inevitably means developing certain features, and that the distinction between tigers and other animals reflects the real structure of the biological world). These beliefs all seem to stem from a core commitment to the idea that categories are determined by an intrinsic category "essence." Yet these various aspects of people's essentialist beliefs are separable from one another, and they are often more highly dissociated for beliefs about social categories than for animal species (Gelman, Heyman, & Legare, 2007). For instance, in thinking about the social world, children might come to think that some category identities are informative about what a person is like (e.g., groups based on age or team memberships) but still not view these categories as stable over time. As another example, for race, children appear to think that a person's racial identity is determined by birth (e.g., at least to the extent that think a person's skin color will match their birth parents; Hirschfeld, 1995), but they do not think that race is an objective way to classify people (Rhodes & Gelman, 2009), or that people of the same race share behaviors or psychological properties (Mandalaywala, Ranger-Murdock, Amodio, & Rhodes, 2019). As these examples illustrate, essentialism is not the product of an on-or-off switch in the mind that is locked in the "on" position for social differences—instead, essentialism reflects a series of interrelated beliefs that can be endorsed for various degrees for different types of social categories.

Second, if the human mind could not help but think of social differences like animal species differences, then we would expect children to think of all social differences in this way, from as soon as they begin to recognize them. But instead children develop essentialist beliefs about social categories slowly, and in a manner that varies across cultural contexts. This pattern suggests that thinking of social differences like animal species (in essentialist terms) is the product of cultural learning, rather than an inevitable feature of the human mind.

To illustrate, consider the development of representations of race. Although by age four children view skin color as inherited (Hirschfeld, 1995), and sometimes they classify by race when prompted in experimental contexts by age five (Rhodes & Gelman, 2009), they do not think of race in strongly essentialist terms. For instance, five-year-old White children in the United States seem unsure about whether race is stable across an individual's life span, especially once that person experiences other types of changes (Kinzler & Dautel, 2012; Roberts & Gelman, 2016, 2017). Children at these ages also do not expect people of the same race to share any physical, psychological, or behavioral features (aside from skin color; Mandalaywala et al., 2019) and view the decision to classify by skin color as subjective and flexible (as reflecting a social construction rather than the objective structure of the world; Rhodes & Gelman, 2009; Diesendruck, Goldfein-Elbaz, Rhodes, Gelman, & Neumark, 2013). Across childhood, particularly between the ages of seven and ten years, children sometimes develop more strongly essentialist views of race, but whether they do and the extent to which they endorse these beliefs depends on their own racial identity, the diversity of their neighborhood, and the social and political ideology of their parents (Kinzler & Dautel, 2012; Mandalaywala et al., 2017; Rhodes & Gelman, 2009). Similar patterns of context-dependent developmental change have been found for other categories as well (e.g., those based on ethnicity, religion, and status) (Birnbaum, Deeb, Segall, Ben-Eliyahu, & Diesendruck, 2010; Deeb, Segall, Birnbaum, Ben-Eliyahu, & Diesendruck, 2011; Smyth, Feeney, Eidson, & Coley, 2017).

Thus, children *learn* to think of certain social differences in essentialist terms—like animal species differences—if those differences are made salient in their environment. Cultural learning plays a fundamental role of in shaping how children think about social differences, including whether they develop problematic representations that tie differences to natural "essences" and feed into prejudice, or more positive representations that allow them to appreciate social diversity without viewing all the differences between people as reflecting the essential structure of the world.

Both of these features of essentialist thought—that it is composed of a set of interrelated but separable beliefs and that these beliefs emerge slowly across childhood for particular social dimensions—can be understood by thinking about the processes underlying conceptual development.

Consideration of these processes also provides guidance about how to prevent the development of essentialist beliefs about particular social differences (and the resulting negative consequences).

For instance, children are particularly likely to develop essentialist beliefs about categories that they hear adults in their community describe with *generic claims* (Gelman, Taylor, & Nguyen, 2004; Gelman, Ware, & Kleinberg, 2010; Rhodes, Leslie, & Tworek, 2012; Segall, Birnbaum, Deeb, & Diesendruck, 2014). Generic descriptions include statements such as "Jews celebrate Passover," "Girls have long hair," or "A boy doesn't cry." Children recognize that generics describe abstract information about what kinds of people are like, and they thus assume that adults use generics to describe categories that are important, coherent, and meaningful (Foster-Hanson, Leslie, & Rhodes, 2019). Indeed, adults are more likely to produce generics in conversation with children (e.g., about gender, ethnicity, or another particular dimension of social difference) when they themselves hold essentialist beliefs about the category (Rhodes et al., 2012; Segall et al., 2014). Hearing generics can then lead children to identify a particular way of classifying people as reflecting an essential "species-like" difference when they would not otherwise view a group in this manner (Gelman et al., 2010; Rhodes et al., 2012; Rhodes, Leslie, Bianchi, & Chalik, 2018). Generics do not *create* essentialist beliefs—they guide children to map essentialist intuitions onto particular culturally relevant distinctions.

Careful analysis of the processes by which essentialist beliefs arise can provide insight into how to prevent the development of racism and prejudice. Thus, the third lesson from developmental science is to talk to children about social differences in ways that promote appreciation of diversity but do not promote the development of essentialist beliefs. The lesson from this is *not* to avoid talking about differences, but instead to keep in mind that children often draw certain assumptions from language that might be different from what the speakers intend. Thus, to avoid promoting essentialism, it can be helpful to talk about specific examples, to explicitly discuss within-group variation, and to provide direct information that group differences do not reflect differences in inherent potential or value. Thus, this research can provide direction on how to have productive conversations with children about differences, which can help them appreciate the importance of diversity without viewing it as indicating that people from different groups are fundamentally distinct kinds of people.

Conclusion

This chapter addressed three common views about the psychological origins of racism and other forms of prejudice, and argued that these views miss the mark—babies are not born racist, people are not predisposed to hate out-groups, and children do not inevitably think of differences between people in the same way as they understand differences between animal species. Yet each of these phenomena can—and often does—develop. That is, people become racist, begin to hate certain out-groups, and adopt views that particular differences between people are as fundamental and natural as animal species. When these beliefs develop, they have pernicious consequences for intergroup relations and the members of stigmatized groups. Therefore, it is critical to take a developmental approach to understanding the origins of these beliefs and attitudes. If we do not ask how these beliefs develop—and instead erroneously believe they are inevitable or there from the start—then we miss the opportunity to consider how they might be preempted or changed. By carefully analyzing the processes that give rise to racism and prejudice, we are better prepared to tackle the question of how these negative phenomena might be prevented in the next generation of children.

Acknowledgments: This chapter was supported by the Beyond Conflict Innovation Lab and the Eunice Kennedy Shriver National Institute of Child Health and Human Development of the National Institutes of Health under Award Number R01HD087672. The content is solely the responsibility of the authors and does not necessarily represent the official views of the National Institutes of Health.

References

Anzures, G., Quinn, P. C., Pascalis, O., Slater, A. M., & Lee, K. (2010). Categorization, categorical perception, and asymmetry in infants' representation of face race. *Developmental Science, 13*(4), 553–564. doi: 10.1111/j.1467-7687.2009.00900.x

Anzures, G., Quinn, P. C., Pascalis, O., Slater, A. M., & Lee, K. (2013). Development of own-race biases. *Visual Cognition, 21*(9–10), 1165–1182. doi: 10.1080/13506285.2013.821428

Bar-Haim, Y., Ziv, T., Lamy, D., & Hodes, R. M. (2006). Nature and nurture in own-race face processing. *Psychological Science, 17*(2), 159–163. doi: 10.1111/j.1467-9280.2006.01679.x

Bastian, B., & Haslam, N. (2006). Psychological essentialism and stereotype endorsement. *Journal of Experimental Social Psychology, 42*(2), 228–235. doi: 10.1016/J.JESP.2005.03.003

Berndt, T. J., & Heller, K. A. (1986). Gender stereotypes and social inferences: A developmental study. *Journal of Personality and Social Psychology, 50*(5), 889–898. doi: 10.1037/0022-3514.50.5.889

Bigler, R. S., & Liben, L. S. (2007). Developmental intergroup theory: Explaining and reducing children's social stereotyping and prejudice. *Current Directions in Psychological Science, 16,* 162–166.

Birnbaum, D., Deeb, I., Segall, G., Ben-Eliyahu, A., & Diesendruck, G. (2010). The development of social essentialism: The case of Israeli children's inferences about Jews and Arabs. *Child Development, 81*(3), 757–777. doi: 10.1111/j.1467-8624.2010.01432.x

Brewer, M. B. (1999). The psychology of prejudice: Ingroup love and outgroup hate? *Journal of Social Issues, 55,* 429–444. doi: 10.1111/0022–4537.00126

Buttelmann, D., & Boehm, R. (2014). The ontogeny of the motivation that underlies in-group bias. *Psychological Science, 25,* 921–927.

Buttelmann, D., Zmyj, N., Daum, M., & Carpenter, M. (2012). Selective imitation of in-group over out-group members in 14-month-old infants. *Child Development, 84*(2), 422–428. doi: 10.1111/j.1467-8624.2012.01860.x

Davoodi, T., Soley, G., Harris, P. L., & Blake, P. R. (2019). Essentialization of social categories across development in two cultures. *Child Development.* Advance online publication. doi: 10.1111/cdev.13209

Deeb, I., Segall, G., Birnbaum, D., Ben-Eliyahu, A., & Diesendruck, G. (2011). Seeing isn't believing: The effect of intergroup exposure on children's essentialist beliefs about ethnic categories. *Journal of Personality and Social Psychology, 101*(6), 1139–1156. doi: 10.1037/a0026107

Diesendruck, G., Goldfein-Elbaz, R., Rhodes, M., Gelman, S., & Neumark, N. (2013). Cross-cultural differences in children's beliefs about the objectivity of social categories. *Child Development, 84*(6), 1906–1917. doi: 10.1111/cdev.12108

Dunham, Y. (2018). Mere membership. *Trends in Cognitive Sciences, 22*(9), 780–793. doi: 10.1016/j.tics.2018.06.004

Dunham, Y., Baron, A. S., & Carey, S. (2011). Consequences of "minimal" group affiliations in children. *Child Development, 82*(3), 793–811. doi: 10.1111/j.1467-8624.2011.01577.x

Ellis, A. E., Xiao, N. G., Lee, K., & Oakes, L. M. (2017). Scanning of own- versus other-race faces in infants from racially diverse or homogenous communities. *Developmental Psychobiology, 59*(5), 613–627. doi: 10.1002/dev.21527

Foster-Hanson, E., Leslie, S.-J., & Rhodes, M. (2019). Speaking of kinds: How generic language shapes the development of category representations.

Gelman, S. A. (2003). *The essential child.* New York: Oxford University Press. doi: 10.1093/acprof:oso/9780195154061.001.0001

Gelman, S. A., Collman, P., & Maccoby, E. E. (1986). Inferring properties from categories versus inferring categories from properties: The case of gender. *Child Development, 57*(2), 396–404.

Gelman, S. A., Heyman, G. D., & Legare, C. H. (2007). Developmental changes in the coherence of essentialist beliefs about psychological characteristics. *Child Development, 78*(3), 757–774. doi: 10.1111/j.1467-8624.2007.01031.x

Gelman, S. A., & Rhodes, M. (2012). "Two-thousand years of stasis": How psychological essentialism impedes evolutionary understanding. In K. S. Rosengren, S. K. Brem, E. M. Evans, & G. M. Sinatra (Eds.) *Evolution challenges: Integrating research and practice in teaching and learning about evolution* (pp. 3–21). Oxford: Oxford University Press. doi: 10.1093/acprof:oso/9780199730421.003.0001

Gelman, S. A., Taylor, M. G., & Nguyen, S. P. (2004). Mother-child conversations about gender: Understanding the acquisition of essentialist beliefs. *Monographs of the Society for Research in Child Development, 69*(1). doi: 10.1111/j.1540-5834.2004.06901002.x

Gelman, S. A., Ware, E. A., & Kleinberg, F. (2010). Effects of generic language on category content and structure. *Cognitive Psychology, 61*(3), 273–301. doi: 10.1016/J.COGPSYCH.2010.06.001

Gelman, S. A., & Wellman, H. M. (1991). Insides and essences: Early understandings of the non-obvious. *Cognition, 38*(38), 213–244.

Gil-White, F. J. (2001). Are ethnic groups biological "species" to the human brain? Essentialism in our cognition of some social categories. *Current Anthropology, 42*(4), 515–554. doi: 10.2307/3596550

Haslam, N. (2006). Dehumanization: An Integrative Review. *Personality and Social Psychology Review, 10*(3), 252–264. doi: 10.1207/s15327957pspr1003_4

Haslam, N., Rothschild, L., & Ernst, D. (2002). Are essentialist beliefs associated with prejudice? *British Journal of Social Psychology, 41*(1), 87–100. doi: 10.1348/014466602165072

Hirschfeld, L. A. (1995). Do children have a theory of race? *Cognition, 54*(2), 209–252. doi: 10.1016/0010-0277(95)91425-R

Hirschfeld, L. A. (1996). *Learning, development, and conceptual change. Race in the making: Cognition, culture, and the child's construction of human kinds.* Cambridge, MA: MIT Press.

Kelly, D. J., Gibson, A., Smith, M., Pascalis, O., Quinn, P. C., Slater, A. M., ... Ge, L. (2005). Three-month-olds, but not newborns, prefer own-race faces. *Developmental Science, 8*(6), 31–36. doi: 10.1111/j.1467-7687.2005.0434a.x

Kelly, D. J., Quinn, P. C., Slater, A. M., Lee, K., Ge, L., & Pascalis, O. (2007). The other-race effect develops during infancy. *Psychological Science, 18*(12), 1084–1089. doi: 10.1111/j.1467-9280.2007.02029.x

Kinzler, K. D., & Dautel, J. B. (2012). Children's essentialist reasoning about language and race. *Developmental Science, 15*(1), 131–138. doi: 10.1111/j.1467-7687.2011.01101.x

Kinzler, K. D., Dupoux, E., & Spelke, E. S. (2007). The native language of social cognition. *Proceedings of the National Academy of Sciences of the United States of America, 104*(30), 12577–12580. doi: 10.1073/pnas.0705345104

Kinzler, K. D., & Spelke, E. S. (2011). Do infants show social preferences for people differing in race? *Cognition, 119*(1), 1–9. doi: 10.1016/j.cognition.2010.10.019

Kluger, J. (2014, April 17). Your baby is a racist—and why you can live with that. *Time.* https://time.com/67092/baby-racists-survival-strategy/

Lee, K., Quinn, P. C., & Pascalis, O. (2017). Face race processing and racial bias in early development: A perceptual-social linkage. *Current Directions in Psychological Science, 26*(3), 256–262. doi: 10.1177/0963721417690276

Leslie, S. J. (2013). Essence and natural kinds: When science meets preschooler intuition. In T. S. Gendler & J. Hawthorne (Eds.), *Oxford studies in epistemology* (Vol. 4, pp. 108–165). New York: Oxford University Press. doi: 10.1093/acprof:oso/9780199672707.003.0005

Leslie, S. J. (2017). The original sin of cognition. *Journal of Philosophy, 114,* 393–421. doi: 10.5840/jphil2017114828

Leslie, S., Cimpian, A., Meyer, M., & Freeland, E. (2015). Expectations of brilliance underlie gender distributions across academic disciplines. *Science, 347*(6219), 262–265.

Liu, S., Xiao, W. S., Xiao, N. G., Quinn, P. C., Zhang, Y., Chen, H., … Lee, K. (2015). Development of visual preference for own-versus other-race faces in infancy. *Developmental Psychology, 51*(4), 500–511. doi: 10.1037/a0038835

LoBue, V., & DeLoache, J. S. (2009). Superior detection of threat-relevant stimuli in infancy. *Developmental Science, 13*(1), 221–228. doi: 10.1111/j.1467-7687.2009.00872.x

Mahajan, N., & Wynn, K. (2012). Origins of "us" versus "them": Prelinguistic infants prefer similar others. *Cognition, 124*(2), 227–233. doi: 10.1016/j.cognition.2012.05.003

Mandalaywala, T. M., Amodio, D. M., & Rhodes, M. (2017). Essentialism promotes racial prejudice by increasing endorsement of social hierarchies. *Social Psychological and Personality Science, 9*(4), 461–469. doi: 10.1177/1948550617707020

Mandalaywala, T. M., Ranger-Murdock, G., Amodio, D. M., & Rhodes, M. (2019). The nature and consequences of essentialist beliefs about race in early childhood. *Child Development, 90*(4), e437–e453. doi: 10.1111/cdev.13008

Nesdale, D. (2004). Social identity processes and children's ethnic prejudice. In M. Bennett & F. Sani (Eds.), *The development of the social self* (pp. 219–245). New York: Psychology Press. doi: 10.4324/9780203391099_chapter_8

Parry, W. (2012, May 4). Racist babies? Nine-month-olds show bias when looking at faces, study shows. *Huffpost.* https://www.huffpost.com/entry/racist-babies-nine -month-olds-bias-faces_n_1477937

Powell, L. J., & Spelke, E. S. (2018). Human infants' understanding of social imitation: Inferences of affiliation from third party observations. *Cognition, 170,* 31–48. doi: 10.1016/j.cognition.2017.09.007

Qian, M. K., Quinn, P. C., Heyman, G. D., Pascalis, O., Fu, G., & Lee, K. (2019). A long-term effect of perceptual individuation training on reducing implicit racial bias in preschool children. *Child Development, 90*(3), e290–e305. doi: 10.1111/cdev.12971

Rakison, D. H., & Derringer, J. (2008). Do infants possess an evolved spider-detection mechanism? *Cognition, 107*(1), 381–393. doi: 10.1016/j.cognition.2007.07.022

Rhodes, M., & Gelman, S. A. (2009). A developmental examination of the conceptual structure of animal, artifact, and human social categories across two cultural contexts. *Cognitive Psychology, 59*(3), 244–274. doi: 10.1016/j.cogpsych.2009.05.001

Rhodes, M., Leslie, S. J., Bianchi, L., & Chalik, L. (2018). The role of generic language in the early development of social categorization. *Child Development, 89*(1), 148–155. doi: 10.1111/cdev.12714

Rhodes, M., Leslie, S. J., & Tworek, C. M. (2012). Cultural transmission of social essentialism. *Proceedings of the National Academy of Sciences of the United States of America, 109*(34), 13526–13531. doi: 10.1073/pnas.1208951109

Roberts, S. O., & Gelman, S. A. (2016). Can white children grow up to be black? Children's reasoning about the stability of emotion and race. *Developmental Psychology, 52*(6), 887–893. doi: 10.1037/dev0000132

Roberts, S. O., & Gelman, S. A. (2017). Now you see race, now you don't: Verbal cues influence children's racial stability judgments. *Cognitive Development, 43,* 129–141. doi: 10.1016/j.cogdev.2017.03.003

Roberts, S. O., Ho, A. K., Rhodes, M., & Gelman, S. A. (2017). Making boundaries great again: Essentialism and support for boundary-enhancing initiatives. *Personality and Social Psychology Bulletin, 43*(12), 1643–1658. doi: 10.1177/0146167217724801

Rothbart, M., & Taylor, M. (1992). Category labels and social reality: Do we view social categories as natural kinds? In G. R. Semin & K. Fiedler (Eds.), *Language, interaction and social cognition* (pp. 11–36). Thousand Oaks, CA: Sage.

Ryan, L. (2017, April). Science says everyone's a little bit racist—even babies. *Parents.* https://www.parents.com/baby/all-about-babies/science-says-everyones-a-little-bit -racist-even-babies

Segall, G., Birnbaum, D., Deeb, I., & Diesendruck, G. (2014). The intergenerational transmission of ethnic essentialism: How parents talk counts the most. *Developmental Science, 18*(4), 543–555.

Shtulman, A., & Schulz, L. (2008). The relation between essentialist beliefs and evolutionary reasoning. *Cognitive Science: A Multidisciplinary Journal, 32*(6), 1049–1062. doi: 10.1080/03640210801897864

Singarajah, A., Chanley, J., Gutierrez, Y., Cordon, Y., Nguyen, B., Burakowski, L., & Johnson, S. P. (2017). Infant attention to same- and other-race faces. *Cognition, 159,* 76–84. doi: 10.1016/j.cognition.2016.11.006

Smyth, K., Feeney, A., Eidson, R. C., & Coley, J. D. (2017). Development of essentialist thinking about religion categories in Northern Ireland (and the United States). *Developmental Psychology, 53*(3), 475–496. doi: 10.1037/dev0000253

Tajfel, H., & Turner, J. C. (2004). The social identity theory of intergroup behavior. In J. T. Jost & J. Sidanius (Eds.), *Key readings in social psychology. Political psychology: Key readings* (pp. 276–293). New York: Psychology Press. http://dx.doi.org/10.4324/9780203505984-16

Taylor, M. G. (1996). The development of children's beliefs about social and biological aspects of gender differences. *Child Development, 67*(4), 1555–1571.

Taylor, M. G., Rhodes, M., & Gelman, S. A. (2009). Boys will be boys; Cows will be cows: Children's essentialist reasoning about gender categories and animal species. *Child Development, 80*(2), 461–481.

Waxman, S. R. (1990). Linguistic biases and the establishment of conceptual hierarchies: Evidence from preschool children. *Cognitive Development, 5*(2), 123–150. doi: 10.1016/0885-2014(90)90023-M

Waxman, S., Medin, D., & Ross, N. (2007). Folkbiological reasoning from a cross-cultural developmental perspective: Early essentialist notions are shaped by cultural beliefs. *Developmental Psychology, 43*(2), 294–308. doi: 10.1037/0012-1649.43.2.294

Xiao, W. S., Fu, G., Quinn, P. C., Qin, J., Tanaka, J. W., Pascalis, O., & Lee, K. (2015). Individuation training with other-race faces reduces preschoolers' implicit racial bias: a link between perceptual and social representation of faces in children. *Developmental Science, 18*(4), 655–663. doi: 10.1111/desc.12241

V   Atypical Social Cognition

18   Toward a Translational Developmental Social Cognitive Neuroscience of Autism

Kevin A. Pelphrey, Jennifer R. Frey, and Michael J. Crowley

Overview

The field of developmental social neuroscience has recently matured to yield the first tentative, biologically meaningful markers of autism. This advancement signals a new era in which advanced neuroimaging analysis could evolve into an integral part of a translational research chain wherein biomarkers will guide personalized therapies. Here we review how these initial autism spectrum disorder biomarkers have arisen from (a) major advances from social neuroscience studies of adults; (b) rapid progress in applying these advances to the study of infants, children, and adolescents; (c) insights gained from comparing groups with and without autism spectrum disorder; and (d) applying artificial intelligence to derive sensitive, reliable brain measures that are informative at the level of the individual. Work on the horizon could provide an empirical realization, in humans, of "probabilistic epigenetic" theories (Gottlieb, 1992) wherein social cognitive development is modeled as emerging from gene ↔ brain ↔ behavior relationships occurring between levels of an active, developing organism in its social environment.

Introduction

Autism spectrum disorder (ASD) is a common neurodevelopmental syndrome characterized by deficits in social communication and the presence of restricted and repetitive patterns of thought and behavior (American Psychiatric Association, 2013). ASD usually emerges in the first three years of life, persisting throughout the individual's life. Under this single syndromic umbrella, we see tremendous variability across affected individuals—well beyond differences in level of intellectual functioning or overall symptom severity. In addition to clinical comorbidities and sex differences, this heterogeneity likely reflects distinct etiologies underlying multiple "autisms" (Geschwind & Levitt, 2007). Although researchers have identified genetic differences associated with individual cases of ASD, each one accounts for only a small number of the actual cases, suggesting that no single genetic cause will apply to the majority of people with ASD.

Historically, a lack of predictive, biologically informed profiles has contributed to the imprecise treatments, misused time and resources, and failures to optimize progress for children and families living with ASD. Fortunately, this situation is rapidly improving. Research within the field of developmental social neuroscience has now advanced sufficiently to provide the first sensitive, quantitative, and biologically meaningful markers of ASD symptoms (e.g., Kaiser et al., 2010). These tentative *biomarkers* are now being evaluated for their utility in measuring and predicting individual responses to evidenced-based interventions (e.g., Yang et al., 2017). As such, the work has advanced our understanding of ASD and may ultimately inform and guide personalized therapies. These promising advances in ASD biomarker discovery emerged from the juncture of four streams of inquiry.

The first tributary was major advances generated through social cognitive neuroscience studies of "neurotypical" (typically developing) adults. Humans are deeply social creatures. For millennia, we have existed in highly collective environments in which each individual depends upon others as well as larger social entities, such as family and kin, peers and neighbors, institutions and society. Social cognition is the term we use to describe abilities to categorize, remember, analyze, reason with, and behave toward other conspecifics. The extent to which such processes work successfully helped determined the fate of individual humans in the past and continues to do so today. Much of the neural circuitry supporting social cognition consists of mechanisms that are relatively old in evolutionary terms. The social world in which the human brain carries out its functions, however, has changed dramatically over a relatively short period. For example, a child born 50,000 years ago would have very similar mental powers as a child today but a decidedly different set of demands, aspirations, educational tools, and opportunities—not to mention rights, responsibilities, chances of survival, and definitions of success. For much of the twentieth century, experimental social psychologists produced theories and a wealth of behavioral evidence concerning the basic building blocks of social cognition. In contrast, a neuroscience of social cognition is a relatively new invention (e.g., Cacioppo et al., 2007).

Traditional invasive neuroscience techniques were applied to the brains of nonhuman animals to study how brains see, hear, feel, move, and remember. Uniquely human cognitive capacities, such as language and social cognition, could not be studied in nonhuman animals. When noninvasive brain imaging techniques, including scalp-recorded electroencephalography (EEG) and functional magnetic resonance imaging (fMRI), were first applied to human adults, the key test of these technologies was replication of functions known from nonhuman animals. Thus, early neuroimaging studies focused on the human homologues of known regions from other primates: early visual cortex, the motion perception region (MT), early sensory, and motor cortices. Building upon these early neuroimaging successes, and armed with a wealth of theories and hypotheses generated within the fields of social psychology, behavioral neurology (the study of the behavioral consequences of localized brain injury), and neuroscientific studies of social behavior in nonhuman animals, a generation of social neuroscientists has discovered that the human brain develops a network of many cortical and subcortical regions, previously unknown or little studied, which have apparently social functions. For instance, the "fusiform face area" (FFA) is involved in perceiving human faces (Kanwisher, McDermott, & Chun, 1997; McCarthy, Puce, Gore, & Allison, 1997; Puce, Allison, Gore, & McCarthy, 1995). The extrastriate body area is involved in perceiving human bodies (Downing, Jiang, Shuman, & Kanwisher, 2001). The right posterior superior temporal sulcus (pSTS) is involved in perceiving and analyzing human actions (Pelphrey, Viola, & McCarthy, 2004). The right temporoparietal junction (RTPJ) is involved in reasoning about people's thoughts (Saxe & Kanwisher, 2003). The medial precuneus, posterior cingulate and medial prefrontal cortices are involved in other aspects of higher level social cognition (Amodio & Frith, 2006). A set of regions involving portions of the parietal cortex, including the inferior and superior parietal lobules, the anterior intraparietal sulcus, and frontal cortical regions such as the inferior frontal gyrus (IFG), has been dubbed the "mirror system" in humans. These regions are involved in both the execution of a motor action and the observation of a motor action performed by another person (Rizzolatti & Fabbri-Destr, 2008). The individual discoveries of these "social brain" regions, combined with knowledge of how they interact, led to a deeper understanding of how humans think about themselves and other minds, mimic other's actions and change, and regulate emotions and actions. In turn, the functioning of the social brain has allowed us to learn about the path toward better and worse mental health outcomes. However, up until the past decade, social neuroscience studies focused on the mature minds of adult humans, almost exclusively.

More recently, with a wave of brain imaging studies conducted by pioneers in the field and a generation of new scientists trained in the methods and theories of developmental social cognitive neuroscience, autism biomarker discovery is now benefiting from a very rich and quickly expanding understanding of how developmental changes in brain structure, neural function and connectivity, and temporal dynamics support the emergence, refinement, and integration of social cognitive abilities. Rapid advances have occurred in what we know about the neural mechanisms supporting multimodal social representations in infants and children and the changes in brain function, connectivity, and temporal dynamics that underlie normative behavioral development in social cognition.

Consider, for example, our laboratory's study of neural responses to speech compared with biological nonspeech sounds in sleeping one- to four-month-old infants using fMRI (Shultz, Vouloumanos, Bennett, & Pelphrey, 2014). Although the behavioral findings indicate that neonates' listening biases are sharpened over the first months of life, with a species-specific preference for speech emerging by three months, the neural substrates underlying this developmental change are unknown. To address this knowledge gap, we did a study where infants heard speech and biological nonspeech sounds, including heterospecific vocalizations and human nonspeech. We observed a left-lateralized response in temporal cortex for speech compared with biological nonspeech sounds, indicating that this region is highly selective for speech by the first month of life. Moreover, this brain region becomes increasingly selective for speech over the next three months, while neural substrates become less responsive to nonspeech sounds. These results reveal specific changes in neural responses during a developmental period characterized by rapid behavioral changes.

This study is particularly significant because it provided us with the methodological and theoretical foundations to study the development of neural specialization for social information processing in a very large, epidemiologically representative sample of infants. By establishing a normative trajectory of social brain development from zero to twelve months, we will provide critical information that is positioned to revolutionize how our society thinks about and facilitates the development of social cognition. Further, this study provides the foundation for work that has the potential to change the ways in which the medical community predicts and treats the emergence of childhood neurodevelopmental disorders, such as ASD. To extend the power of our approach, our future work will produce "growth charts" of social brain development and use them to develop and validate a highly sensitive, reliable test that can reveal early neural systems markers of the development of ASD.

Similarly, in preschool age children, we have used fMRI to examine the typical development of brain mechanisms for the visual processing of social categories (e.g., faces) and nonsocial cultural artifacts (e.g., words and numbers). It was well known that in adults category-based specializations manifest as greater neural responses in visual regions of the brain (e.g., the fusiform gyrus) to some categories over others. However, few studies examine how these specializations originate in the brains of young children. Moreover, it was unknown whether the development of visual specializations hinges upon "increases" in the responses to the preferred categories, "decreases" in the responses to nonpreferred categories, or both. This question is relevant to a long-standing debate concerning whether neural development is driven by building up or pruning back representations.

We measured patterns of visually evoked brain activity in four-year-old children for four categories (faces, letters, numbers, and shoes) (Cantlon, Pinel, Dehaene, & Pelphrey, 2011). We reported two key findings regarding the development of visual categories in the brain: (a) the categories "faces" and "symbols" doubly dissociate in the fusiform gyrus before children can read, and (b) the development of category-specific responses in young children depends on cortical responses to nonpreferred categories that decrease as preferred category knowledge is acquired. Our results indicate that children's increasing skill in face and symbol recognition is accompanied by decreased responses to nonpreferred stimuli (pruning) as opposed to an increased response to the preferred category.

The notion that learning proceeds by selection, attrition, or a "use it or lose it" principle has long been proposed at the theoretical level and has received empirical support in domains such as bird song acquisition or speech perception that exhibit perceptual narrowing over development. Our data suggest a similar selectionist principle in the development of category-selective occipitotemporal cortex in human children.

Cross-sectional and snapshots of development studies like those just discussed are useful for sketching out the broad outline of developmental processes. But ultimately, we look forward to the time when we have enough findings in hand to construct normative developmental curves or "growth charts"—for the functioning of circuits supporting integration of visual, auditory, and tactile social cues in the service of social perception and social cognition. These normative curves would document the changes in brain function, connectivity, and temporal dynamics as a function of the infant/young child's age and sex and also detect changes in the circuitry underlying developments in these domains during critical periods of childhood before, during, and after major developmental transitions. The availability of these normative data would facilitate efforts to characterize atypical developmental pathways, advancing the development of more effective interventions for social deficits across diagnostic categories, including the autism spectrum.

Because autism is a developmental disorder, it is particularly important to diagnose and treat ASD early in life. Early deficits in attention to other's actions, for instance (what we call biological motion), derail subsequent experiences in attending to higher level social information, thereby driving development toward more severe dysfunction and stimulating deficits in additional domains of functioning, such as language development. The lack of reliable predictors of ASD during the first year of life has been a major impediment to its effective treatment. Without early predictors, and in the absence of a firm diagnosis until behavioral symptoms emerge, treatment is often delayed for two or more years, eclipsing a crucial period in which intervention may be particularly successful in ameliorating some of the social and communicative impairments seen in ASD.

In response to the great need for sensitive (able to identify subtle cases) and specific (able to distinguish autism from other disorders) early indicators of ASD, such as biomarkers, many research teams from around the world have been studying patterns of infant development using prospective longitudinal studies of infant siblings of children with ASD and compared to infant siblings without familial risks. Such designs gather longitudinal information about developmental trajectories across the first three years of life, followed by clinical diagnosis at approximately thirty-six months. Biobehavioral markers—especially eye-tracking studies—of multidimensional social processes are beginning to prove useful in early identification of atypical developmental processes among children at increased genetic risk for developing difficulties in social information processing (e.g., Jones & Klin, 2013).

Studies pursing early biobehavioral markers are potentially problematic in that many of the social features of autism do not emerge in typical development until after twelve months of age. It is not certain that these symptom features will manifest during the limited periods of observation involved in clinical evaluations or in pediatricians' offices. Moreover, across development, but especially during infancy, behavior is widely variable and often unreliable. At present, longitudinal behavioral observation is the only means to detect the emergence of ASD and to predict a diagnosis (Jones & Klin, 2013).

However, measuring the brain activity associated with social perception can detect differences not appearing behaviorally until much later. The identification of biomarkers using the imaging methods we have described offers promise for earlier detection of atypical social development. ERP measures of brain response predict subsequent development of autism in infants as young as six months old who showed normal patterns of visual fixation (as measured by eye tracking) (Elsabbagh et al., 2012). These types studies illustrate the great promise of brain imaging for earlier recognition of ASD. With earlier detection, treatments could move from addressing existing symptoms to preventing their emergence by altering the course of abnormal brain development and steering it toward compensation or normality.

To the extent that research can further elucidate developmental trajectories of the neural circuitry supporting pivotal, early developing social abilities, it might inform the design of more effective programs for identifying and remediating risk for difficulties in these areas. It is generally accepted that earlier educational interventions are more effective for treating a variety of behavioral and academic childhood problems and neurodevelopmental disorders. Therefore, early identification of children with difficulties in social cognition is pivotal to optimizing individual intervention outcomes.

A neurobiological marker for individual differences in multidimensional social representation abilities would then not solely be important for improving early identification, but also could offer advantages for earlier interventions. Moreover, it could be that the neurobiological marker would relate to the severity of specific deficits, helping us to better understand the heterogeneity characteristic of neuropsychiatric disorders. With this information, more targeted treatments could be developed, on a child-by-child basis, and implemented early in ontogeny. Early targeted intervention then might guarantee the most effective course of intervention possible, also improving efficiency and standardization. Further, functional neuroimaging techniques might actually provide a means to better quantify treatment effectiveness and reveal whether behavioral improvements correspond to compensatory changes in brain function or normalization of developmental pathways.

Early and longitudinal study will be critical in defining brain phenotypes because the shape of developmental trajectories of brain functioning in specific circuits will provide more detail on the nature of the abnormalities than will analysis of brain phenotypes in adulthood. Despite the progress, there is still much to learn about the early longitudinal changes in brain connectivity, function, and temporal dynamics that support the development of the ability to integrate a broad array of emotional and social cues from multiple sensory modalities (e.g., vision, touch, audition) in the service of social cognition. Likewise, the neurobiological basis of individual differences in multimodal, multidimensional social cognitive abilities remains poorly understood.

There are straightforward and compelling methodological reasons to adopt a longitudinal design. Neuroimaging data are inherently noisy because individual brains are different from one another. A longitudinal design is the only way to study developmental processes coupled with the power of within subject statistics. It also makes it more likely that we will be able to detect relationships between different developmental changes. For example, just knowing that both task A and task B change between four and six months of age tells us almost nothing about those tasks, but individual differences in the response to task A at time 1 or the change in task A performance are good predictors of individual differences in the magnitude of change in task B will yield much stronger inferential leverage upon which to build lasting theoretical contributions.

There is currently no single biological test for ASD. The diagnostic process involves a combination of parental report and clinical observation.

Children with significant impairments across the social/communication domain, who also exhibit repetitive behaviors, can qualify for the ASD diagnosis. As discussed earlier, there is wide variability in the precise symptom profile an individual may exhibit. Since Kanner first described ASD in 1943, important commonalities in symptom presentation have been used to compile criteria for an ASD diagnosis. These diagnostic criteria have evolved during the past seventy-six years and continue to evolve, yet impaired social functioning remains a required symptom for an ASD diagnosis.

Deficits in social functioning are present in varying degrees for simple behaviors such as eye contact and complex behaviors like navigating the give and take of a group conversation for individuals of all functioning levels (i.e., high or low IQ). Moreover, difficulties with social information processing occur in both visual (e.g., Pelphrey et al., 2004) and auditory (e.g., Dawson, Meltzoff, Osterling, Rinaldi, & Brown, 1998) sensory modalities. While repetitive behaviors or language deficits are seen in other disorders (e.g., obsessive-compulsive disorder and specific language impairment, respectively), basic social deficits of this nature are unique to ASD. Onset of the social deficits appears to precede difficulties in other domains (Osterling, Dawson, & Munson, 2002) and may emerge as early as six months of age (Maestro et al., 2002).

This focus on social cognition and the importance of the social brain led us to hypothesize that factors contributing to the expression of ASD exert their effects through a circumscribed set of neuroanatomical structures, so the simplest and potentially most powerful signatures of ASD will be found at the level of brain systems. Such "neural signatures" of ASD may serve as critical endophenotypes to facilitate the study of the pathophysiological mechanisms. Endophenotypes, or characteristics that are not immediately available to observation but reflect an underlying genetic liability for disease, expose the most basic components of a complex psychiatric disorder and are more stable across the life span than observable behavior (Shields & Gottesman, 1973).

In a study pursuing the neural signatures of ASD, we assessed three groups of children with fMRI. These were children diagnosed with ASD, unaffected siblings (US) of children with ASD who were typically developing, and typically developing (TD) children without a relative with ASD. The three groups of participants were matched on chronological age and were of similar cognitive ability, all within the average range. Notably, the

US and TD groups were matched on measures of social responsiveness and adaptive behavior. This rigorous matching ensured that both groups were unaffected by ASD and demonstrated equivalent levels of social responsiveness. In addition, strict exclusion criteria were used for the TD and US groups to rule out other developmental disorders and the "broader autism phenotype" (BAP) in each participant, as well as in first- and second-degree relatives of the US participants.

We measured brain responses in the children using fMRI while they viewed socially meaningful biological motion (movements of other people) to reveal three types of neural signatures: (a) state activity related to having ASD that characterizes the nature of disruption in brain circuitry; (b) trait activity reflecting shared areas of dysfunction in US and children with ASD, thereby providing a promising neuroendophenotype to facilitate efforts to bridge genomic complexity and disorder heterogeneity; and (c) compensatory activity, unique to US, suggesting a neural systems–level mechanism by which US might compensate for an increased genetic risk for developing ASD.

The identification of state activity extends previous research implicating the right amygdala, right pSTS, bilateral fusiform gyrus (FG), left ventrolateral prefrontal cortex (vlPFC), and ventromedial prefrontal cortex (vmPFC) in adults with ASD, by showing that dysfunction in these regions is already present in school-age children with ASD (Castelli, Frith, Happé, & Frith, 2002; Gilbert, Meuwese, Towgood, Frith, & Burgess, 2009; Schultz et al., 2000). This finding was an important advance in the field, given that previous reports of atypical neural response to biological motion included only adult subjects.

In addition, activity in the state-defined right pSTS was associated with the severity of social deficits in individuals with ASD. Individuals with higher social responsiveness scale (SRS) scores (Constantino, 2013) exhibited less activation to biological motion within the right pSTS. This finding suggests that activity in the pSTS might serve as a biological marker to subdivide the autism spectrum on the basis of severity. Furthermore, activity in the state-defined region of the left vlPFC was found to reflect the level of social responsiveness of the TD children, indicating a coupling of social behavior and brain mechanisms for social perception.

The evidence of dysfunction in brain mechanisms for social perception in young children with ASD explains previous behavioral findings of disrupted biological motion perception (Klin, Lin, Gorrindo, Ramsay, & Jones, 2009). Given that social interaction relies on the accurate perception of other people's actions, state activity indicating regions of dysfunction associated with the manifestation of ASD provides a significant step toward more fully characterizing the biological underpinnings of this neurodevelopmental disorder.

In accordance with Gottesman and Gould's (2003) characterization of endophenotypes, trait activations—including those in the left dorsolateral prefrontal cortex (dlPFC), right inferior temporal gyrus, and bilateral FG—were shared between affected individuals (ASD group) and first-degree relatives (US group). These findings are particularly noteworthy because we explicitly ruled out the BAP in the US group. This implies that our neuroimaging paradigm offers a remarkable level of sensitivity that transcends clinical evaluation. Although the US group was indistinguishable from the TD group at the behavioral level, the trait activity findings revealed similar neural signatures in the US and ASD groups. Consistent with this interpretation, social responsiveness was associated with overall trait activity in the US group and with trait-defined left dlPFC in the TD group. Furthermore, whereas the state regions could arise as an effect of having ASD, the trait activity could not be explained in this way; rather, this trait activity likely reflects the genetic vulnerability to develop ASD.

The trait activity findings provided a functional neuroendophenotype that has helped bridge the gene–behavior gap and accelerated the search for pathophysiological mechanisms. The key implication of our trait activity findings is that we provide a functional neuroendophenotype that should help bridge the gene–behavior gap, thereby accelerating the search for pathophysiological mechanisms.

The US group exhibited unique areas of activation in the vmPFC and the right pSTS, regions previously implicated in aspects of social perception and social cognition (Adolphs, 1999). These regions might reflect the absence of additional genetic or environmental factors that confer risk for ASD. Alternatively, they could represent a process through which brain function was altered over development to compensate for an increased genetic risk to develop ASD. We found that the activity in these regions did not vary with chronological age. Thus, it is possible that the compensatory regions reflect the outcome of a process occurring earlier in development, during a sensitive period for the development of brain mechanisms for social perception.

This process might be likely, given that autism is a developmental disorder that emerges during the first years of life, well before age four years (the youngest age studied in this sample).

Nonetheless, we cannot yet draw firm conclusions regarding the compensatory activity. Indeed, longitudinal research in younger children is critically needed to better understand the origins of this compensatory activity, which likely has both genetic and environmental influences. Future studies are needed to compare the activity in these regions in US participants with and without BAP to determine the function and etiology of this brain response to biological motion. The implication of these findings is that these regions could represent important targets for treatment and provide a measure of the effectiveness of intervention as well as a better understanding of the mechanisms through which successful treatments function. The US group exhibited unique areas of activation in regions previously implicated in aspects of social perception and social cognition. This might reflect the absence of additional genetic or environmental factors that confer risk for ASD. Alternatively, it could represent a process through which brain function is altered over development to compensate for an increased genetic risk to develop ASD.

Our study and many other group-based comparative studies reveal important clues about the neurobiological mechanisms that give rise to ASD symptomatology. Yet treating ASD as a unitary condition—whereby individuals with ASD are grouped together and compared with "neurotypical" (TD) people—undermines the potential of translational research to contribute to "precision medicine" (Insel, 2014) in ASD. Because of the limited quality of the behavioral methods used to diagnose ASD and current clinical diagnostic practice, which permits similar diagnoses despite distinct symptom profiles (McPartland, Webb, Keehn, & Dawson, 2011), it is possible that the group of children currently referred to as having ASD may actually represent different syndromes with distinct causes.

The ability to integrate a broad array of social cues from multiple sensory domains is impaired in many neuropsychiatric disorders. The spectrum from mental health to mental illness is continuous and not categorical—Mother Nature has not yet read the DSM-5. Thus, a dimensional, individual-differences approach is crucial to understanding multimodal social cognitive abilities and their development. Individuals with autism are currently defined solely on the basis of behavioral indicators. Undoubtedly this approach lumps together individuals with common behavioral phenotypes but possibly quite different underlying etiologies. By defining functional brain phenotypes based on neurofunctional/behavioral developmental pathways and activation patterns, fMRI studies of children have the potential to dissect the heterogeneity present in these disorders. Functional neuroimaging studies could reveal different brain phenotypes in the circuitry involved in social cognition. This approach may allow us to partition individuals with autism—a complex, etiologically heterogeneous disorder—into more homogenous subgroups (e.g., Yang, Sukhodolsky, Lei, Dayan, Pelphrey & Ventola, 2017). These profiles, in turn, may inform treatment of ASD by helping us to match specific treatments to specific profiles.

The challenge of quantifying brain profiles in ASD is now being addressed via the application to brain imaging data of artificial intelligence (AI) theories and analytic techniques in order to derive sensitive, reliable brain measures that are informative at the level of the individual (e.g., Björnsdotter, Wang, Pelphrey, & Kaiser, 2016; Yang et al., 2016; Yang et al., 2017). Consider, for instance, our use of fMRI and machine learning, multivariate pattern analysis (MVPA) techniques to identify profiles of activation in young children with ASD, which predicts responses to sixteen weeks of an evidence-based behavioral treatment (Yang et al., 2016)—pivotal response treatment (PRT). Neural predictors were identified in the pretreatment levels of activity in response to biological versus scrambled motion in the neural circuits that support social information processing (superior temporal sulcus, fusiform gyrus, amygdala, inferior parietal cortex, and superior parietal lobule) and social motivation/reward (orbitofrontal cortex, insula, putamen, pallidum, and ventral striatum). The predictive value of our findings for individual children with ASD was supported by a MVPA with cross validation.

Predicting who will respond to a particular treatment for ASD, these findings marked the very first evidence of prediction/stratification biomarkers in young children with ASD. In MVPA, the samples were divided into training and testing data sets, which has constituted a *cross validation framework* in which the predictive model is first trained with the training set and then used to predict the regression labels of the sample in the testing set. This type of cross validation provides approximately unbiased estimates of effects, generalizable to new samples, helping to minimize the likelihood that the results overfit the data.

Our findings move the field toward targeted, personalized treatment for individuals with ASD. The knowledge gained can be used in future work to tailor individualized treatment, refine PRT, and develop novel interventions. This study adds to the understanding of the pretreatment neural underpinnings of successful behavioral response to PRT. In the future, our results may drive the construction of algorithms to predict which, among several treatments, is most likely to benefit a given person. In addition, PRT is a multicomponent treatment; hence future studies might use dismantling designs to isolate treatment components and their association with the neuropredictive targets identified here. This line of work could inform the development of treatment strategies that would target specific patterns of neural strengths and vulnerabilities within an individual—consistent with the priority of creating individually tailored interventions, customized to the characteristics of a given person.

The predictive biomarkers identified in this study can be interpreted as the pretreatment neurobiological readiness to respond to a specific treatment, PRT. It should be noted that the brain regions where activity before treatment correlated with SRS scores before treatment did not overlap with the neuropredictive network described here, which indicates that the neuropredictive network is specific to change in severity in young children with ASD. As such, our findings offer the hope that the pretreatments or concurrent treatments (whether pharmacological, direct stimulation, neurofeedback, or behaviorally based) that improve the functioning of the neuropredictive markers identified here may increase the effectiveness of evidenced-based behavioral treatments for core deficits in children with ASD.

Our findings are also particularly important for those children who would otherwise be the least likely to benefit from these expensive and time-consuming forms of treatment. For example, in a randomized, double-blind, cross-over functional fMRI study (Gordon et al., 2013), we reported that intranasal oxytocin administered to children with ASD increased activity during social versus nonsocial judgments in several of the same brain regions identified as predictive in the present study (e.g., the amygdala, orbitofrontal cortex, superior temporal sulcus region, and ventral striatum). These findings, coupled with those in the current report, raise the provocative hypothesis that the administration of intranasal oxytocin, by priming key neural circuits for social motivation and social perception, may serve to enhance the effectiveness of interventions like PRT in the very children who might be less biologically ready to respond.

This work and similar research developments indicate a very new approach, and a new period in which advanced neuroimage analysis will evolve into an integral part of a translational research chain. Novel behavioral treatment and pharmacotherapies for ASD may be further developed in young children with the tremendous benefit of directly and more precisely assessing impairment and change in targeted neural circuits. Neuroimaging-derived biological markers could be used from the beginning to make treatment decisions related to dose, duration, intensity, and specific behavioral treatment approach as well as related to the use of concurrent pharmacological intervention.

As the field moves forward in conducting brain imaging studies that leverage AI in the context of behavioral and pharmacological treatments for ASD, we anticipate the empirical realization of a "transactional" approach in humans to the study of mechanisms for developmental changes in social cognition (Gottlieb, 1997; Sameroff & Chandler, 1975). As illustrated in figure 18.1, such a perspective emphasizes the necessity to characterize the development of social cognition as an emergent property reflecting transactions occurring across levels of an active, developing organism in its environment. The goal of theory building is then to specify mechanisms through the identification of critical transactions between two, three, or more levels of analysis (e.g., gene $\leftrightarrow$ brain $\leftrightarrow$ behavior transactions over a specific developmental period).

We embrace this approach as a way to generate exciting and deeply influential theoretical advances in our field, identifying the neural systems required

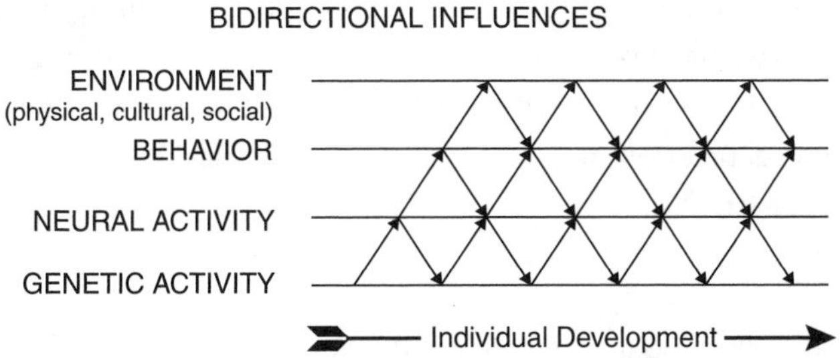

BIDIRECTIONAL INFLUENCES

ENVIRONMENT
(physical, cultural, social)
BEHAVIOR

NEURAL ACTIVITY

GENETIC ACTIVITY

Individual Development

Figure 18.1
A probabilistic epigenetics theory of development. Adapted from Gottlieb, 1998. Mechanisms of change are understood as arising from transactions between two, three, or more levels of analysis (e.g., gene $\leftrightarrow$ brain $\leftrightarrow$ behavior transactions over a specific developmental period).

for social success. Though the relative influence of biology and environment have often been considered as independent, a probabilistic epigenetic framework provides the foundation for a unified model of development that offers powerful explanatory capabilities for our understanding of the complexities of social development and the interaction of biology with environment.

Working with TD infants and children, our colleagues at the University of Virginia are currently testing the central hypothesis that epigenetic mechanisms provide an interface between biological capabilities and developmental experience, shaping the developing brain and establishing a wide spectrum of behavioral and cognitive outcomes—which, in turn, continue to shape the developing brain in a bidirectional fashion over ontogeny. To empirically test this theory in humans, they are studying the endogenous oxytocin system—a candidate biological system that is under epigenetic regulation. Their foundational work has established that DNA methylation along the promoter region of the oxytocin receptor (OXTRm) is a regulatory marker of the endogenous oxytocin system and strongly associated with several social phenotypes and neuroendophenotypes. By dissecting the function of this "nodal" system, they are making specific predictions and providing a mechanistic, biobehavioral explanation of the role played by the interplay among levels of analysis from regulation of gene expression, to brain activity and connectivity, to behavioral and cognitive development in driving human social communicative development. This mechanistic account of the bidirectional, transactional nature of development will serve as a model basis for deciphering the many such systems that interact to drive the full spectrum of variability in human development. This unique program of multidisciplinary research combines techniques from molecular biology, genetics, and epigenetics with tools from developmental cognitive and social neuroscience, which will allow for unprecedented training opportunities the next generation of developmental social cognitive neuroscientists.

References

Adolphs, R. (1999). Social cognition and the human brain. *Trends in Cognitive Sciences, 3*(12), 469–479.

American Psychiatric Association. (2013). *Diagnostic and statistical manual of mental disorders* (5th ed.). Arlington, VA: American Psychiatric Association.

Amodio, D. M., & Frith, C. D. (2006). Meeting of minds: the medial frontal cortex and social cognition. *Nature Reviews Neuroscience, 7*(4), 268.

Björnsdotter, M., Wang, N., Pelphrey, K., & Kaiser, M. D. (2016). Evaluation of quantified social perception circuit activity as a neurobiological marker of autism spectrum disorder. *JAMA Psychiatry, 73*(6), 614–621.

Cacioppo, J. T., Amaral, D. G., Blanchard, J. J., Cameron, J. L., Carter, C. S., Crews, D.,…Levenson, R. W. (2007). Social neuroscience: Progress and implications for mental health. *Perspectives on Psychological Science, 2*(2), 99–123.

Cantlon, J. F., Pinel, P., Dehaene, S., & Pelphrey, K. A. (2011). Cortical representations of symbols, objects, and faces are pruned back during early childhood. *Cerebral Cortex, 21,* 191–199.

Castelli, F., Frith, C., Happé, F., & Frith, U. (2002). Autism, Asperger syndrome and brain mechanisms for the attribution of mental states to animated shapes. *Brain, 125*(8), 1839–1849.

Chong, T. T., Cunnington, R., Williams, M. A., Kanwisher, N., & Mattingley, J. B. (2008). fMRI adaptation reveals mirror neurons in human inferior parietal cortex. *Current Biology, 18*(20), 1576–1580. doi: 10.1016/j.cub.2008.08.068

Constantino, J. N. (2013). Social responsiveness scale. In Volkmar F. R. (Ed.), *Encyclopedia of autism spectrum disorders* (pp. 2919–2929). New York: Springer.

Dawson, G., Meltzoff, A. N., Osterling, J., Rinaldi, J., & Brown E. (1998). Children with autism fail to orient to naturally occurring social stimuli. *Journal of Autism and Developmental Disorders, 28*(6), 479–485.

Downing, P. E., Jiang, Y., Shuman, M., & Kanwisher, N. (2001). A cortical area selective for visual processing of the human body. *Science, 293*(5539), 2470–2473.

Elsabbagh, M., Mercure, E., Hudry, K., Chandler, S., Pasco, G., Charman, T.,…& BASIS Team. (2012). Infant neural sensitivity to dynamic eye gaze is associated with later emerging autism. *Current Biology, 22*(4), 338–342.

Geschwind, D. H., & Levitt, P. (2007). Autism spectrum disorders: Developmental disconnection syndromes. *Current Opinion in Neurobiology, 17*(1), 103–111.

Gilbert, S. J., Meuwese, J. D., Towgood, K. J., Frith, C. D., & Burgess, P. W. (2009). Abnormal functional specialization within medial prefrontal cortex in high-functioning autism: A multi-voxel similarity analysis. *Brain, 132*(4), 869–878.

Gordon, I., Vander Wyk, B. C., Bennett, R. H., Cordeaux, C., Lucas, M. V., Eilbott, J. A.,…Pelphrey, K. A. (2013). Oxytocin enhances brain function in children with autism. *Proceedings of the National Academy of Sciences of the United States of America, 110*(52), 20953–20958.

Gottesman, I. I., & Gould, T. D. (2003). The endophenotype concept in psychiatry: Etymology and strategic intentions. *American Journal of Psychiatry, 160*(4), 636–645.

Gottlieb, G. (1998). Normally occurring environmental and behavioral influences on gene activity: From central dogma to probabilistic epigenesis. *Psychological Review, 105*(4), 792.

Insel, T. R. (2014). The NIMH research domain criteria (RDoC) project: Precision medicine for psychiatry. *American Journal of Psychiatry, 171*(4), 395–397

Jones, W., & Klin, A. (2013). Attention to eyes is present but in decline in 2–6-month-old infants later diagnosed with autism. *Nature, 504*(7480), 427.

Kaiser, M. D., Hudac, C. M., Shultz, S., Lee, S., Cheung, C., Berken, A. M., … Pelphrey, K. A. (2010). Neural signatures of autism. *Proceedings of the National Academy of Sciences of the United States of America, 107*, 21223–21228.

Kanner, L. (1943). Autistic disturbances of affective contact. *Nervous Child, 2,* 217–250.

Kanwisher, N., McDermott, J., & Chun, M. M. (1997). The fusiform face area: A module in human extrastriate cortex specialized for face perception. *Journal of Neuroscience, 17*(11), 4302–4311.

Klin, A., Lin, D. J., Gorrindo, P., Ramsay, G., & Jones, W. (2009). Two-year-olds with autism orient to non-social contingencies rather than biological motion. *Nature, 459*(7244), 257.

McCarthy, G., Puce, A., Gore, J. C., & Allison, T. (1997). Face-specific processing in the human fusiform gyrus. *Journal of Cognitive Neuroscience, 9*(5), 605–610.

McPartland, J. C., Webb, S. J., Keehn, B, & Dawson, G. (2011). Patterns of visual attention to faces and objects in autism spectrum disorder. *Journal of Autism and Developmental Disorders, 41*(2), 148–157.

Osterling, J. A., Dawson, G., & Munson, J.A. (2002). Early recognition of 1-year-old infants with autism spectrum disorder versus mental retardation. *Development and Psychopathology, 14*(2), 239–251.

Pelphrey, K. A., Viola, R. J., & McCarthy, G. (2004). When strangers pass: Processing of mutual and averted social gaze in the superior temporal sulcus. *Psychological Science, 15*(9), 598–603.

Puce, A., Allison, T., Gore, J. C., & McCarthy, G. (1995). Face-sensitive regions in human extrastriate cortex studied by functional MRI. *Journal of Neurophysiology, 74*(3), 1192–1199.

Rizzolatti, G., & Fabbri-Destro, M. (2008). The mirror system and its role in social cognition. *Current Opinion in Neurobiology, 18*(2), 179–184.

Saint-Georges, C., Cassel, R. S., Cohen, D., Chetouani, M., Laznik, M. C., Maestro, S., & Muratori, F. (2010). What studies of family home movies can teach us about autistic infants: A literature review. *Research in Autism Spectrum Disorders, 4*(3), 355–366.

Sameroff, A. J., & Chandler, M. J. (1975). Reproductive risk and the continuum of caretaking casualty. In F. D. Horowitz, E. M. Hetherington, S. Scarr-Salapatek, G. M. Siegel (Ed.), *Review of child development research* (Vol. 4, pp. 187–244). Chicago: University of Chicago Press.

Saxe, R., & Kanwisher, N. (2003). People thinking about thinking people: the role of the temporo-parietal junction in "theory of mind". *Neuroimage, 19*(4), 1835–1842.

Schultz, R. T., Gauthier, I., Klin, A., Fulbright, R. K., Anderson, A. W., Volkmar, F., …Gore, J. C. (2000). Abnormal ventral temporal cortical activity during face discrimination among individuals with autism and Asperger syndrome. *Archives of General Psychiatry, 57*(4), 331–340.

Shields, J., & Gottesman, I. I. (1973). Genetic studies of schizophrenia as signposts to biochemistry. In L. L. Iversen & S. P. R. Rose (Eds.), Biochemistry and mental illness (pp. 165–174). London: Biochemical Society.

Shultz, S., Vouloumanos, A., Bennett, R. H., Pelphrey, K. A. (2014). Neural specialization for speech in the first months of life. *Developmental Science, 17*(5), 766–774.

Yang, D., Pelphrey, K. A., Sukhodolsky, D. G., Crowley, M. J., Dayan, E., Dvornek, N. C., …Ventola, P. (2016). Brain responses to biological motion predict treatment outcome in young children with autism. *Translational Psychiatry, 6*(11), e948.

Yang, Y. D., Allen, T., Abdullahi, S. M., Pelphrey, K. A., Volkmar, F. R., & Chapman, S. B. (2017). Brain responses to biological motion predict treatment outcome in young adults with autism receiving virtual reality social cognition training: Preliminary findings. *Behaviour Research and Therapy, 93*, 55–66.

Yang, Y. J. D., Sukhodolsky, D. G., Lei, J., Dayan, E., Pelphrey, K. A., & Ventola, P. (2017). Distinct neural bases of disruptive behavior and autism symptom severity in boys with autism spectrum disorder. *Journal of Neurodevelopmental Disorders, 9*, Article ID 1. doi: 10.1186/s11689-017-9183-z

19 Developmental Origins of Psychopathy

Essi Viding, Eamon McCrory, and Ruth Roberts

Overview

Psychopathy is a devastating personality disorder associated with considerable financial and social cost. Individuals with psychopathy lack empathy and guilt and fail to care for those around them. The primary focus of this chapter is to consider the developmental origins of psychopathy. There has been progress in charting the social brain abnormalities associated with adult psychopathy and many of the neurocognitive hallmarks of the disorder are also seen in children at risk of developing the disorder. Twin and adoption data indicate that both genetic risk and environmental factors play a role in the development of psychopathy, but we know less about the precise identity of genetic and environmental risk factors or how the atypical social cognition emerges in psychopathy. We argue that carefully designed genetically informative and neurocognitive studies have the potential to elucidate how developmental risk for psychopathy unfolds and help us understand social interactions that characterize psychopathy in a new way.

Social Brain in Psychopathy

Tom's mother remembers how his behavior as a child used to scare her. Tom could be violent toward his peers and cruel to animals. He would also often do something hurtful or destructive, but deny his involvement—even when it was clear that he had been the culprit. As Tom got older he often manipulated other children to do things for him so that he could blame them if they got caught. Tom showed no evidence of caring about other people's feelings, he simply did not seem to mind if his behavior made others upset. All of his relationships seemed dispensable. He could be charming as long as he was getting what he wanted, but then would turn on people if they did not do his bidding. He did not seem to feel any genuine remorse for his actions. As Tom entered adolescence, he started to engage in criminal activities. He was good at implicating other people and did not often get caught. His "friendships" were very transactional, and he had multiple girlfriends on the go, all of whom he treated badly. Eventually Tom was convicted of a violent, premeditated robbery that resulted in severe and life-changing injuries to the victim. There were eye witnesses to the crime, whose testimony helped convict Tom. Tom did not show any sympathy for his victim. Instead he claimed that it was him who was a victim of a grave miscarriage of justice.

As illustrated by this fictional case study, individuals with psychopathy are characterized by lack of empathy and remorse, manipulation of other people, the ability to engage in premeditated, cold, and calculated aggression to achieve their goals, and an impoverished capacity to take care of their responsibilities and make good decisions (Hare & Neumann, 2008; Patrick & Drislane, 2015). Decades of experimental and neurocognitive research have focused on understanding why individuals with psychopathy do not readily empathize with others or why they make poor decisions (Baskin-Sommers, Stuppy-Sullivan & Buckholtz, 2016; Blair, Leibenluft & Pine, 2014; Hosking et al., 2017). These studies have demonstrated atypical structure and function in a network of emotion and reward processing areas that are thought to facilitate emotional resonance, empathy, moral processing, and decision-making guided by reinforcement information—such as the amygdala, anterior insula, anterior cingulate cortex, ventral striatum, and different regions of the prefrontal cortex (Viding & McCrory, 2018).

Not only do we see atypical brain structure and function in adults with psychopathy, but studies of children and youth at risk of developing psychopathy (as measured by standard instruments charting psychopathic personality features and behavior in children and young people) have, on the whole, produced comparable findings to those seen in adults with psychopathy (Blair et al., 2014; Viding & McCrory, 2018). Reduced amygdala activity to fearful faces has been reported in several studies that have compared children at risk of developing psychopathy with their peers (typically developing children, those with attention deficit/hyperactivity disorder, or those exhibiting disruptive behaviors without psychopathic traits) (Marsh et al., 2008; Jones, Laurens, Herba, Barker, & Viding, 2009; Viding et al., 2012; Blair et al., 2014). One such study found that the association between psychopathic personality features and proactive aggression is partially mediated by low amygdala reactivity to fearful faces (Lozier, Cardinale, VanMeter, & Marsh, 2014).

Reduced amygdala and insula activity in children at risk of developing psychopathy are also seen when these children engage in more complex forms of social judgment regarding other people's distress, such as categorization of legal and illegal behaviors in moral judgment tasks (Marsh et al., 2011), making decisions about appropriate responses to the distress of others (Sebastian et al., 2012), or making decisions about whether to benefit self by harming others (Sakai et al., 2017). Finally, five recent studies of children at risk of developing psychopathy, (four involving functional magnetic resonance imaging, and the one using event-related potential measurement) have reported atypical neural reactivity to other people's pain (Cheng, Hung, & Decety, 2012; Lockwood et al., 2013; Marsh et al., 2013; Michalska, Zeffiro, & Decety, 2016; Yoder, Lahey, & Decety, 2016). Collectively, these studies implicate reduced activity and altered connectivity in children at risk of developing psychopathy in a network of brain areas known to be associated with empathy for other people's pain in healthy individuals. This network encompasses a number of brain regions including the anterior insula, posterior insula, anterior cingulate cortex, and the amygdala. Importantly, this profile of reduced neural reactivity to expressions of pain is not coupled with difficulty in understanding intentionality on the part of others (Cheng et al., 2012).

An atypical ventromedial prefrontal cortex and orbitofrontal cortex response to punishment and reward in children at risk of developing psychopathy has also been reported in several studies, although this presentation seems to be common to most children with disruptive behavior (Finger et al., 2008, 2011; White et al., 2013, 2014) and relates to the levels of conduct disorder symptoms (White et al., 2016a, 2016b). However, poor decision making may be particularly toxic for social functioning when combined with deficient empathy, which is what we see in those at risk of developing psychopathy.

Collectively, the imaging studies have considerably advanced our understanding of the neurocognitive presentation that accompanies psychopathic features and helps explain why individuals with these features appear relatively unaffected by other people's distress, do not feel guilt, and often make repeated disadvantageous decisions. By contrast what leads to atypical social affiliation and lack of prosocial behaviors in psychopathy has received remarkably little attention, despite the fact that individuals with psychopathy appear to have reduced motivation and capacity to develop social relationships founded on an enjoyment of prosocial interactions or genuine love and concern for others' well-being (Hare & Neumann, 2008).

Only a handful of studies to date have probed the potential mechanisms of impaired affiliation and social connectedness in those with or at risk of developing psychopathy. A recent study our group investigated how children at risk of developing psychopathy process laughter (O'Nions et al., 2017). Laughter is a universal expression of emotion used to maintain social bonds (Scott, Lavan, Chen, & McGettigan, 2014; Gervais & Wilson, 2005; Sauter, Eisner, Ekman, & Scott, 2010; Provine, 2001). It is a highly contagious behavior: it can be primed simply by listening to others' laughter (Provine, 1992). Such emotional contagion has been posited as a mechanism for facilitating the coupling of emotions and behavior within groups, increasing cooperation, cohesiveness, and social connectedness (Scott et al., 2014; Gervais & Wilson, 2005; Warren et al., 2006). Laughter also plays a role in the vicarious experience of positive emotions, and it triggers the endogenous opioid system, argued to be key for prosocial communication and social bonding in primates and other mammals (Gervais & Wilson, 2005; Nummenmaa et al., 2016; Manninen et al., 2017).

Our work has shown that children at risk of developing psychopathy report a reduced desire to join in when they hear clips of others' genuine laughter, compared with their typically developing peers matched on ability and socioeconomic status. Neuroimaging studies of typical individuals demonstrate that listening to laughter automatically recruits motor and premotor regions involved in the production of emotional expressions and empathy (Fonagy, Gergely, & Target, 2007), including the precentral gyrus, supplementary motor area, inferior frontal gyrus, and anterior insula (Warren et al., 2006; McGettigan et al., 2013; Szameitat et al., 2010; Lima et al., 2015). The preparatory motor response associated with laughter production is thought to facilitate joining in with others' positive vocalizations during social behavior, representing a neural mechanism for experiencing these emotions vicariously and promoting social connectedness (Gervais & Wilson, 2005; Warren et al., 2006). Compared with typically developing boys, those at risk of developing psychopathy displayed reduced neural response to laughter in the supplementary motor area and anterior insula. The reduced anterior insula response in part accounted for the reduced desire to join in with others' laughter in the group at risk of developing psychopathy.

These findings open up the prospect of a new and interesting avenue of research to explore the role of affective signals that facilitate social affiliation, and promote and maintain social bonds, in individuals with or at risk of developing psychopathy. They add to the more established evidence base regarding the atypical processing of other people's distress and development of empathy in this population.

What Do We Know about the Origins of Psychopathy?

The brief overview of the studies focusing of neurocognitive profile of individuals with or at risk of developing psychopathy indicates that they see the world around them differently from the rest of us. Unsurprisingly their brains are different too, with structural and functional atypicalities in areas that support information processing critical for empathy, morality, social affiliation, and planning—as we would expect from people who feel and choose differently. Why and how have they developed this way? Is psychopathy a genetic condition? Or do people become psychopaths because they have been maltreated?

Is It in Their Genes?

We have robust evidence of genetic risk for psychopathy from twin and adoption studies (Viding & McCrory 2015, 2018). However, it is important to note that any heritability estimate from such studies reflects the extent of genetic influences on individual differences or group differences and does not provide an estimate that tells us about the origins of a single individual's personality or behavior. There is always a degree of error in these estimates (they are not precise), and the relative importance of genetic influences may differ between populations. (To illustrate, if adequate nutrition is available, individual differences in height in the population are almost exclusively explained by genetic differences between individuals; if parts of the population suffer from malnutrition, then individual differences in height will in part be explained by environmental factors—namely, availability of nutrition.)

Because the neurocognitive functions and some behaviors associated with psychopathy are at least partially distinct from those associated with antisocial behavior in general, researchers think that the risk genes for psychopathy may not always be the same as the risk genes for antisocial behavior in the absence of psychopathic traits (Viding & McCrory, 2015). We can make reasonable guesses regarding the risk genes—for example, those genes that have an impact on the range of emotional reactivity, empathy, and capacity for attachment are likely to be important. The problem is, most of these genes are yet to be found. Not just in relation to psychopathy, but in general. Only a handful of candidate gene molecular genetic investigations have focused on psychopathic traits, and those have identified particular serotonin and oxytocin system genes—thought to contribute to emotional reactivity and capacity for attachment—as increasing risk of psychopathy (for a review, see Viding & McCrory, 2018). Such findings are exciting, but they need to be replicated in larger samples to evaluate whether they represent robust associations. Genome-wide association studies have not produced any promising insights either (Viding & McCrory, 2018).

Why has it been so difficult to find the genes that increase the risk of developing psychopathy when the twin and adoption studies indicate that psychopathy has a substantial heritable component? There are several reasons. The genetic risk for most disorders, even those that are strongly heritable, is "polygenic." In other words, we are very unlikely to be searching for one or few devastating genes that push certain individuals to a risk group. Psychopathy is not a single-gene disorder, unlike Huntington's disease. Instead we are hunting for multiple genes of small effect size that probabilistically increase the risk of developing the disorder (Knopik, Neiderhiser, DeFries, & Plomin, 2016). These genes may confer advantages as well as disadvantages, depending on the environmental context and the full set of other genes that an individual has. This of course makes our task of finding the genes a lot more difficult.

Finding genes that have a small effect requires very large sample sizes, and unfortunately the molecular genetic studies of psychopathy have had small sample sizes to date. We do not yet have large international scientific consortia to focus on the genetic underpinnings of psychopathy. Most existing large-scale studies do not include good measures of psychopathic personality. It is also very likely that we need to become better at understanding how genes interact with other genes as well as the environmental risks if we want to get a better sense of how the genetic risk for psychopathy impacts brain development and manifests across development. Although a person's genome (in other words, their whole set of genes) likely limits the "range for phenotypic expression," it does not prespecify how an individual will turn out. The specific developmental trajectory of any individual is determined by a complex interplay between genetic propensities and environmental factors that constrain how those genetic propensities are expressed across development.

Exciting new methodological developments provide some basis for optimism in relation to ongoing genetic research, which is likely to advance greatly in the coming decade. Notably, studies using novel epigenetic approaches may help uncover mechanisms of gene–environment interaction. Genetic researchers are also now focusing on finding rare genetic variants that may have more substantial effects but may only affect a small subset of population. Molecular genetic studies of psychopathy to date have not looked for rare variants.

Is Psychopathy the Result of Environmental Adversity?

Both cross-sectional and longitudinal studies identify maltreatment, harsh discipline, negative parental emotions, disorganized attachment with the caregiver, and disrupted family functioning as risk factors associated with antisocial behavior and psychopathic features (Viding & Kimonis, 2018; Waller, Gardner, & Hyde, 2013; Pasalich, Dadds, Hawes, & Brennan, 2012; Roberts, McCrory, Joffe, De Lima, & Viding, 2018). In contrast, warm and consistent parenting has been associated with a reduced risk of antisocial behavior and psychopathy (Waller et al., 2013; Clark & Frick, 2016).

What should we make of these associations? Several risk factors that are commonly thought to be "environmental" may in part reflect genetic predispositions of people who are part of that environment—a phenomenon known as *gene environment correlation* (Knopik, Neiderhiser, DeFries, & Plomin, 2016). Children, as well as the adults who interact with them, have substantial, in part heritable individual differences in their social information processing capacities and behavior. Not all children are equally cooperative, empathetic, loving, or easy to manage. And not all adults are equally responsive to social cues, able to regulate their emotions, or capable of planning ahead. Individual differences in a variety of social-cognitive abilities clearly impact how the children behave and what they best respond to, as well as how the adults behave and how they respond to the child or meet his or her needs. Just as we are not going to find one gene that will explain why someone becomes a psychopath, it is also helpful to assume that "it's complicated" when we seek to understand environmental risk.

In typical families where family members are biologically related to each other, the parents and children share genetic endowments and information processing styles. This is likely to constrain the range of 'inputs' and learning outcomes that are probable for a particular child and may explain why behaviors develop in certain way in particular family ecologies. Parents with genetic risk factors for psychopathy and antisocial behavior are likely to display parenting behaviors in line with these risks (e.g., harsh and inconsistent parenting) and may also pass on some of their risk genes to their offspring, which can increase the chance of developing disruptive behaviors and a lack of empathy. This means that part of the association between less harsh parenting and psychopathic traits in the child may represent a genetic confound. Put differently, we cannot assume pure environmental causality when the association may in part be due to shared genetic risk between parents and offspring and may not entirely reflect a pure impact of parenting on child behavior.

We also know that some of the differences in the reactions that different children evoke in their caregivers are in part driven by heritable dispositions. Children at risk of developing psychopathy are extremely challenging to parent. They typically show diminished empathy for others and less remorse, and they tend to manipulate others, engage in self-serving behaviors, be impulsive, and show little interest in being nice to other people. It is therefore likely that they evoke different parenting reactions than less-challenging children. Research has suggested that this is indeed the case.

Our recent study found compromised family functioning in families with a child who is at risk of developing psychopathy (Roberts et al., 2018). The families had difficulty with affective responsiveness, only showing interest in each other if was of instrumental benefit. Qualitatively, the parents of children at risk of developing psychopathy described their child as being volatile and able to turn on the charm when they wanted something. Parents also seemed to have less rapport with their child, focusing on their child's behaviors rather than thinking about the child's personality or mental state, and they described challenges in family life that were in line with the child's self-interest and lack of empathy for others. Without genetically informative longitudinal studies it is not possible to tell to what extent an association between family functioning, harsh parenting, and psychopathic features purely reflects environmental risk.

We now have a handful of such studies that have investigated parenting and development of psychopathic traits. Longitudinal twin data from our group indicate that part of the association between harsh and negative parenting and higher levels of psychopathic traits in children reflects genetic vulnerability within biological families (Viding, Fontaine, Oliver, & Plomin, 2009). Very encouraging adoption data have shown that positive reinforcement by an adoptive mother can buffer the effects of heritable risk for psychopathic traits (Hyde et al., 2016). Children who have experienced this kind of adoptive parenting develop less psychopathic features than their adopted away peers who also have a biological risk for antisocial behavior and psychopathy (as indexed by the biological mother's disposition), but whose adoptive mothers are not able to provide consistent positive reinforcement. This study demonstrates a purely environmental effect of positive parenting on child outcome.

It is tempting to conclude from the adoption data that if we just had parenting interventions that curb harsh parenting and promote warm parent–child interactions, then we could stop the development of psychopathy. It may not be quite so straightforward, and the phenomenon of gene–environment correlation is almost certainly likely to introduce a complicating factor. Parents in adoptive families are typically very motivated to undertake the challenges of parenting, are often relatively well-resourced, and do not, as a rule, share the same constitutional vulnerabilities than the children they adopt. By contrast, in biological families, parents of children with psychopathic traits are more likely to have a host of genetic and contextual risk factors (including socioeconomic disadvantage), which can make it more difficult to deliver interventions that seek to promote prosocial behavior. We also know that not all adoptive parents manage to parent the children with difficult temperaments in a way that promotes prosocial development. Some children are extremely difficult and evoke a host of negative feelings in their parents, no matter how motivated the parents are and how genuinely they want the best for their children. Some parents even describe feelings akin to parental "burn out" in the face of repeated difficult interactions (Roberts et al., 2018). We need to better understand how both the child and parent vulnerabilities impact the social fabric of the

families and try to offer the right kind of support. We might try to understand what would help the parent to regulate their emotions while dealing
with a very difficult child or how the parent could best motivate prosocial
behavior when typical parenting practices do not work.

It is also important to note in this context that accumulating evidence base
indicates that some individuals may develop psychopathic features following
childhood maltreatment (Viding & Kimonis, 2018). These individuals appear
callous and unemotional and display high levels of antisocial behavior, but
intriguingly also have internalizing problems and a distinct neurocognitive
profile that is in line with extreme threat reactivity (Viding & Kimonis, 2018;
Kimonis, Goulter, Hawes, Wilber, & Groer 2017; Kimonis, Fanti, Goulter, &
Hall, 2017). We can think of them as "behavioral phenocopy" of classic "primary" psychopathy, but with a distinct developmental route to and neurocognitive profile associated with their psychopathic behavioral profile. We
need to systematically research neurocognitive processes related to empathy
and social affiliation in this group and investigate how their social cognition
develops. In other words, we must be open to investigating different developmental pathways into psychopathic presentation.

Translational Implications

Psychopathy incurs significant financial and human costs for society. Current treatment approaches for adults with psychopathy have modest effectiveness (Kiehl & Hoffman, 2011). The evidence for efficacy of treatment
for children and young people at risk of developing psychopathy is mixed
but more hopeful (Waller et al., 2013; Hawes, Price, & Dadds, 2014; Wilkinson, Waller, & Viding, 2016). We want to raise two important considerations in relation to prevention and treatment in the context of children
at risk of developing psychopathy. First, we propose that it is not sufficient
to focus on behavior modification and teaching caregivers and educators
behavior management techniques. It is also important to understand why
it can be more challenging to deliver traditional systemic approaches with
this population. A conceptual framework is needed to help inform alternative interpretations of behavioral difficulty and motivate the rationale for
effectively supporting those caring for these children. Second, we outline a
number of reasons why it may be particularly fruitful to develop techniques
for motivating affiliative, prosocial behavior in children at risk.

Interventions for children with conduct problems (including those at risk of developing psychopathy) predominantly draw on systemic principles, focusing on the relationship between the child, their peers, and the adults around them (e.g., parents, carers, teachers, and social workers) (National Institute for Heath and Care Excellence, 2013). Yet many aspects of establishing a mutual and balanced reciprocal relationship are contingent on prosocial and affiliative processes that function quite differently in children at risk for psychopathy. Atypical affect processing and a reduced drive to affiliate with others is likely to contribute to a distinct pattern of socialization difficulties. Currently we have a poor understanding of precisely how such atypical affiliative processing and behavior could inform the formulation of a child's presenting problems and guide approaches to change. The substantial variability in how children with conduct problems respond to interventions may in part derive from the impact of these specific information processing biases in how they process social/affiliative stimuli.

Social learning principles used in therapeutic programs emphasize the ways in which adult behavior can impact on the child outcome. However, children also play a key role in shaping the responses of adults around them, and in this case often evoking particularly negative reactions. Furthermore, parents of these children may share some of the vulnerabilities of their child, augmenting the challenge of delivering a systemic intervention. Helping parents, carers, and teachers reframe a child's behavior (including in relation to affiliative behavior) in the context of a profile of dispositional strengths and weaknesses that the child presents with may change how the adults around them respond. Moreover, having systems in place to ensure that adults caring for the child themselves receive support and a space to process their relationship with the child is a prerequisite for providing sustained, predictable support.

A more precise understanding the neurocognitive processes that contribute to the atypical affiliation could help sharpen the clinical formulation. Some previous work with adults with and children at risk of developing psychopathy has focused on lack of empathy (as opposed to affiliation) and trialed effortful strategies to modify how negative/fearful stimuli are processed. For example, participants have been asked to up-regulate their emotional response or direct attention to the relevant features of the face (Meffert, Gazzola, Den Boer, Bartels, & Keysers, 2013; Dadds et al., 2006). While neurocognitive changes toward more typical presentation have been observed in these laboratory studies, there is no empirical or clinical evidence that individuals with psychopathic features are then motivated to apply such strategies in everyday life. This is consistent with the long recognized persistent deficits in victim empathy in this population (Blair, 2013). Consequently, exploiting low-level automatic processes, such as conditioning and attentional bias modification, may represent more effective ways to modify behavior. However, modification of how negative stimuli (such as another's pain or distress) are processed using implicit strategies is inherently ethically problematic. It is not at all clear that aversive conditioning or attentional cuing to such stimuli would engender victim empathy and elicit desired behavioral outcomes; indeed, using such approaches may simply produce heightened arousal and behavioral unpredictability. By contrast, promoting responses to positive affect by modifying automatic/implicit processing—such as by pairing social affiliative stimuli with stimuli that the child finds rewarding—has the potential to make a child more receptive to adult affect/feedback/behavior modification, thus offering a potential to scaffold existing intervention approaches.

Conclusion

Both genetic and environmental risk factors play a role in predisposing a child to the development of psychopathy. A number of brain areas associated with processing of socially relevant information—such as stimuli that index distress or invitation to join in with others—show atypical structure and function in children at risk of developing psychopathy, but we need more longitudinal research on how maladaptive cognitions develop. Such research has particular promise in helping us understand how maladaptive social interactions between children at risk and their parents, teachers, or peers develop. This in turn will offer clues regarding how we might better promote prosocial development and prevent emergence of psychopathy in vulnerable individuals.

Acknowledgments: Essi Viding and Eamon McCrory were supported by grants from the Economic and Social Research Council (MR/N017749/1; principal investigator Viding, co-principal investigator McCrory) and the ESRC/NSPCC (principal investigator McCrory, co-principal investigator Viding) during the writing of this chapter.

References

Baskin-Sommers, A., Stuppy-Sullivan, A. M., & Buckholtz, J. W. (2016). Psychopathic individuals exhibit but do not avoid regret during counterfactual decision making. *Proceedings of the National Academy of Sciences of the United States of America, 113*(50), 14438–14443.

Blair R. J. (2013). The neurobiology of psychopathic traits in youths. *Nature Reviews Neuroscience, 14*(11), 786–799.

Blair, R. J., Leibenluft, E., & Pine, D. S. (2014). Conduct disorder and callous–unemotional traits in youth. *New England Journal of Medicine, 371*(23), 2207–2216.

Cheng, Y., Hung, A. Y., & Decety, J. (2012). Dissociation between affective sharing and emotion understanding in juvenile psychopaths. *Development and Psychopathology, 24*(2), 623–636.

Clark, J. E., & Frick, P. J. (2016). Positive parenting and callous-unemotional traits: Their association with school behavior problems in young children. *Journal of Clinical Child and Adolescent Psychology, 47*(Suppl. 1), S242–S254. doi: 10.1080/15374416.2016.1253016.

Dadds, M. R., Perry, Y., Hawes, D. J., Merz, S., Riddell, A. C., Haines, D. J., ... Abeygunawardane, A. I. (2006). Attention to the eyes and fear-recognition deficits in child psychopathy. *British Journal of Psychiatry, 189*(3), 280–281.

Finger, E. C., Marsh, A. A., Blair, K. S., Reid, M. E., Sims, C., Ng, P., ... Blair, R. J. R. (2011). Disrupted reinforcement signaling in the orbitofrontal cortex and caudate in youths with conduct disorder or oppositional defiant disorder and a high level of psychopathic traits. *American Journal of Psychiatry, 168*(2), 152–162.

Finger, E. C., Marsh, A. A., Mitchell, D. G., Reid, M. E., Sims, C., Budhani, S., ... Blair, R. J. R. (2008). Abnormal ventromedial prefrontal cortex function in children with psychopathic traits during reversal learning. *Archives of General Psychiatry, 65*(5), 586–594.

Fonagy, P., Gergely, G., & Target, M. (2007). The parent–infant dyad and the construction of the subjective self. *Journal of Child Psychology and Psychiatry, 48*(3–4), 288–328.

Gervais, M., & Wilson, D. S. (2005). The evolution and functions of laughter and humor: A synthetic approach. *Quarterly Review of Biology, 80*(4), 395–430.

Hare, R. D., & Neumann, C. S. (2008). Psychopathy as a clinical and empirical construct. *Annual Review of Clinical Psychology, 4*(1), 217–246.

Hawes, D. J., Price, M. J., & Dadds, M. R. (2014). Callous-unemotional traits and the treatment of conduct problems in childhood and adolescence: A comprehensive review. *Clinical Child and Family Psychology Review, 17*(3), 248–267.

Hosking, J. G., Kastman, E. K., Dorfman, H. M., Samanez-Larkin, G. R., Baskin-Sommers, A., Kiehl, K. A.,…Buckholtz, JW. (2017). Disrupted prefrontal regulation of striatal subjective value signals in psychopathy. *Neuron, 95*(1), 221–231.

Hyde, L. W., Waller, R., Trentacosta, C. J., Shaw, D. S., Neiderhiser, J. M., Ganiban, J. M.,…Leve, L. D. (2016). Heritable and nonheritable pathways to early callous-unemotional behaviors. *American Journal of Psychiatry, 173*(9), 903–910.

Jones, A. P., Laurens, K. R., Herba, C. M., Barker, G. J., & Viding, E. (2009). Amygdala hypoactivity to fearful faces in boys with conduct problems and callous-unemotional traits. *American Journal of Psychiatry, 166*(1), 95–102.

Kiehl, K. A., & Hoffman, M. B. (2011). The criminal psychopath: History, neuroscience, treatment, and economics. *Jurimetrics, 51,* 355–397.

Kimonis, E. R., Fanti, K. A., Goulter, N., & Hall, J. (2017). Affective startle potentiation differentiates primary and secondary variants of juvenile psychopathy. *Development and Psychopathology, 29*(4), 1149–1160.

Kimonis, E. R., Goulter, N., Hawes, D. J., Wilbur, R. R., & Groer, M. W. (2017). Neuroendocrine factors distinguish juvenile psychopathy variants. *Developmental Psychobiology, 59*(2), 161–173

Knopik, V., Neiderhiser, J., DeFries, J., & Plomin, R. (2016). *Behavioral genetics* (7th ed.). New York: Worth/Macmillian Learning.

Lima, C. F., Lavan, N., Evans, S., Agnew, Z., Halpern, A. R., Shanmugalingam, P.,…Warren, J. E. (2015). Feel the noise: Relating individual differences in auditory imagery to the structure and function of sensorimotor systems. *Cerebral Cortex, 25*(11), 4638–4650.

Lockwood, P. L., Sebastian, C. L., McCrory, E. J., Hyde, Z. H., Gu, X., De Brito, S. A., & Viding, E. (2013). Association of callous traits with reduced neural response to others' pain in children with conduct problems. *Current Biology, 23*(10), 901–905.

Lozier, L. M., Cardinale, E. M., VanMeter, J. W., & Marsh, A. A. (2014). Mediation of the relationship between callous-unemotional traits and proactive aggression by amygdala response to fear among children with conduct problems. *JAMA Psychiatry, 71*(6), 627–636.

Manninen, S., Tuominen, L., Dunbar, R., Karjalainen, T., Hirvonen, J., Arponen, E.,…Nummenmaa, L. (2017). Social laughter triggers endogenous opioid release in humans. *Journal of Neuroscience, 37*(25), 6125–6131.

Marsh, A. A., Finger, E. C., Fowler, K. A., Adalio, C. J., Jurkowitz, I. T., Schechter, J. C.,…Blair, R. J. R. (2013). Empathic responsiveness in amygdala and anterior cingulate cortex in youths with psychopathic traits. *Journal of Child Psychology and Psychiatry, 54*(8), 900–910.

Marsh, A. A., Finger, E. C., Mitchell, D. G., Reid, M. E., Sims, C., Kosson, D. S.,… Blair, R. J. (2008). Reduced amygdala response to fearful expressions in children and adolescents with callous-unemotional traits and disruptive behavior disorders. *American Journal of Psychiatry, 165*(6), 712–720.

Marsh, A. A., Finger, E. C., Schechter, J. C., Jurkowitz, I. T., Reid, M. E., & Blair, R. J. R. (2011). Adolescents with psychopathic traits report reductions in physiological responses to fear. *Journal of Child Psychology and Psychiatry, 52*(8), 834–841.

McGettigan, C., Walsh, E., Jessop, R., Agnew, Z. K., Sauter, D. A., Warren, J. E., & Scott, S. K. (2013). Individual differences in laughter perception reveal roles for mentalizing and sensorimotor systems in the evaluation of emotional authenticity. *Cerebral Cortex, 25*(1), 246–57.

Meffert, H., Gazzola, V., Den Boer, J. A., Bartels, A. A., & Keysers, C. (2013). Reduced spontaneous but relatively normal deliberate vicarious representations in psychopathy. *Brain, 136*(8), 2550–2562.

Michalska, K. J., Zeffiro, T. A., & Decety, J. (2016). Brain response to viewing others being harmed in children with conduct disorder symptoms. *Journal of Child Psychology and Psychiatry, 57*(4), 510–519.

National Institute for Heath and Care Excellence. (2013). *Antisocial behavior and conduct disorders in children and young people: Recognition, intervention and management* (Clinical guideline 158). Retrieved from https://www.nice.org.uk/guidance/cg158.

Nummenmaa, L., Tuominen, L., Dunbar, R., Hirvonen, J., Manninen, S., Arponen, E.,…Sams, M. (2016). Social touch modulates endogenous μ-opioid system activity in humans. *Neuroimage, 138*, 242–247.

O'Nions, E., Lima, C. F., Scott, S. K., Roberts, R., McCrory, E. J., & Viding, E. (2017). Reduced laughter contagion in boys at risk for psychopathy. *Current Biology, 27*(19), 3049–3055.

Pasalich, D. D., Dadds, M. R., Hawes, D. J., & Brennan, J. (2012). Attachment and callous-unemotional traits in children with early-onset conduct problems. *Journal of Child Psychology and Psychiatry, 53*(8), 838–845.

Patrick, C. J., & Drislane, L. E. (2015). Triarchic model of psychopathy: Origins, operationalizations, and observed linkages with personality and general psychopathology. *Journal of Personality, 83*(6), 627–643.

Provine, R. R. (1992). Contagious laughter: Laughter is a sufficient stimulus for laughs and smiles. *Bulletin of the Psychonomic Society, 30*(1), 1–4.

Provine, R. R. (2001). *Laughter: A scientific investigation.* New York: Penguin.

Roberts, R., McCrory, E., Joffe, H., De Lima, N., & Viding, E. (2018). Living with conduct problem youth: Family functioning and parental perceptions of their child. *European Child and Adolescent Psychiatry, 27*(5), 595–604.

Sakai, J. T., Dalwani, M. S., Mikulich-Gilbertson, S. K., Raymond, K., McWilliams, S., Tanabe, … Crowley, T. J. (2017). Imaging decision about whether to, or not to, benefit self by harming others: Adolescents with conduct and substance problems, with or without callous-unemotionality, or developing typically. *Psychiatry Research: Neuroimaging, 263,* 103–112.

Sauter, D. A., Eisner, F., Ekman, P., & Scott, S. K. (2010). Cross-cultural recognition of basic emotions through nonverbal emotional vocalizations. *Proceedings of the National Academy of Sciences of the United States of America, 107*(6), 2408–2412.

Scott, S. K., Lavan, N., Chen, S., & McGettigan, C. (2014). The social life of laughter. *Trends in Cognitive Sciences, 18*(12), 618–620.

Sebastian, C. L., McCrory, E. J., Cecil, C. A., Lockwood, P. L., De Brito, S. A., Fontaine, N. M., & Viding, E. (2012). Neural responses to affective and cognitive theory of mind in children with conduct problems and varying levels of callous-unemotional traits. *Archives of General Psychiatry, 69*(8), 814–822.

Szameitat, D. P., Kreifelts, B., Alter, K., Szameitat, A. J., Sterr, A., Grodd, W., & Wildgruber, D. (2010). It is not always tickling: Distinct cerebral responses during perception of different laughter types. *Neuroimage, 53*(4), 1264–1271.

Viding, E., Fontaine, N. M. G., Oliver, B. R., & Plomin, R. (2009). Negative parental discipline, conduct problems and callous-unemotional traits: Monozygotic twin differences study. *British Journal of Psychiatry, 195*(5), 414–419.

Viding, E., & Kimonis, E. R. (2018). Callous-unemotional traits. In C. J. Patrick (Ed.), *Handbook of psychopathy* (2nd ed., pp. 144–164). New York: Guildford Press.

Viding, E., & McCrory, E. J. (2015). Developmental risk for psychopathy. In A. Thapar, D. S. Pine, J. F. Leckman, S. Scott, M. J. Snowling, & E. Taylor (Eds.), *Rutter's child and adolescent psychiatry* (6th ed., pp. 966–980). New York: John Wiley & Sons.

Viding, E., & McCrory, E. (2018). Understanding the development of psychopathy: Progress and challenges. *Psychological Medicine, 48*(4), 566–577.

Viding, E., Sebastian, C. L., Dadds, M. R., Lockwood, P. L., Cecil, C. A., De Brito, S. A., & McCrory, E. J. (2012). Amygdala response to preattentive masked fear in children with conduct problems: The role of callous-unemotional traits. *American Journal of Psychiatry, 169*(10), 1109–1116.

Waller, R., Gardner, F., & Hyde, L. W. (2013). What are the associations between parenting, callous–unemotional traits, and antisocial behavior in youth? A systematic review of evidence. *Clinical Psychology Review, 33*(4), 593–608.

Warren, J. E., Sauter, D. A., Eisner, F., Wiland, J., Dresner, M. A., Wise, R. J., ... Scott, S. K. (2006). Positive emotions preferentially engage an auditory–motor "mirror" system. *Journal of Neuroscience, 26*(50), 13067–13075.

White, S. F., Brislin, S., Sinclair, S., Fowler, K. A., Pope, K., & Blair, R. J. R. (2013). The relationship between large cavum septum pellucidum and antisocial behavior, callous-unemotional traits and psychopathy in adolescents. *Journal of Child Psychology and Psychiatry, 54*(5), 575–581.

White, S. F., Fowler, K. A., Sinclair, S., Schechter, J. C., Majestic, C. M., Pine, D. S., & Blair, R. J. R. (2014). Disrupted expected value signaling in youth with disruptive behavior disorders to environmental reinforcers. *Journal of American Academy of Child and Adolescent Psychiatry, 53*(5), 579–588.

White, S. F., Briggs-Gowan, M. J., Voss, J. L., Petitclerc, A., McCarthy, K. R., Blair, R. J. R., & Wakschlag, L. S. (2016a). Can the fear recognition deficits associated with callous-unemotional traits be identified in early childhood? *Journal of Clinical and Experimental Neuropsychology, 38*(6), 672–684.

White, S. F., Tyler, P. M., Erway, A. K., Botkin, M. L., Kolli, V., Meffert, H., Pope, K., & Blair, R. J. R. (2016b). Dysfunctional representation of expected value is associated with reinforcement-based decision-making deficits in adolescents with conduct problems. *Journal of Child Psychology and Psychiatry, 57*(8), 938–946.

Wilkinson, S., Waller, R., & Viding, E. (2016). Practitioner review: Involving young people with callous unemotional traits in treatment—does it work? A systematic review. *Journal of Child Psychology and Psychiatry, 57*(5), 552–65.

Yoder, K. J., Lahey, B. B., & Decety, J. (2016). Callous traits in children with and without conduct problems predict reduced connectivity when viewing harm to others. *Scientific Reports, 6,* 20216.

20   Morals, Money, and Risk-Taking from Childhood to Adulthood: The Neurodevelopmental Framework of Fuzzy Trace Theory

Valerie F. Reyna and Christos Panagiotopoulos

Overview

Fuzzy trace theory (FTT) explains how cognitive representations of moral and monetary decisions, along with reward motivation and social values, are essential for understanding the adaptive social brain. What this means is that the way people think—whether they focus on exact details (called "verbatim" thinking) or the simple meaning behind those details (called "gist" thinking)—determines social behavior, such as risk-taking or committing crimes against others. Children, and adolescents who take unhealthy risks, rely more on verbatim thinking, but neurotypical adults progress to gist-based intuition, which is reflected in differences in the brain. Adults who do not develop properly continue engaging in unhealthy risk-taking and criminal behavior. We explain how FTT accounts for these developmental disorders. FTT's neurodevelopmental framework distinguishes autism from adult psychopathy, predicting and explaining paradoxes such as how these disorders are associated with fewer thinking biases in the laboratory but worse decisions in life.

Introduction

In this chapter, we provide a framework for understanding how people make moral and monetary decisions and how thinking about decisions changes with development from childhood to adulthood. We also apply this framework to explore developmental disorders such as autism and psychopathy. Our framework is grounded in experiments, mathematical models, neuroscientific, observational, and interventional studies (i.e., designing programs that support healthy and socially adaptive decision-making) on FTT (Blalock & Reyna, 2016; Reyna, 2012).

FTT's core assumption is that people mentally represent decision information in two basic ways: verbatim (the literal details) and gist (simple bottom-line meaning). For example, a decision to take a plea deal (accept a lighter sentence rather than risk going to trial) can be thought about in terms of details about potential outcomes, such as the number of years in prison, and their probabilities (the certain option offered by the prosecutor to serve fewer years in prison versus the risky option of going to trial and possibly serve more years in prison). Alternatively, an innocent person might think about the decision in much simpler terms: never plead guilty to something you did not do (Helm & Reyna, 2017).

How people mentally represent their decision options has a tremendous influence on the choices they make. Moreover, mental representation—the degree to which decision-making is based on verbatim details or gist meaning—changes with age and experience. In concert with changes in the brain, reward motivation, and socialization of values, cognitive representations determine whether people take unhealthy and antisocial risks (for a brief review of brain evidence, see Reyna, 2018).

Background

As we discuss below, we argue that neurotypical adults make most moral, monetary, and other reward-related decisions by relying on simple gist, ignoring numerical magnitudes and trade-offs. To be sure, most adults process numbers and trade-offs, and these exert some influence on decisions, but our central point is that simple gist tends to dominate. For example, asked whether you would push an innocent bystander off a bridge in order to divert a trolley so that it would kill one person rather than five people, most adults say "no" even though five is more than one (Bartels & Pizarro, 2011). Such decisions reflect combined influences of cognitive representations (thinking about both verbatim numbers and bottom-line gist), reward motivations (if money or other rewards are involved), and affective social and moral values (money is good; killing people is bad; Broniatowski & Reyna, 2018). (The ability to inhibit rash impulses and delay gratification are also factors but not ones we discuss here; see Reyna & Wilhelms, 2017; Romer, Reyna, & Pardo, 2016).

To preview, research indicates that children lean toward processing verbatim details, trading off objective levels of risk and reward if those are made clear to them (Schlottmann & Wilkening, 2011). Adolescents engage in a mixture of cognitive styles because they are in transition from childhood to adulthood, but those who take unhealthy and antisocial risks tend to engage in thinking about risks that is closer to the verbatim than gist side of processing (Kwak, Payne, Cohen, & Huettel, 2015; Reyna, Estrada et al., 2011). For example, adolescents might attend a party with illegal drugs because the benefits of having fun at the party are perceived to be high while the probability of getting caught (and going to jail) is perceived to be low. Surprisingly, teens often consider details, such as whether the amount of fun offsets the amount of risk, but these teens underemphasize the life-altering gist that they are risking a felony conviction. Thus, with respect to their cognitive representations, adolescent risk-takers frequently think in a younger way because they focus on details rather than the bigger picture (i.e., they are developmentally delayed).

The reason for this unhealthy risk-taking is not that gist thinking is inherently risk-discouraging and verbatim thinking is inherently risk-promoting. Rather, it is because many unhealthy risks are characterized by large rewards with low probabilities of bad consequences for a single act of engagement (Reyna & Farley, 2006). For example, many crimes go unpunished, addiction does not routinely occur with the first use of illicit drugs, and human immunodeficiency virus (the virus that causes AIDS) is unlikely to be transmitted even with unprotected sex. Engaging in these behaviors once is unlikely to be punished. Therefore, thinking about magnitudes of risk and reward, and trading these off as economists recommend, is likely to promote unhealthy and antisocial risk-taking. People who think this way about crime, addiction, and AIDS tend to have bad outcomes. If rewards were low (e.g., small amounts of money were at stake) and the probability of bad consequences were high (e.g., high chances of getting caught for committing a crime such as stealing the small amount of money), verbatim thinking would *discourage* risk-taking. Commonly, however, verbatim thinking is about high rewards and low probabilities of bad consequences, which encourages risk-taking. In fact, many crimes seem to reflect such technically "rational" considerations of risk and reward (though other crimes are impulsive; Matsueda, 2013).

Building on this foundation of research on neurotypical development, we argue that autism is characterized, in part, by a greater reliance on verbatim as opposed to gist thinking, which generally characterizes younger children, but without the heightened reward motivation that characterizes adolescents or psychopaths (Reyna & Brainerd, 2011). That is, those with autism would take the plea deal if the risk-reward ratio were favorable, regardless of factual guilt or innocence. Adolescents would be unduly influenced by the prospect of freedom even if going to trial was risky. Adolescence, in addition to being a period of cognitive transition, is characterized by an increase in sensation seeking or reward motivation (e.g., for fun, freedom, and other rewards), which combines with cognitive representations to further promote risk-taking for rewards, including social rewards (e.g., impressing peers; Steinberg, 2008). Thus, risk-taking and antisocial behavior are normative for teenagers in the sense that these behaviors increase during normal adolescence for both cognitive and social motivational reasons. For most individuals, these behaviors decline in adulthood, hence the term "adolescent-limited" antisocial behavior (Moffitt, 1993).

Adolescent-limited is contrasted with life-course persistent antisocial behavior (Moffitt, 1993). The latter individuals who engage in persistent antisocial behavior are characterized by continued high sensation seeking (beyond adolescence) and impulsivity, among other traits. This unhealthy risk-taking may progress in severity to *criminal* antisocial risk-taking in adulthood because of high reward sensitivity (e.g., characterized by higher activation in reward centers, such as the ventral striatum, in decision-making tasks) and/or inability to control reward-related impulses (e.g., deficits in executive processes that reflect trauma or developmental disorders in the operation of frontal control networks in the brain) (Bjork & Pardini, 2015; Glenn & Raine, 2014). Thus, FTT anticipates two routes to risk-taking in adolescence and adulthood: (a) a "hot" kind in which emotion, temptation, and passion dominate (risks that, on reflection, people often regret and would not want to take again) and (b) a "cold" one in which decision makers take "rational" risks to gain rewards, calculated risks they would want to take again, even if they turned out badly, as long as the odds and outcomes were favorable (Reyna & Farley, 2006).

When risk-taking occurs in the absence of empathy for the feelings of others, it can be particularly dangerous to society. Life-course persistent antisocial risk-takers include psychopaths, who also lack empathic caring, the ability to care about the feelings of others (Decety & Yoder, 2016). There is some neural evidence that the brains of adult psychopaths resemble those of younger people (Shannon et al., 2011), suggesting a developmental disorder related to reward motivation and, we speculate, to developmentally inappropriate reliance on verbatim representations. An important lynchpin in this argument is Bartels and Pizarro's (2011) finding that utilitarian thinking in the trolley problem mentioned above is related to psychopathy. Utilitarian thinking is, by definition, maximizing the risk-reward ratio as stipulated in economics; in our earlier example, the utilitarian solution is to save more people by murdering one. In other words, both options entail little risk (the options are presented as involving sure outcomes), but one option saves more lives and thus is preferred according to utilitarianism. The developmentally normative response for neurotypical adults, however, is to reject quantitative comparisons of numbers of lives in favor of qualitative, categorical thinking: no amount of lives can compensate for murder of an innocent person (Reyna & Casillas, 2009).

As we will discuss, similar noncompensatory thinking—in which tradeoffs are rejected in favor of qualitative, categorical thinking—also characterizes developmental differences in the singularity effect. The singularity effect is, for example, donating more money to one identifiable victim (e.g., the victim of a disease who needs expensive therapy) than to a group of eight identifiable people that includes the same one victim (Kogut & Ritov, 2005). When these two scenarios are presented together, people donate more to eight victims than one victim; they know that eight is more than one. However, when the scenarios are not presented together, adults give more to the single victim, responding based on qualitative gist rather than verbatim quantitative details.

In contrast to adults, younger children give more of their candies to more children compared to one child, contrary to the singularity effect. Children gradually reverse their donations as they get older and move toward adulthood; they become less technically rational, less utilitarian, and less verbatim in their thinking—eventually exhibiting the singularity effect as adults (Kogut & Slovic, 2016). As we will discuss, this developmental reversal echoes other results predicted by FTT in which literal verbatim thinking, which is more objective and reflects reality, is gradually replaced by gist-based thinking that foments specific cognitive biases and turns quantitative comparisons on their head.

In both the trolley car example and the singularity effect, the role of emotion seems obvious and explanatory: psychopaths feel little emotion for others (i.e., empathic caring) and so they are willing to murder one person whereas neurotypical adults feel emotion more intensely for one identifiable victim than for eight (Kogut & Ritov, 2005). However, although the effects of emotion are probably real, they do not fully explain the results. Why do people feel more emotion for one than for eight? Donating more to one person compared to eight (or saving one person compared to five) cannot be explained by saying that this is just a failure to think deliberatively—why would that kind of thinking *increase* from childhood to adulthood, as shown in myriad studies (Weldon, Corbin, & Reyna, 2014)? Why does longer deliberation produce greater biases under specific circumstances (Duke, Goldsmith, & Amir, 2018), and worse decision-making, compared with gist-based reasoning (Abadie, Waroquier, & Terrier, 2013)? These puzzles and paradoxes, from the perspectives of standard theories, are explained by FTT.

There are three reasons we know that this kind of processing and decision-making has a cognitive representational component rather than *only* an emotion-versus-control component. First, gist and verbatim thinking independently predict real-world self-reported risk-taking when sensation seeking is controlled for statistically (Reyna, Estrada et al., 2011). In other words, gist and verbatim thinking predict unique variance in risk-taking in addition to what sensation seeking predicts. Second, behavioral and neural differences that correlate with risk-taking are elicited when the evidence indicates that decision makers are processing risky options in a more verbatim as opposed to gist way (Reyna, Helm et al., 2018); these comparisons are within-subjects, meaning that the same people are compared with themselves under different conditions that elicit different kinds of cognitive processes. Within-subjects comparisons control for differences in types of people, which rules out a host of alternative explanations, such as that differences across people in sensation seeking or emotionality wholly explain the results. Third, differences in verbatim and gist processing that replicate behavioral patterns of risky choice can be induced with purely cognitive manipulations (again, within the same experimental session and within the same people) that do not involve any variation in sensation seeking or any manipulation of emotion. Neurotypical adults can be induced to make choices that resemble those of children or adolescents simply by using manipulations predicted by FTT to change cognitive representations (Kühberger & Tanner, 2009; Reyna, Chick, Corbin, & Hsia, 2014).

In summary, we implement three constructs from FTT: (a) verbatim/gist cognitive representations; (b) individual and developmental differences in reward motivation (i.e., sensation seeking); and (c) affective valences associated with social and moral principles stored in long-term memory. Verbatim and gist representations underpin processing that varies from a focus on precise details (verbatim-based analysis) to one on overall simple gist-based intuition (see table 20.1). Although FTT assumes that both kinds of representations are usually processed, the balance of these representations in decision-making varies developmentally. FTT is the only theory that predicts that reliance on intuition *increases* from childhood to adulthood, causing gist-based biases to increase. This developmental trend has been predicted and found in diverse domains of cognitive development such as false memory (remembering events that are consistent with the gist of what happened although those events never actually happened) and many other cognitive biases. Next, we explain how FTT predicts such biases in children versus adults, which shapes risk preferences. We then extend FTT to explain predictions for those with autism, adolescents, adult criminal risk-takers, and psychopaths.

Decision Biases in Children and Adults: Framing and Singularity Effects

Framing biases occur when people make inconsistent choices when confronted with two different versions of the same problem. One version presents the decision dilemma in a gain frame, while the other presents it in a loss frame. An example of the framing task is Tversky and Kahneman's (1986) dread disease problem, in which participants are informed of a disease that is expected to kill 600 people and then are presented with a gain and a loss frame dilemma about saving these people. In the gain frame, they choose between saving 200 people for sure or one-third chance of saving all 600 people and two-thirds chance of saving none. In the loss frame, they choose between either 400 people dying for sure, or a one-third chance of no deaths and a two-thirds chance of 600 deaths (see table 20.1 for a similar money problem). When confronted with such a problem, people will typically choose the *safe* option in the gain frame—saving 200 people for sure—but choose the *risky* option in the loss frame—gambling on the chance of no deaths—thus being inconsistent with themselves. These inconsistencies are described as "cognitive biases" because the number of people saved in the end is the same in the gain and loss versions (e.g., 600 lives minus 400 who die equals 200 saved).

According to FTT, this gain–loss framing bias is predicted because people use simple gist to reason and make decisions. Therefore, they mentally

Table 20.1

Verbatim and Categorical Gist Representations of Gain–Loss Framing Problems and Associated Social and Moral Principles (Affective Values)

Decision Problem	Verbatim Representation	Gist Representation	Affective Values
Gain Frame			
Option A: Saving 200 people for sure	Option A: 200 people saved	Option A: Saving some people for sure	Saving people is good.
Option B: 1/3 chance of saving 600 people and 2/3 chance of saving none	Option B: 1/3 × 600 = 200 people saved Preference: Indifference because expected values are equal	Option B: Saving some people or saving none Preference: Option A because saving some is better than saving none	
Loss Frame			
Option C: 400 people dying for sure	Option C: 400 people dying	Option C: Some people dying for sure	People dying is bad.
Option D: 2/3 chance of 600 people dying and 1/3 chance of none dying	Option D: 2/3 × 600 = 400 dying Preference: Indifference because expected values are equal	Option D: Some people dying or none dying Preference: Option D because none dying is better than some dying	
Gain Frame			
Option A: Winning $200 for sure	Option A: Win $200	Option A: Winning some money for sure	Gaining money is good.
Option B: 1/3 chance of winning $600 and 2/3 chance of winning none	Option B: 1/3 × $600 = Win $200 Preference: Indifference because expected values are equal	Option B: Winning some money or none Preference: Option A because some money is better than none	
Option C: Losing $400 for sure	Option C: Lose $400	Option C: Losing some money for sure	Losing money is bad.
Option D: 2/3 chance of losing $600 and 1/3 chance of losing none	Option D: 2/3 × $600 = Lose $400 Preference: Indifference because expected values are equal	Option D: Losing some money or none Preference: Option D because losing no money is better than losing some	

Note: Loss framing problems are usually preceded by a preamble (e.g., 600 people are expected to die or $600 has already been won) such that the gain and loss net outcomes are equivalent. When these decisions are presented to children and young adolescents, the outcomes are prizes (e.g., stickers) and the probabilities are represented by colored areas of spinners.

compare saving some people to saving none, favoring saving some, and none dying to some dying, favoring none. (This prediction is supported by separate evidence from separate tasks about how people represent information in their minds.) Gist supports "fuzzy" intuition, which is argued to be an advanced form of thought in FTT because it captures the meaning of information, not just memorized meaningless words or numbers (Reyna, 2012).

In a critical test of framing biases in children, FTT also predicted that intuitive gist-processing increases with age. Early work (reviewed in Reyna & Farley, 2006) showed that children are indifferent when making decisions in a framing task. The youngest children, preschoolers, did not have framing biases. Children focused on the final outcomes of the decision tasks, regardless of whether the problem was presented in a gain or loss frame. Unlike adults, their responses were not biased. Children were paying attention; for example, they gambled less as the risk (chances of gaining none or losing some) increased, regardless of gains or losses (see examples of similar decisions in table 20.1). Children's thinking showcases what most theories would regard as rational decision-making, not exhibiting gist-based biases in their decision-making process—therefore, technically behaving more like rational actors than adults do.

From childhood to adolescence, a different pattern of preferences emerges: reverse framing, or framing-inconsistent behavior when outcomes are large (Reyna & Farley, 2006). In contrast to standard framing effects in adults, or no framing in children, reverse framing consists of a preference for the *risky* option in the gain frame and the *sure* option in the loss frame. For example, in the money problem in table 20.1, adolescents are more likely to prefer the risky option that offers the chance to win $600 compared with winning $200 for sure. They are also more likely to prefer losing $400 for sure compared with taking a chance and losing $600. This pattern of preferences occurs during adolescence, especially for large quantitative differences in rewards, such as $600 versus $200 as opposed to $6 versus $1 (Reyna, Estrada et al., 2011). Reverse framing reveals a more precise analysis of a decision problem in which quantitative differences in outcomes matter, rather than the simple gist that adults rely on (see gist in table 20.1). In terms of complexity of quantitative processing, this reverse-framing processing lies between processing both risk and reward quantities, as young children do, and processing simple gist, as adults do, and involves processing mainly reward.

Thus, reverse framing is choosing the higher magnitude of outcome in the gamble for the gain frame and the lower magnitude in the sure loss for the loss frame. This more precise kind of thinking, compared with gist thinking, is called "verbatim" processing because it relies on precise words or quantities (the latter if quantities are presented). Many scholars assume that adolescence is a period when reward sensitivity and not-yet-fully-developed inhibition lead to risk-taking (Reyna & Farley, 2006). We agree. However, FTT also predicts this reverse-framing pattern for adolescence, which cannot be predicted by just higher reward sensitivity and lower inhibition. Furthermore, this pattern of reverse framing is associated with greater real-life risk-taking (Reyna, Estrada et al., 2011). Thus, ideas about cognitive representations (verbatim and gist) are required to understand the development of risky choices.

FTT explains that adolescents approach risky decision dilemmas in a way that is closer to verbatim than adults do, concentrating on the literal facts instead of reaching for the gist, the bottom-line meaning (Mills, Reyna, & Estrada, 2008; Reyna & Farley, 2006). When confronted with a decision, they trade off risks and rewards to determine which choice is more attractive to them. In other words, they engage in an implicit cost–benefit analysis of their options, relying on the representations of the superficial information as known or presented to them. Presented with a framing task, which involves risky decision dilemmas, adolescents who take risks reverse frame more than adults do. Most adolescents are not reverse framers, but this varies cross individuals; they show less standard framing than adults do. Note that reverse framing does not equal more risk-taking overall, but more risk-taking when rewards (gain outcomes) are higher and risk avoidance when loss outcomes are lower, all else being equal.

Hence, adolescent risk-taking is a multifactorial process. Sensation seeking (reward sensitivity) and behavioral inhibition, both contributing factors, have been shown to be related to age. Inhibition gradually rises throughout adolescence; sensation seeking that draws teens to rewards rises and then falls (Reyna, Estrada et al., 2011; Steinberg, 2008). When these age-related factors are controlled for, gist and verbatim processing were shown to still predict unique variance in adolescent risk-taking. FTT indicates that how young people think, as assessed by their framing patterns, explains crucial aspects of adolescent risk-taking.

This thinking is particularly evident in sexual risk-taking, a domain in which adolescents might be assumed—falsely—to underestimate the risks, even though data show that they are very well aware of and overestimate risks (e.g., for sexually transmitted infections; Reyna & Farley, 2006; Reyna & Mills, 2014). Non–risk-taking adults are more likely to categorically refuse to have unprotected sex even if the chances of being infected by the AIDS virus are very low, relying on their gist representations (no risk is better than some risk of AIDS) which cue core values (getting AIDS is bad). In a similar scenario, adolescents would attempt to trade off risks and benefits. For example, having sex is an important reward that might outweigh the risks of possible sexually transmitted infection (STI), especially since statistically the probability of an STI is low. In other words, in their decision-making, adolescents are less likely to use the categorical distinction "no risk is better than some risk of AIDS" or "it only takes once" to be infected by an STI (Reyna, Estrada, et al., 2011). Because they weigh risks and rewards, they might lean toward the much-sought reward of having (unprotected) sex—the slight chance of getting infected by an STI is not downplayed, but its effect is swamped by the magnitude of benefits. This rational process, which is distinct from after-the-fact rationalization (Cushman, 2019) is objective, but would ultimately lead adolescents to risky behavior, with potentially severe repercussions for their long-term health (Reyna & Mills, 2014).

Similar to framing biases, the singularity effect emerges and becomes stronger with age, from childhood to adulthood, consistent with FTT. Research has shown that adults donate more money to, and are more affectively moved by, one identifiable victim in distress compared with a group of people who includes that same one victim (Kogut & Ritov, 2005). This preference toward the identifiable unity versus the more numerous group is called the *singularity effect*, and like the framing effect it is a cognitive bias. Contrary to the assumption that people would consistently help a group of people more than they would help one individual, the singularity effect indicates an increased insensitivity toward quantitative magnitudes—to *greater numbers* of people—and a preference to help one identifiable individual.

The economically rational prediction that, given the choice between helping one individual or a group of eight, most people would choose to help the group because it consists of more people—it is quantitatively larger—is inconsistent with the experimental observations of the singularity effect. This rational vision implies that people will process this dilemma in a verbatim way, processing one versus eight; in other words, that they will cognitively represent the single individual also as a number. Research has shown that most adults tend to approach similar moral dilemmas in a gist-based fashion and thus show a consistent preference to help the one identifiable person in lieu of the larger number, the group. Most adults would not think one as opposed to eight; instead, when they see a single person they think categorically about personhood, meaning they would represent the unit as a categorical gist. Eight people invites more exact calibration of the number of people with the number of candies or dollars (Reyna, 2012). Focusing on such numerical details supports verbatim-based analysis.

Conversely, as Kogut and Slovic (2016) showed, younger children donated or shared less candy with one identifiable child compared with six children who included that one. However, this tendency begins to reverse as children get older, when their donations become more consistent with the singularity effect; they share more with the single individual than the group. The single individual stands for a category; it is qualitative and not a mere number like "6," which is a quantitative representation. As with framing, the singularity effect becomes greater with age, from childhood to adulthood, as people mature cognitively and become less sensitive to quantitative differences in outcomes (less scope sensitivity). This is another example of developmental reversals in cognitive representations from childhood to adulthood, indicating an increase in reliance on gisty categorical thinking as age progresses.

These effects are not due to an overall tendency to share or to be selfish. The overall tendency to help or share with other people also increases with age. The youngest children share less of their candy than their older counterparts overall—whether with the single individual or the group (Kogut & Slovic, 2016) and are therefore more selfish, extreme in psychopaths. However, selfishness alone does not explain why the singularity effect emerges and increases with age. Cognitive representations, and a preference for gisty intuitions versus verbatim quantitative thinking, explains why the singularity effect becomes more common.

Criminal Risk-Taking, Autism, and Psychopathy

This FTT theoretical framework contrasting gist and verbatim representations has been extended to nontypical development—namely, to criminal risk-taking and autism (e.g., Reyna & Brainerd, 2011). Regarding criminal risk-taking, research suggests that there are two kinds of risk-taking, with overlapping but distinguishable brain substrates (Reyna, Helm et al., 2018).

Noncriminal risk-taking behavior, linked to impulsivity and reward sensitivity, was associated with more activation in the amygdala and striatal areas, areas of emotional processing and reward motivation. Criminal risk-taking was associated with these kinds of areas but also with activation in temporal and parietal cortices, their junction, and insula, areas related to moral cognition, risk preference, and numerical processing.

Moreover, all of this activation was detectable when adults engaged in reverse framing—that is, choosing risky gambles in a gain frame and sure losses in the loss frame (Reyna, Helm et al., 2018). Thus, when adults displayed risk preferences in framing problems that were similar to those of risk-taking adolescents, the extent of brain activation in these areas correlated with the extent of their risk-taking activities. More noncriminal risk-taking was associated with more activation in reward and emotion areas ("hot" cognition), and more criminal risk-taking was also associated with activation in cognitive areas including numerical processing areas ("cold" cognition; Dehaene, Piazza, Pinel, & Cohen, 2003). These results suggest that risk-taking in adults, a non-normative behavior for this age group, may reflect developmentally delayed cognitive representational, as well as emotional/motivational, processing.

FTT indicates that both the hot and cold routes to risk-taking characterize adolescence. In particular, the second type of risk-taking, the cold route, involves verbatim analytical thinking about risk-reward trade-offs; when reward sensitivity and reward magnitudes are high, this produces reverse-framing decisional patterns. Individuals with autism also appear to process information in a more verbatim or literal way than neurotypical adults do (e.g., De Martino et al., 2008; Wojcik et al., 2018). We would characterize their information-processing style as high verbatim but low gist, which contributes to less comprehension of metaphors, lower levels of false memory (falsely remembering information that is gist-consistent but verbatim-inconsistent—that is, never presented), and lower likelihood of inferring implicit semantic connections in narratives (e.g., inferring that the bird is under the table from reading that the bird was in the cage and the cage was under the table), as predicted by FTT (Reyna & Brainerd, 2011). (See also the FTT research on mental representation of metaphor, false memory, and inference; Reyna, 2012.)

However, people with autism do not necessarily exhibit higher levels of sensation seeking. Therefore, to the degree that people have autism, FTT predicts that they should be less likely to show standard framing effects. Given a high-verbatim, low-gist cognitive style, they would treat gains and losses more similarly than neurotypical adults and generally be less subject to other gist-based cognitive biases, as has been observed (see Reyna & Brainerd, 2011).

However, FTT would not expect that those with autism would show reverse framing. That is, people with autism appear to be more technically rational but not reward-sensitive (e.g., not drawn to higher rewards in the gamble). In addition, although people with autism have some difficulty inferring the feelings of others (cognitive empathy), they are not less likely to experience empathic caring. Thus, the affective valence of their social and moral values should not necessarily differ from those of neurotypical individuals (cf. Shah, Catmur, & Bird, 2016). FTT would therefore predict attenuation of framing among autistic individuals for both the money and lives dilemmas shown in table 20.1.

In contrast to people with autism, psychopaths are higher in sensation seeking or reward sensitivity and lower in empathic caring, compared with neurotypical individuals (Buckholtz et al., 2010; Glenn & Raine, 2014). Adult risk-takers in the study by Reyna and colleagues (2018) were also higher in sensation seeking (as well as being less likely to show standard framing). Psychopathy is also distinct from conduct disorder in adolescence (risk-taking and aggression) or impulsive antisocial behavior in adulthood. Traits of adult psychopathy, especially callous-unemotional traits, are detectable in childhood, and such traits during childhood predict adult psychopathy (Frick & Viding, 2009; Lynam, Caspi, Moffitt, Loeber, & Stouthamer-Loeber, 2007). *Criminal* risk-taking (e.g., drunk driving) in adulthood represents a more extreme form of antisocial behavior compared with *noncriminal* risk-taking (e.g., getting drunk). However, it could be argued that psychopathy represents the most extreme form of adult antisocial behavior because it involves intentional manipulation of others to obtain rewards (e.g., money) without empathic caring (Decety, Skelly, & Kiehl, 2013), as opposed to acting on impulse because of tempting rewards.

Based on behavioral and brain evidence, it appears that psychopathy represents a severe form of developmental delay in both reward sensitivity and cognitive processing (Buckholtz et al., 2010; Shannon et al., 2011). Crucially, Bartels and Pizarro (2011) showed that psychopathy is characterized by utilitarian thinking—which is, by definition, trading off risk and reward to determine the rationally superior option, what FTT calls verbatim thinking. Thus, psychopaths would be expected to show reverse framing for monetary dilemmas, because they involve valued rewards (see table 20.1), but merely attenuated framing for lives problems. To the degree that psychopaths fail to endorse the moral value that saving lives is good, the lives dilemmas become just math problems governed by the verbatim numbers (cf. Bloom, 2017). Hence, risky choices of psychopaths would be a product of reward sensitivity (i.e., maximizing personal gains), cognitive representations that favor verbatim thinking, and lower levels of endorsement of moral principles such as saving lives is good.

For similar reasons, psychopaths would be expected to process the trolley problem analytically (five people killed is more than one person killed), although they would be less affectively or emotionally responsive to this dilemma (see Glenn & Raine, 2014, for evidence about lower levels of affective responsiveness; Patil, 2015). Like children, they should donate less in the singularity tasks than neurotypical adults to both one and six victims (because they experience lower affective valence or empathic caring), but again they would be more likely to consider these to be math problems because of verbatim thinking (Kogut & Slovic, 2016). Thus, if they donated, it would be more to six victims than one.

Note that, like autism, psychopathy is a matter of degree and varies across individuals. Presumably *pure* psychopaths would not donate anything if they could avoid social sanctions for doing so. Those with psychopathic *tendencies* would give less than other adults without such tendencies but would give more to six than one. Although we can piece together theoretical expectations based on prior work (e.g., showing reverse framing associated with adult risk-taking, including self-reported criminal behavior), there is as yet no published evidence that supports the latter predictions for psychopaths.

Summary and Conclusions

FTT attributes decision-making to three kinds of causes: how people think (focusing on literal details or simple gist), how responsive they are to rewards (sensation seeking), and their experienced affective valences associated with social and moral values (their internalization of social and moral norms). Thus, typical adults will choose to have some money for sure rather than take a risk and possibly have no money, even when they probably could get more money by gambling (e.g., in the Allais problem in which a sure option of lower numerical value is preferred to a gamble of higher numerical value; see Reyna & Brainerd, 2011). Scientific tests show that simple gist representations of information (get some money versus get none) determine risk preferences, in concert with values such as "money is good," as illustrated in table 20.1. These simple gist representations also create cognitive biases such as framing effects in which the same person wants to avoid risks for gains but seek risks for objectively equivalent losses.

Counter to other theories of development, FTT predicts that children should be more objective than adults because reliance on verbatim analysis of surface details goes down and gist-based intuition grows from childhood to adulthood. Ironically, this means that children are less likely than adults to show a variety of gist-based biases, such as the framing effect, singularity effect (donating more money to one victim than eight), gist-based false memory, and other cognitive illusions. However, for the same developmental reasons, younger people are more likely to think about risk and reward analytically, which encourages risk-taking when rewards are high and probabilities of bad consequences are low. This vulnerability to risk-taking is worse in adolescence because reward sensitivity or motivation also increases and inhibition is not yet fully developed. Adolescent risk-takers are more likely to show a pattern of risk preferences that is rare in adults—reverse framing—preferring sure losses but seeking risky gains. Thus, with respect to their cognitive representations, adolescent risk-takers are often developmentally delayed. Reward motivation, less inhibition, and the thinking characterized by reverse framing all combine to predict greater real-life risk-taking (Reyna, Estrada et al., 2011).

Drawing on behavioral and brain evidence, FTT predicts that adult risk-takers are also more likely to show reverse framing. The degree to which they engage in criminal and noncriminal risk-taking is correlated with both reward motivation (i.e., they are higher in sensation seeking) and reverse framing (Reyna, Helm et al., 2018). Brain differences emerge for both types of risk-takers that further support a hot (related to reward and emotion areas, such as the amygdala) and a cold (related to the temporal and parietal cortex) route to risk-taking. These differences are detectable when the adults make choices consistent with reverse framing; when adults choose sure losses and risky gains, their brain activation covaries with the level of self-reported unhealthy risk-taking. Adult risk-takers, especially those who engage in criminal risk-taking, appear to be developmentally delayed with respect to both motivation and cognition.

Building on other research, FTT suggests that disorders such as autism and psychopathy are characterized by reliance on verbatim analysis of risks and rewards rather than gist—again, with respect to their cognitive representations, exhibiting developmental delay. Psychopathy differs from autism, however, in also showing more responsiveness to rewards and less affective (emotional) responsiveness to social and moral values. Consistent with this framework, those with autism are less likely to demonstrate cognitive biases—they are *technically* more rational than typical adults in a variety of laboratory tasks because they base their responses on verbatim reality, such as objective outcomes and their probabilities, rather than being biased by subjective gist. Psychopaths would be expected to be similar to adolescent risk-takers and adult criminal risk-takers (who may include psychopaths or those with psychopathic tendencies) in showing reverse framing and differences in other cognitive biases, too.

FTT suggests that effects of impulsive reward seeking can be distinguished from such cognitive effects in accounting for developmental disorders such as psychopathy. The utilitarian approach of psychopaths is non-normative for adults, who typically respond more to intuitive gist than to precise numbers, undergirding social and moral development. In other words, morally, saving human life or sharing resources is a categorical good, as opposed to being a math problem involving distinguishing the number of lives saved or dollars shared. Thinking about moral choices in simple gist terms highlights categorical values, which helps most people make moral choices.

In summary, developmentally advanced gist intuitions are central to social and moral development. People who rely more on their gist intuitions when making decisions tend to make healthier socially adaptive choices and avoid unnecessary risks by thinking simply and categorically. Conversely, adolescents—and adults with developmentally immature cognition—think in a more literal, verbatim way, and this can manifest in atypical social development, including utilitarian approaches to moral dilemmas, unhealthy risk-taking, antisocial behavior, and even criminal activity. Ironically, irrational biases that promote healthy and moral choices seem to be a hallmark of the adaptive social brain.

Acknowledgments:	Preparation of this article was supported in part by grants from the National Institutes of Health (National Institute of Nursing Research R21NR016905) and the National Institute of Food and Agriculture (NYC-321407).

References

Abadie, M., Waroquier, L., & Terrier, P. (2013). Gist memory in the unconscious-thought effect. *Psychological Science, 24*(7), 1253–1259. doi: 10.1177/0956797612470958

Bartels, D. M., & Pizarro, D. A. (2011). The mismeasure of morals: Antisocial personality traits predict utilitarian responses to moral dilemmas. *Cognition, 121,* 154–161. doi: 10.1016/j.cognition.2011.05.010

Bjork, J. M., & Pardini, D. A. (2015). Who are those risk-taking adolescents: Individual differences in developmental neuroimaging research. *Developmental Cognitive Neuroscience, 11,* 56–64. doi: 10.1016/j.dcn.2014.07.008

Blalock, S. J., & Reyna, V. F. (2016). Using fuzzy-trace theory to understand and improve health judgments, decisions, and behaviors: A literature review. *Health Psychology, 35*(8), 781–792. doi: 10.1037/hea0000384

Bloom, P. (2017). Empathy and its discontents. *Trends in Cognitive Science, 21*(1), 24–31.

Broniatowski, D. A., & Reyna, V. F. (2018). A formal model of fuzzy-trace theory: Variations on framing effects and the Allais Paradox. *Decision, 5*(4), 205–252. doi: 10.1037/dec0000083

Buckholtz, J. W., Treadway, M. T., Cowan, R. L., Woodward, N. D., Benning, S. D., Li, R.,…Zald, D. H. (2010). Mesolimbic dopamine reward system hypersensitivity in individuals with psychopathic traits. *Nature Neuroscience, 13*(4), 419–421. doi: 10.1038/nn.2510

Cushman, F. (2019, May 28). Rationalization is rational. *Behavioral and Brain Sciences.* Advance online publication. doi: 10.1017/S0140525X19001730

Decety J., Skelly L. R., & Kiehl K. A. (2013). Brain response to empathy-eliciting scenarios involving pain in incarcerated individuals with psychopathy. *JAMA Psychiatry, 70*(6), 638–645.

Decety, J., & Yoder, K. J. (2016). Empathy and motivation for justice: Cognitive empathy and concern, but not emotional empathy, predict sensitivity to injustice for others. *Social Neuroscience, 11*(1), 1–14. doi: 10.1080/17470919.2015.1029593

Dehaene, S., Piazza, M., Pinel, P., & Cohen, L. (2003). Three parietal circuits for number processing. *Cognitive Neuropsychology, 20*(3–6), 487–506. doi: 10.1080 /02643290244000239

De Martino, B., Harrison, N. A., Knafo, S., Bird, G., & Dolan, R. J. (2008). Explaining enhanced logical consistency during decision making in autism. *The Journal of Neuroscience, 28*, 10746–10750. http://dx.doi.org/10.1523/JNEUROSCI.2895-08.2008

Duke, K., Goldsmith, K., & Amir, O. (2018). Is the preference for certainty always so certain? *Journal of the Association for Consumer Research, 3*(1), 63–80.

Frick, P. J., & Viding, E. (2009). Antisocial behavior from a developmental psychopathology perspective. *Development and Psychopathology, 21*, 1111–1131. doi: 10.1017/S0954579409990071

Glenn, A. L., & Raine, A. (2014). Neurocriminology: Implications for the punishment, prediction and prevention of criminal behaviour. *Nature Reviews Neuroscience, 15*, 54–63. doi: 10.1038/nrn3640

Helm, R. K., & Reyna, V. F. (2017). Logical but incompetent plea decisions: A new approach to plea bargaining grounded in cognitive theory. *Psychology, Public Policy, and Law, 23*, 367–380. doi: 10.1037/law0000125

Kogut, T., & Ritov, I. (2005). The "identified victim" effect: An identified group, or just a single individual? *Journal of Behavioral Decision Making, 18*(3), 157–167. doi: 10.1002/bdm.492

Kogut, T., & Slovic, P. (2016). The development of scope insensitivity in sharing behavior. *Journal of Experimental Psychology: Learning, Memory, and Cognition, 42*(12), 1972–1981. doi: 10.1037/xlm0000296

Kühberger, A., & Tanner, C. (2009). Risky choice framing: Task versions and a comparison of prospect theory and fuzzy-trace theory. *Journal of Behavioral Decision Making, 23*(3), 314–329. doi: 10.1002/bdm.656

Kwak, Y., Payne, J. W., Cohen, A. L., & Huettel, S. A. (2015). The rational adolescent: Strategic information processing during decision making revealed by eye tracking. *Cognitive Development, 36*, 20–30. doi: 10.1016/j.cogdev.2015.08.001

Lynam, D. R., Caspi, A., Moffitt, T. E., Loeber, R., & Stouthamer-Loeber, M. (2007). Longitudinal evidence that psychopathy scores in early adolescence predict adult psychopathy. *Journal of Abnormal Psychology, 116*(1), 155–165. doi: 10.1037/0021-843X.116.1.155

Matsueda, R. L. (2013). Rational choice research in criminology: A multilevel framework. In R. Wittek, T. Snijders, & V. Nee (Eds.), *The handbook of rational choice social research* (pp. 283–321). Palo Alto, CA: Stanford University Press.

Mills, B., Reyna, V. F., & Estrada, S. (2008). Explaining contradictory relations between risk perception and risk taking. *Psychological Science, 19*(5), 429–433. doi: 10.1111/j.1467-9280.2008.02104.x

Moffitt, T. E. (1993). Adolescent-limited and life-course-persistent antisocial behaviour: A developmental taxonomy. *Psychological Review, 100*(4), 674–701.

Patil, I. (2015). Trait psychopathy and utilitarian moral judgement: The mediating role of action aversion. *Journal of Cognitive Psychology, 27.* doi: 10.1080 /20445911.2015.1004334

Reyna, V. F. (2012). A new intuitionism: Meaning, memory, and development in fuzzy-trace theory. *Judgment and Decision Making, 7*(3), 332–359. doi: 10.1017/ CBO9781107415324.004

Reyna, V. F. (2018). Neurobiological models of risky decision-making and adolescent substance use. *Current Addiction Reports, 5*(2), 128–133. http://dx.doi.org/10.1007 /s40429-018-0193-z

Reyna, V. F., & Brainerd, C. J. (2011). Dual processes in decision making and developmental neuroscience: A fuzzy-trace model. *Developmental Review, 31*(2–3), 180– 206. doi: 10.1016/j.dr.2011.07.004

Reyna, V. F., & Casillas, W. (2009). Development and dual processes in moral reasoning: A fuzzy-trace theory approach. In D. M. Bartels, C. W. Bauman, L. J. Skitka, & D. L. Medin (Eds.), *The psychology of learning and motivation: Vol. 50, Moral judgment and decision making* (pp. 207–236). San Diego, CA: Elsevier. doi: 10.1016/ S0079-7421(08)00407-6

Reyna, V. F., Chick, C. F., Corbin, J. C., & Hsia, A. N. (2014). Developmental reversals in risky decision making: Intelligence agents show larger decision biases than college students. *Psychological Science, 25*(1), 76–84. doi: 10.1177/0956797613497022

Reyna, V. F., Estrada, S. M., DeMarinis, J. A., Myers, R. M., Stanisz, J. M., & Mills, B. A. (2011). Neurobiological and memory models of risky decision making in adolescents versus young adults. *Journal of Experimental Psychology: Learning, Memory, and Cognition, 37*(5), 1125–1142. doi: 10.1037/a0023943

Reyna, V. F., & Farley, F. (2006). Risk and rationality in adolescent decision making— Implications for theory, practice, and public policy. *Psychological Science, 7*(1), 1–44. doi: 10.1145/1142680.1142682

Reyna, V. F., Helm, R. K., Weldon, R. B., Shah, P. D., Turpin, A. G., & Govindgari, S. (2018). Brain activation covaries with reported criminal behaviors when making risky choices: A fuzzy-trace theory approach. *Journal of Experimental Psychology: General, 147*(7), 1094–1109. doi: 10.1037/xge0000434

Reyna, V. F., & Mills, B. A. (2014). Theoretically motivated interventions for reducing sexual risk taking in adolescence: A randomized controlled experiment applying fuzzy-trace theory. *Journal of Experimental Psychology: General, 143*(4), 1627–1648. doi: 10.1037/a0036717

Reyna, V. F., & Wilhelms, E. A. (2017). The gist of delay of gratification: Understanding and predicting problem behaviors. *Journal of Behavioral Decision Making, 30*(2), 610–625. doi: 10.1002/bdm.1977

Romer, A. L., Reyna, V. F., & Pardo, S. T. (2016). Are rash impulsive and reward sensitive traits distinguishable? A test in young adults. *Personality and Individual Differences, 99,* 308–312. doi: 10.1016/j.paid.2016.05.027

Schlottmann, A., & Wilkening, F. (2011). Judgment and decision making in young children: Probability, expected value, belief updating, heuristics and biases. In A. Schlottmann, M. Dhami, & M. Waldmann (Eds.), *Judgement and decision making as a skill* (pp. 55–83). Cambridge: Cambridge University Press.

Shah, P., Catmur, C. & Bird, G. (2016). Emotional decision-making in autism spectrum disorder: The roles of interoception and alexithymia. *Molecular Autism, 7,* 43, doi:10.1186/s13229-016-0104-x

Shannon, B. J., Raichle, M. E., Snyder, A. Z., Fair, D. A., Mills, K. L., Zhang, D., … Kiehl, K. A. (2011). Premotor functional connectivity predicts impulsivity in juvenile offenders. *Proceedings of the National Academy of Sciences of the United States of America, 108*(27), 11241–11245. doi: 10.1073/pnas.1108241108

Steinberg, L. (2008). A social neuroscience perspective on adolescent risk-taking. *Developmental Review, 28*(1), 78–106. doi: 10.1016/j.dr.2007.08.002

Tversky, A. & Kahneman, D. (1986). Rational choice and the framing of decisions. *Journal of Business, 59,* S251–S278.

Weldon, R. B., Corbin, J. C., & Reyna, V. F. (2014). Gist processing in judgment and decision making: Developmental reversals predicted by fuzzy-trace theory. In H. Markovits (Ed.), *Current issues in thinking and reasoning: The developmental psychology of reasoning and decision making* (pp. 36–62). New York: Psychology Press.

Wojcik, D. Z., Díez, E., Alonso, M. A., Martín-Cilleros, M. V., Guisuraga-Fernández, Z., Fernández, M., … Fernandez, A. (2018). Diminished false memory in adults with autism spectrum disorder: Evidence of identify-to-reject mechanism impairment. *Research in Autism Spectrum Disorders, 45,* 51–57.

Contributors

Lior Abramson   Hebrew University of Jerusalem

Renée Baillargeon   University of Illinois at Urbana-Champaign

Pascal Belin   Aix-Marseille Université, CNRS

Frances Buttelman   Friedrich Schiller University Jena

Melody Buyukozer Dawkins   University of Illinois at Urbana-Champaign

Sofia Cardenas   University of Southern California

Michael J. Crowley   Yale University

Fabrice Damon   Université Grenoble-Alpes, CNRS

Michelle de Haan   University College London

Jean Decety   University of Chicago

Ghislaine Dehaene-Lambertz   University Paris Saclay

Xiao Pan Ding   National University of Singapore

Kristen A. Dunfield   Concordia University

Rachel D. Fine   University of Michigan

Ana Fló   Scuola Internazionale Superiore di Studi Avanzati

Jennifer R. Frey   The George Washington University

Susan A. Gelman   University of Michigan

Diane Goldenberg   University of Southern California

Marie-Hélène Grosbras   Aix-Marseille Université, CNRS

Tobias Grossmann   University of Virginia

Caitlin M. Hudac   University of Washington

Dora Kampis University of Copenhagen

Tara A. Karasewich Queen's University

Ariel Knafo-Noam Hebrew University of Jerusalem

Tehila Kogut Ben-Gurion University

Ágnes Melinda Kovács Central European University

Valerie A. Kuhlmeier Queen's University

Kang Lee University of Toronto

Narcis Marshall University of Southern California

Eamon McCrory University College London

David Méary Université Grenoble-Alpes, CNRS

Christos Panagiotopoulos Cornell University

Olivier Pascalis Université Grenoble-Alpes, CNRS

Markus Paulus Ludwig-Maximilians-Universität München

Kevin A. Pelphrey University of Virginia

Marcela Peña Pontificia Universidad Católica de Chile

Valerie F. Reyna Cornell University

Marjorie Rhodes New York University

Ruth Roberts University College London

Hagit Sabato Ben-Gurion University

Darby Saxbe University of Southern California

Virginia Slaughter University of Queensland

Jessica A. Sommerville University of Toronto

Maayan Stavans Central European University, Budapest

Nikolaus Steinbeis University College London

Fransisca Ting University of Illinois at Urbana-Champaign

Florina Uzefovsky Hebrew University of Jerusalem

Essi Viding University College London

Index

Note: Page numbers followed by f and t indicate figures and tables, respectively.